The Pollution Biology of Aquatic Oligochaetes

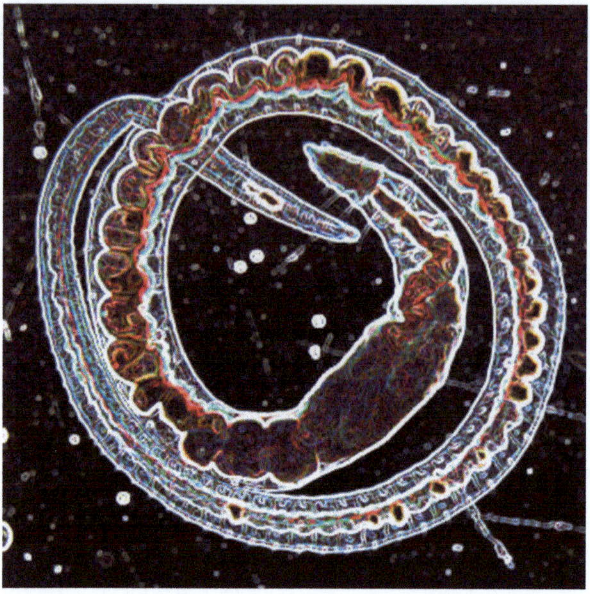

A specimen of *Tubifex tubifex* from the culture in the laboratory of Animal Ecotoxicology and Water Quality at the University of Basque Country. Photo: Pilar Rodriguez.

Pilar Rodriguez • Trefor B. Reynoldson

The Pollution Biology of Aquatic Oligochaetes

Springer

Pilar Rodriguez
Department of Zoology
and Animal Cell Biology
Faculty of Science and Technology
University of the Basque Country
P.O. Box 644 Bilbao 48080
Spain
pilar.rodriguez@ehu.es

Trefor B. Reynoldson
Acadia Centre for Estuarine Research
National Water Research Institute
Environment Canada
Acadia University
P.O. Box 115
Wolfville, NS B4P 2R6
Canada
trefor.reynoldson@acadiau.ca

ISBN 978-94-007-1717-6 e-ISBN 978-94-007-1718-3
DOI 10.1007/978-94-007-1718-3
Springer Dordrecht Heidelberg London New York

Library of Congress Control Number: 2011934967

Printed on acid-free paper

Springer is part of Springer Science+Business Media (www.springer.com)

Prologue

Some 30 years ago, Ralph Brinkhurst thought that it would be appropriate to prepare a volume pulling together the current state of knowledge on the biology of aquatic oligochaetes. This led to a meeting of those involved in oligochaete research and a subsequent volume which was the first Proceedings of the International Symposia of Aquatic Oligochaeta, held in Sidney, British Columbia in 1979 (Brinkhurst and Cook 1980) and was not a review but rather a summary of the state of the art of the studies on biology of aquatic oligochaetes at that time, including the biology of pollution. That first meeting in British Columbia initiated a triennial series of meetings and proceedings including ecotoxicology and pollution. In 1997, from these beginnings, Ralph suggested that we continue work that he had already initiated and over the ensuing period we have gradually completed this project both individually, in the rare periods that teaching or research projects has allowed us time, and also together. We have had to convince our family to meet in Canada, Spain, or Wales, for a few weeks every year from 1997, to move the work forward. For that we are very grateful for the indulgence of our families, for their patient and unconditional support over these years.

As the work has taken longer than expected, the additions to and re-organisation of the original manuscript have been substantial and, while Ralph is not a co-author on the volume, we wish to acknowledge his role in both developing the original manuscript and the fact that without his first contribution the book would probably have never seen the light of day. Thus, we wish to express our gratitude to Ralph Brinkhurst, colleague and good friend, for his key role in the development of this book. We also wish to express our gratitude to an array of colleagues which have helped with this volume. Particularly, Philipp Egeler for his review and valuable comments on the draft manuscript on bioaccumulation. Our gratitude also to Ruth Collado, Steve Fend, David Fogarty, Enrique Navarro, Adrian Pinder, Tarmo Timm, and Mark Wetzel for answering an array of questions.

The first author wants to express thanks for the financial support for the acquisition of part of the bibliographic material as well as part of work invested in this book, by the projects CGL2005-04943, CGL2008-04502 (Spanish Government) and IT405-10 (Basque Government). The second author would also like to acknowledge Environment Canada who supported this work before his retirement. We also wish to thank the various publishers who have given permission to reproduce published data and especially to Springer who have financed the publication of the book. We hope that it will be useful in your research and teaching, this will have made our effort worthwhile.

Finally we would like to dedicate this book to our respective fathers, to D. Primitivo Rodriguez who from childhood instilled a thirst for knowledge and scientific discovery and Professor Thomas B. Reynoldson who led the way in working with oligochaete worms.

February 2011

Pilar Rodriguez
Trefor B. Reynoldson

Contents

Chapter 1
Introduction

Abstract The first chapter provides a general introduction and describes the layout and objectives of the book. This includes a review of literature from the middle of the twentieth century, much of which is not easily accessible. The book addresses three main themes related to oligochaete biology in pollution assessment studies. First, ecological and field studies using composition and structure of oligochaete communities (Chap. 3). This section is devoted to the studies performed in applied ecology in aquatic systems, where aquatic oligochaete communities and indicator species exposed to contaminants in field sediments have had a core role in the environmental assessment of water quality. Second, toxicology and laboratory studies (Chap. 4) and is organized into water-only vs. sediment toxicity tests, and acute vs. chronic toxicity tests. The chapter also deals with cellular and molecular biomarkers and their relationship with toxicological effects. Third, bioaccumulation studies in aquatic oligochaete species (Chap. 5). Both Chaps. 4 and 5 have appendices where toxicological parameters (LC_{50} and EC_{50}) and several bioaccumulation data (bioaccumulation factors, biological half-life, toxicokinetic coefficients) are reported for different oligochaete species. The remaining chapters of the book provide additional information that may be helpful for those readers that will use this book as an aid to working with aquatic oligochaetes in pollution assessment. Chapter 2 addresses the taxonomy of the aquatic oligochaete families, providing references to key taxonomic studies on the identification of organisms. The final chapter addresses a few selected methodological issues which will hopefully be useful to the reader when making decisions on field sampling and laboratory work.

1.1 Background on Aquatic Oligochaetes in Pollution Biology

The term pollution is commonly used to describe the introduction by humans of chemicals and wastes or the alteration of physical conditions (temperature, pH, etc) that result in deleterious effects to living organisms. The operational definition of pollution is when a site fails to meet those characteristics of a site or sites

P. Rodriguez and T.B. Reynoldson, *The Pollution Biology of Aquatic Oligochaetes*,
DOI 10.1007/978-94-007-1718-3_1, © Springer Science+Business Media B.V. 2011

deemed to be in reference condition, where reference is defined by a set of biological conditions (either a numeric or narrative description of a species or community assemblage) in unaltered locations. Accordingly, reference conditions are associated with conditions suitable for maintaining species or community assemblages. Therefore, the first question in the biology of pollution is what are the natural qualities of unpolluted waters to which other sites can be compared. This question was previously posed by Hynes (1960) in his classic monograph on the *Biology of Polluted Waters*. However, it was not until the 1980s that the pollution assessment of water bodies used a Reference Condition Approach (Wright et al. 1984). This change in perspective for studies on water pollution marks a clear separation between old and new approaches in aquatic pollution studies.

In the case of aquatic oligochaetes, there is no unified theoretical body of knowledge on the pollution biology of the group. The last summary of the response of the group to disturbance was in Brinkhurst and Jamieson's book on Aquatic oligochaetes (1971) and in Brinkhurst and Cook's (1974) chapter on aquatic earthworms. Despite a large volume of work since those reviews, the group still tends to be treated as a taxonomic unit in many pollution studies, and in general most practitioners tend to think of the group as indicative of degraded conditions. In this book we have tried to present and synthesise both the laboratory and field studies and research on oligochaete worms to show that this group of organisms, which occur in almost all aquatic habitats and frequently in large numbers, do have great utility in the study of pollution biology. As usually happens in science, the study of pollution biology of aquatic organisms has not been treated as a single topic and much of the research has been conducted separately in laboratory and field studies. This has had the consequence that there is no coherent view of the relationships between worm species and environmental variables. Most recently work in ecotoxicology and bioaccumulation tends to be a component of the Ecological Risk Assessment methodology which does provide a mechanism for relating community studies and the probability of effects, and will hopefully generate important advances in the near future.

The ability of indicator species to identify organic pollution was observed many centuries ago. The Greek philosopher Aristotle (384–322 BC) noted the small red threads that grew out of foul mud (Hynes 1960). Initial research in pollution biology examined the composition and structure of field macroinvertebrate communities and in particular identified indicator species that responded to organic pollution. From the 1930s to the early 1980s there was a general consideration that aquatic oligochaetes had a single response and were very tolerant of degraded water quality. With the development of identification keys, it was possible to identify taxa to species level, but surveys at that time were mainly descriptive, reporting associations between the degree of environmental disturbance and the species that occurred. The first specific examination, at the community level, of the behaviour of oligochaetes to gross pollution was conducted by Brinkhurst in a series of quantitative surveys, beginning in 1958. Downstream of a major effluent, the macrobenthos in grossly polluted rivers were composed of a few tubificine oligochaetes and some chironomid species (Brinkhurst 1965). Several studies in rivers revealed that *Tubifex tubifex* and *Limnodrilus hoffmeisteri* were the most abundant oligochaetes in grossly polluted sites. Improvement of water quality by installation of sewage treatment plants or the reduction of waste discharges was commonly reported as resulting in a dramatic change in species composition. After recovery, the macrobenthic river fauna included not only members of many more animal phyla but also many more oligochaete species. In Scandinavia, Milbrink (1973) began a series of studies of the larger Swedish lakes and their oligochaete fauna in relation to pollution. He described overlapping distributions of species along a trophic gradient, and proposed an indicator community approach as opposed to using indicator species. Through the 1970s and 1980s, various indices using species richness and abundance were proposed in Europe and North America, illustrating the trend of improving the assessment approach from simple descriptions of species occurrence and association with trophic condition to a more formalized quantitative description. Values for the indices provided a generally good description of the gradient from oligotrophic to eutrophic (or from oligosaprobic to polysaprobic) conditions.

Other approaches have used tolerance values for the species. This approach has two main problems, first that some species are tolerant of a range of conditions but are poor competitors (*e.g. T. tubifex*) and, thus, can be abundant at both ends of a trophic stress gradient; and second, that tolerance values and indices need be developed for each type of stress, and most systems generally ascribe tolerance values based on organic enrichment, which in fact is a reflexion of their ability to regulate oxygen.

In the last 20 years or so, the use of multivariate statistical approaches to describing worm assemblages and their relationship with the environment has resulted in the development of the predictive modelling used in various Reference Condition Approaches (RCA), which form the basis of freshwater bioassessment programmes in Australia, Canada, USA, and Europe. Under requirements of the European Water Framework Directive, Verdonschot (2006) has recently combined the two approaches, and used multivariate analyses to calculate tolerance values in a RCA. He has demonstrated that, at the species level, oligochaetes can discriminate the range of ecological quality described across Europe and the number of oligochaete species occurring across the ecological quality classes was comparable to other invertebrate groups, although there was a trend to increased relative abundance of oligochaetes in poorer quality waters.

The role of sediments as a secondary source of contaminants in water bodies is well known and widely recognized. The relevance of aquatic oligochaetes in sediment quality assessment is largely related to their benthic and detritivorous life habit. Worms interact with sediments through pumping pore water and actively transport particulate matter to surficial layers by feeding activity. The usefulness of oligochaetes in environmental risk assessment has been highlighted by Chapman (2001) and Egeler and Römbke (2007), and these organisms play an important ecological role in sediments where they may occur at high densities and are responsible for bioturbation through their burrowing activity. Aristotle was probably performing the first rudimentary testing placing "blood worms" in salt

Table 1.1 Proceedings of the International Symposia on Aquatic Oligochaete Biology (ISAOB)

Publication year	Editor(s)	Date and symposium location
1980	Ralph O. Brinkhurst and David G. Cook	1979 Sidney (British Columbia, Canada)
1987	Giuliano Bonomi and Christer Erséus	1982 Pallanza (Italy)
1984	Ralph O. Brinkhurst and Robert. J. Diaz	1985 Hamburg (Germany)
1989	Jerry L. Kaster	1988 Baton Rouge (Louisiana, USA)
1994	Trefor B. Reynoldson and Kathryn A. Coates	1991 Tallin (Estonia)
1996	Kathryn A. Coates, Trefor B. Reynoldson and Thomas B. Reynoldson	1994 Strömstat (Sweden)
1999	Brenda M. Healy, Trefor B. Reynoldson and Kathryn A. Coates	1997 Presque Isle (Maine, USA)
2001	Pilar Rodríguez and Piet F. M. Verdonschot	2000 Bilbao (Spain)
2006	Piet F. M. Verdonschot, Hong-Zhu Wang, Adrian Pinder and Rebi Nijboer	2003 Wageningen (The Netherlands)
2007	Hong-Zhu Wang, Mark J. Wetzel, Adrian M. Pinder, Piet F.M. Verdonschot and Naime Arslan	2006 Wuhang (China)
2011	Naime Arslan and Ercument Colak	2009 Antalya (Turkey)

water and observing their responses (Chapman 2001). With better understanding, worms have become more accessible as study organisms and their use more popular in environmental risk assessment of sediments. An extensive database on aquatic oligochaetes has demonstrated their sensitivity to a range of chemicals (metals, pesticides, PAHs, etc). There are also reasonably standardised protocols for laboratory and field research in sediment ecotoxicity and bioaccumulation testing using oligochaetes (USEPA 2000; ASTM 2005; OECD 2007, 2008) which allow comparisons among substances and species, and mapping toxicity risk across different geographic areas.

This book is designed to be of special interest to researchers and postgraduate lecturers who work in the fields of ecotoxicology and aquatic ecology, and it addresses a current need related to the wider use of oligochaetes as test organisms in sediment risk assessment. Further information is also available from the publications from a series of international symposia dedicated to the study of aquatic oligochaetes, the first of which was held in Sidney (British Columbia, Canada) in 1979. To date, eleven international symposia on Aquatic oligochaetes have been convened, and their proceedings contain much information on taxonomy and on the use of oligochaete species in aquatic pollution biology (Table 1.1). Information on past and future symposia can be obtained from the page of the *International Symposia on Aquatic Oligochaete Biology* (ISAOB: http://www.inhs.uiuc.edu/~mjwetzel/ISAOBdir.html).

1.2 Purpose and Contents of the Book

The main objective of this book is to present the current state of the art in the pollution biology of aquatic oligochaetes, with particular emphasis on ecotoxicological studies, both in the laboratory and in the field. The book also includes a review of literature from the middle of the twentieth century, much of which is not easily accessible. The book addresses three main themes related to oligochaete biology in pollution assessment studies:

First, ecological and field studies using composition and structure of oligochaete communities (Chap. 3). This section is devoted to the studies performed in applied ecology in aquatic systems, where aquatic oligochaete communities and indicator species exposed to contaminants in field sediments have had a core role in the environmental assessment of water quality.

Second, toxicology and laboratory studies (Chap. 4). This chapter is dedicated to the field of ecotoxicology where different oligochaete species have been used in bioassays under different environmental conditions, in either short- or long-term exposure periods to water or sediments. The main objective of these studies was to assess effective concentrations of different chemicals (metals and organics). The chapter has been organized into water-only vs. sediment toxicity tests, and acute vs. chronic toxicity tests. As acute effects we have included information on short-term toxicity tests (typically ≤96-h exposure) that usually measure mortality as the endpoint, although other sublethal measurements are also described, including respiration and behaviour. Chronic toxicity tests commonly report the effects of longer exposures (usually more than 10 days up to several months) and the endpoints include mortality as well as sublethal effects related to growth and reproduction, and less frequently behavioural responses. Oligochaete worms are not uniformly more tolerant to contaminants than other test organisms and comparative data for worms and other benthic invertebrates show that responses are species-specific and also contaminant-specific. Both acclimation and genetic adaptation of test organisms are discussed as an explanation of differences in sensitivity between populations used in the toxicological assessment. Finally, this chapter deals with cellular and molecular biomarkers and their relationship with toxicological effects. There has been relatively little work on biomarkers in aquatic oligochaetes, which purportedly have the advantage of occurring at lower doses than effects at the organism scale.

Third, bioaccumulation studies in aquatic oligochaete species (Chap. 5). Bioaccumulation constitutes a separate line of evidence in the environmental risk assessment approach and in this chapter we address three main topics: a description of the bioaccumulation process of contaminants (uptake and elimination rates and pathways), the relationships between body burden and the concentration of the substance in the bulk sediment or in the pore water, and the assessment of risk associated with transfer of contaminants from sediment through aquatic oligochaetes to higher trophic levels. The relationships between bioaccumulation and toxic effects in oligochaetes exposed to pollutants in the field or in laboratory bioassays have been examined under the Critical Body Residue concept. Both Chaps. 4 and 5 have appendices where toxicological parameters (LC_{50}, EC_{50}) and several bioaccumulation data (bioaccumulation factors, biological half-life, toxicokinetic coefficients) are reported for different oligochaete species.

The remaining chapters of the book provide additional information that may be helpful for those readers that will use this book as an aid to working with aquatic oligochaetes in pollution assessment. This is the information and knowledge that we wish we had before working on this interesting group of organisms, and so we hope new workers in the field find this of value. Thus, Chap. 2 addresses the taxonomy of the aquatic oligochaete families, providing references to key taxonomic work for the identification of organisms. It also summarises information on the most recent phylogenetic and taxonomic studies to make researchers aware of some of the taxonomic debates which affect the nomenclature of common taxa used in the field of pollution biology. The final chapter addresses a few selected methodological issues which will hopefully be useful to the reader when making decisions on field and laboratory work, topics include advice on sieving, subsampling, biomass estimation, worm culture and the preparation of worms for identification. Specific details on performing toxicity and bioaccumulation tests are also provided, including issues such as food supplement, aspects of water and sediment characterisation, the use of artificial sediments, and the control of worm age and density. Some examples are provided on the required control of environmental variables (*e.g.* ammonia, temperature) so as to avoid confounding factors in bioassays. We have also made an effort in all the chapters to include recommendations on methodological issues based on the authors' experience.

References

Arslan N, Ercument C (eds) (2011 in press) Aquatic oligochaete biology XI. In: Proceedings of the XI international symposium on aquatic Oligochaeta, held in Alanya, Turkey, October 2009. Turkish J Zool 35(3)

ASTM (2005) Standard guide for conducting sediment toxicity tests with freshwater invertebrates, ASTM E1706-05, Philadelphia

Bonomi G, Erséus C (eds) (1984) Aquatic Oligochaeta. In: Proceedings of the second international symposium on aquatic oligochaete biology, held in Pallanza, Italy, September 1982. Developments in Hydrobiology, Hydrobiologia 155, Dr W. Junk Publ

Brinkhurst RO (1965) Observations on the recovery of a British river from gross organic pollution. Hydrobiologia 25:9–51

Brinkhurst RO, Cook DG (1974) Aquatic earth worms (Annelida: Oligochaeta). In: Hart CW, Fuller SLH (eds) Pollution ecology of freshwater invertebrates. Academic, New York, pp 143–156

Brinkhurst RO, Cook DG (eds) (1980) Aquatic oligochaete biology. Plenum Press, New York, p 529

Brinkhurst RO, Diaz RJ (eds) (1987) Aquatic oligochaete biology III. Proceedings of the third international symposium on aquatic Oligochaeta, held in Hamburg, Germany, 1985. Developments in Hydrobiology, Hydrobiologia 155, Dr W. Junk Publ

Brinkhurst RO, Jamieson BGM (1971) The aquatic Oligochaeta of the world. Oliver & Boyd, Edinburgh, p 860

Chapman PM (2001) Utility and relevance of aquatic oligochaetes in ecological risk assessment. Hydrobiologia 463:149–169

Coates KA, Reynoldson TB, Reynoldson TB (eds) (1996) Aquatic oligochaete biology VI. In: Proceedings of the sixth international symposium on aquatic oligochaetes, held in Strömstat, Sweden, September 1994. Developments in Hydrobiology, Hydrobiologia 334, Kluwer Academic Publ

Egeler P, Römbke J (2007) Oligochaeta (microdrile) worms in the environmental risk assessment of pesticides in the European Union. Acta Hydrobiologica Sinica 31(Suppl):151–162

Healy BM, Reynoldson TB, Coates KA (eds) (1999) Aquatic oligochaetes. Proceedings seventh international symposium on aquatic oligochaetes, held in Presque Isle, Maine, August 1997. Developments in Hydrobiology, Hydrobiologia 406, Kluwer Academic Publ

Hynes HBN (1960) The biology of polluted waters. Liverpool University Press, Liverpool, p 202

Kaster JL (eds) (1989) Aquatic oligochaete biology VII. In: Proceedings of the fourth international symposium on aquatic oligochaete biology, held in Baton Rouge, Louisiana, 1988. Developments in Hydrobiology, Hydrobiologia 180, Kluwer Academic Publ

Milbrink G (1973) On the use of indicator communities of Tubificidae and some Lumbriculidae in the assessment of water pollution in Swedish lakes. Zoon 1:125–139

OECD (2007) Sediment-water lumbriculus toxicity test using spiked sediment. OECD guideline for the testing of chemicals no. 225

OECD (2008) Bioaccumulation in sediment-dwelling benthic oligochaetes. OECD guideline for the testing of chemicals no. 315

Reynoldson TB, Coates KA (eds) (1994) Aquatic oligochaete biology V. Proceedings of the 5th Oligochaete Symposium, held in Tallinn, Estonia, 1991. Developments in Hydrobiology, Hydrobiologia 278, Kluwer Academic Publ

Rodriguez P, Verdonschot PFM (eds) (2001) Aquatic oligochaete biology VIII. Proceedings eighth international symposium on aquatic oligochaete biology, held in Bilbao, Spain, July 2000. Developments in Hydrobiology, Hydrobiologia 463, Kluwer Academic Publ

USEPA (2000) Methods for measuring the toxicity and bioaccumulation of sediment-associated contaminants with freshwater invertebrates, 2nd edn. EPA 600/R-99/064

Verdonschot PFM (2006) Beyond masses and blooms: the indicative value of oligochaetes. Hydrobiologia 564:127–142

Verdonschot PFM, Wang HZ, Pinder A, Nijboer R (eds) (2006) Aquatic oligochaete biology IX. Proceedings ninth international symposium on aquatic Oligochaeta, held in Wageningen, The Netherlands, 2003. Developments in Hydrobiology, Hydrobiologia 564, Springer

Wang HZ, Wetzel M, Pinder A, Verdonschot PFM, Arslan N (eds) (2007) Aquatic oligochaete biology X. In: Proceedings of the tenth international symposium on aquatic oligochaete biology, held in Wuhan, China, 2006. Acta Hydrobiologica Sinica 31

Wright JF, Moss D, Armitage PD, Furse MT (1984) A preliminary classification of running-water sites in Great Britain based on macro-invertebrate species and the prediction of community type using environmental data. Freshwater Biol 14:221–256

Chapter 2
Taxonomy of Aquatic Oligochaetes

Abstract This chapter provides the reader and those studying oligochaetes some general background on the basic anatomy, morphology and taxonomy of the group. The first part of the chapter describes the taxonomic context for the group within the phylum Annelida. It gives the reader the latest taxonomic status of the group and also explains some of the issues on nomenclature and classification. The classification of the Clitellata at the family level is in a state of flux. A recent proposal of synonymy of the families Naididae and Tubificidae affects the taxonomic classification of most species commonly used in pollution biology. Thus, it has been formally proposed that the former families of Naididae and Tubificidae be merged as synonyms in the family Naididae. The growing evidence from molecular data of an increasing number of oligochaete species suggests that this interpretation may change again in the near future. Several oligochaete families contain a small number of species and have either a reduced or scattered geographic distribution and rare occurrence. For example, the families Tiguassidae, Narapidae, Biwadrilidae and Lutodrilidae are monospecific, and Capilloventridae, Dorydrilidae, Opistocystidae, Parvidrilidae, and Propappidae have less than seven species each, with geographical distributions that are often local and/or very scattered. The family Haplotaxidae contains about 20 species and has a cosmopolitan distribution, but species are scarce. For these reasons, members of these ten families are rarely reported in field community studies or water quality assessments and have not been used in laboratory bioassays. Most field aquatic pollution studies only require knowledge of the most common tubificid and naidid oligochaete species, together with the less frequently occurring lumbriculids and enchytraeids. The rest of the chapter describes for the reader general keys or papers on the taxonomy of the commonest families.

P. Rodriguez and T.B. Reynoldson, *The Pollution Biology of Aquatic Oligochaetes*,
DOI 10.1007/978-94-007-1718-3_2, © Springer Science+Business Media B.V. 2011

2.1 What Is an Oligochaete Worm?

The oligochaete worms belong to the class Clitellata (phylum Annelida). They are typically segmented worms of bilateral symmetry, hermaphroditic, with a spacious coelom and a clitellum. The clitellum is a characteristic reproductive organ that appears during sexual maturity. It forms a ring or dorsal glandular epithelial thickening over a few body segments located around and/or behind the female pores. Oligochaetes also possess a pre-oral lobe, the prostomium, an anterior ventral mouth, a posterior anus in the pygidium (Fig. 2.1a), and most commonly possess four bundles of chitinous chaetae segmentally arranged along the greater part of the body, but in contrast to the polychaetes these are not located on parapodia. There are usually one or two pairs of male and female gonads, situated in the anterior part of the body, the testes are anterior to the ovaries, and the genital products (gametes together with accompanying material) are discharged through special ducts.

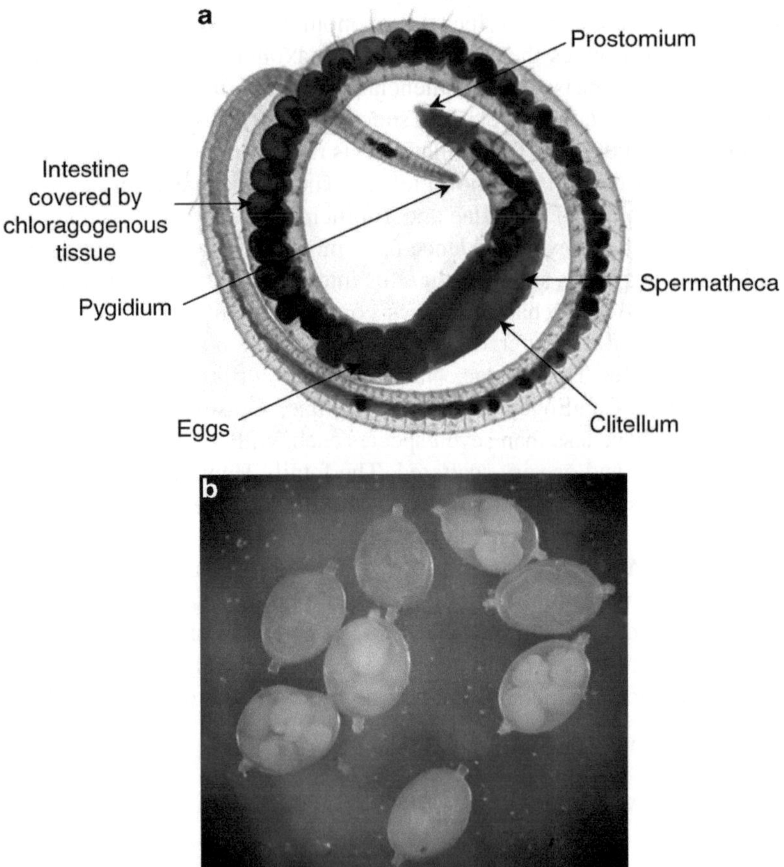

Fig. 2.1 (a) Anatomy of an oligochaete worm. (b) Cocoons

The clitellum secretes a cocoon where eggs are laid and fertilized, and provides both the external cover and the internal content necessary for the direct development of embryos (Fig. 2.1b). Oligochaetes have a specific structure and position of the male ducts and of the sperm receptacula (commonly called spermathecae) relative to the male pores, which is usually constant within a family. The male ducts may have specialised and variously complex structures (*e.g.* vas deferens, atrium, penis or pseudopenis, prostatic glands) that are useful characters for taxonomic classification.

Oligochaete worms have colonized a wide range of aquatic and semi-aquatic habitats and freshwater, brackish and marine benthic communities commonly include several characteristic oligochaete species. These occupy a range of microhabitats in sediments, as well as in aquatic vegetation and decomposed organic matter. Most species are detritivorous sediment-dwellers, selectively feeding on bacteria, algae and mineral particles rich in organic matter (Coler et al. 1967; Harper et al. 1981a, b; Wavre and Brinkhurst 1971). Other, mostly naidines, swim and graze in aquatic vegetation, or are ecto- or endo-symbiotic (*e.g. Chaetogaster limnaei or Allodero* spp., see Gelder 1980), while a few are predaceous (*Chaetogaster, Agriodrilus, Phagodrilus, Haplotaxis*). Among the meiofauna inhabiting the interstitial environment there are many oligochaete species and they can be a major portion of both the freshwater and marine meiobenthic community. Several freshwater oligochaete species also occur in the meiobenthic fauna of the profundal zone of lakes and can be used as indicators of pollution (Särkkä 1996). Ecological studies on the distribution of marine oligochaetes are scarce, but small interstitial oligochaetes can be the dominant taxa in the shallower coarse sand areas of the Middle Atlantic Bight (psammic species), whereas burrowing species predominate in deeper waters or in finer grained sediments (Diaz et al. 1987).

2.2 Classification of the Clitellata

There is general agreement among oligochaetologists that the class Clitellata forms a monophyletic group that includes the oligochaete families along with leech-like worms and leeches (branchiobdellids, acanthobdellids and euhirudineans). The latter three groups are distinguished from the oligochaete worms by the presence of suckers and by having a fixed number of segments, among other characters. Several studies using cladistic analysis based on both morphological and molecular characters (Jamieson 1988; Jamieson et al. 2002; Purschke et al. 1993; Ferraguti and Erséus 1999; Martin 2001; Siddall et al. 2001; Erséus and Källersjö 2004) have revised the systematic position of oligochaetes within the Clitellata. These studies unanimously regard the oligochaetes as a paraphyletic group, *i.e.* the clade derived from a most recent common oligochaete-like ancestor includes not only the oligochaete families but also the branchiobdellids and leeches, and thus oligochaetes do not constitute a natural taxon in a cladistic sense (Fig. 2.2). Therefore, the

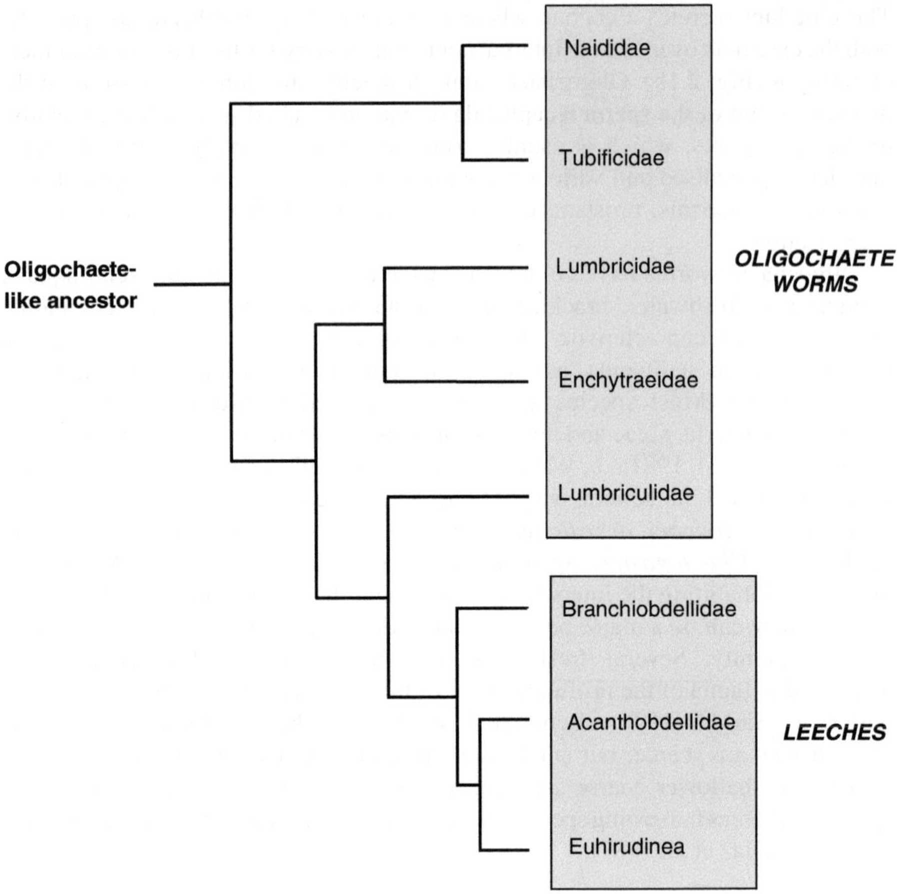

Fig. 2.2 Clitellate relationships based on 18S rRNA sequence (Note: both Naididae and Tubificidae have recently been proposed to merge as synonyms in the single family Naididae, see text) (Redrawn and simplified from Martin 2001, Fig. 3, reproduced with permission of the Royal Society)

oligochaetes are referred to as members of the class Clitellata, although some authors persist in using the old nomenclature as Oligochaeta *sensu stricto* (Martin et al. 2008), while others have used Oligochaeta as synonym of Clitellata (*e.g.* Siddall et al. 2001). The use of Clitellata is preferred in our opinion as it avoids the confusion for non-specialists that would result from the use of the term Oligochaeta when referring to leeches.

Most characters that define the Clitellata relate to their reproductive biology, *e.g.* the presence of a clitellum, hermaphroditism, the organization of reproductive organs (generally, plesiopora male ducts), and the ultrastructure of the spermatozoa (Rouse and Fauchald 1995). Several other morphological characters, such as, the reduction of the prostomium and the lack of prostomial appendages (Purschke et al. 1993), and the backward position of the supra-oesophageal ganglion (Westheide 1997) are also distinct characteristics of the Clitellata that separate them from other annelids. The absence of a trocophore larva, copulatory organs and even hermaphroditism have also been described in some interstitial polychaetes (Westheide 1971) and therefore cannot be considered as attributes restricted to the Clitellata.

Species of the family Aeolosomatidae are also often found in samples from aquatic habitats together with oligochaetes. Species in this family were previously classified as oligochaetes due to their hermaphroditism, the presence of copulatory epidermal glands (although ventral) in some segments, the lack of parapodia, and relatively simple chaetae, but nowadays molecular data support the exclusion of aeolosomatids from the class Clitellata (Martin et al. 2000; Martin 2001; Struck et al. 2002). These annelid worms have been classified as members of the class *Aphanoneura* (Timm 1981; Brinkhurst 1982), which has also been proposed to include the family Potamodrilidae (Struck and Purschke 2005; Glasby and Timm 2008). These are related to the Polychaeta and are therefore not included in this book.

There are general treatises that contain detailed descriptions on the external and internal anatomy of oligochaetes (Stephenson 1930; Avel 1959; Brinkhurst and Jamieson 1971; Adiyodi 1988) and provide detailed descriptions of the morphological characters of the group. Microscopical anatomy has been described in detail by Jamieson (1981).

The division of oligochaetes into *microdriles* and *megadriles* based on their relative size was established more than a century ago by Benham (1890), and this nomenclature is still sometimes used in general monographs and studies on oligochaetes. Microdriles, also loosely known as *limicolae*, are small oligochaetes, usually less than 25 mm long, living in freshwater, brackish and marine habitats, and less often in wet soils and a total of 13 aquatic microdrile families are known. They have a single-layered clitellum, and produce a relatively small number of yolky eggs. Most megadriles, generically termed *earthworms*, are classified into the Crassiclitellata (Jamieson 1988) characterized by having a multi-layered clitellum, which is much more obvious than the single-layered one of the microdriles. However, there are seven megadrile families that contain almost 100 aquatic species (Martin et al. 2008). Approximately 30% of the described oligochaete species are aquatic or semi-aquatic, with about 1,100 freshwater species and 600 marine or estuarine species (Erséus 2005; Martin et al. 2008). Microdrile families of oligochaetes and aquatic megadriles are reported in Table 2.1, with their geographical distribution and selected key papers that are useful for species identification.

The classification of the Clitellata at the family level is in a state of flux. A recent proposal of synonymy of the families Naididae and Tubificidae affects the taxonomic classification of most species commonly used in pollution biology. Erséus and Gustavsson (2002) who reviewed the taxonomic status of the Naididae formally proposed that the family be considered a subfamily of the family Tubificidae based on a phylogenetic assessment using 18S rDNA data, and this

Table 2.1 Aquatic oligochaete families, their geographic distribution and, number of species, and some taxonomic key works

Family	Subfamily	Geographic area	Number of spp.[a]	Reference
Capilloventridae		Southern hemisphere	3	Erséus (1993), Pinder and Brinkhurst (1997b)
Haplotaxidae		Cosmopolitan	21	Omodeo (1987), Brinkhurst (1988)
Tiguassidae		South America	1	Righi et al. (1978)
Narapidae		South America	1	Righi and Varela (1983)
Opistocystidae		America and Africa	7	Harman and Loden (1978)
Parvidrilidae		North America and Europe	2	Erséus (1999), Martinez-Ansemil et al. (2002b)
Dorydrilidae		European	3	Brinkhurst and Jamieson (1971)
Phreodrilidae	Phreodrilinae	Southern hemisphere	50	Brinkhurst (1991), Pinder and Brinkhurst (1997a)
	Phreodriloidinae			
Naididae (syn. Tubificidae)[b]	Tubificinae	Cosmopolitan	350	Brinkhurst and Jamieson (1971), Brinkhurst (1981), Baker and Brinkhurst (1981), Erséus (1982, 1992), Giani et al. (1984), Holmquist (1974, 1985), Martin et al. (2010)
	Phallodrilinae		315	
	Limnodriloidinae		90	
	Telmatodrilinae		8	
	Rhyacodrilinae		155	
	Rhyacodriloidinae		3	
	Naidinae		180	
	Pristininae		40	
			In total > 1,000	
Lumbriculidae		Holarctic (one peregrine species)	204	Brinkhurst and Jamieson (1971), Rodriguez and Giani (1994), Fend and Brinkhurst (2000), Fend (2005) (see text for other references)
Propappidae		Palearctic	3	Coates (1986), Coates et al. (2006)
Enchytraeidae		Cosmopolitan	136 [c]	Nielsen and Christensen (1959, 1961, 1963), Graefe and Römbke (1985), Römbke (1992), Schoch-Bösken and Römbke (1993), Römbke and Dozsa-Farkas (1996), Schmelz and Römbke (1999), Schmelz (2003), Coates et al. (2006)
Alluroididae		Africa and South America	11	Omodeo (1996), Omodeo and Coates (2001)

[a] Number of species for the families taken from Martin et al. (2008). For the subfamilies the number of species is approximate

[b] It has recently been proposed to rename the family Tubificidae to family Naididae which would then include the subfamilies Tubificinae, Naidinae, Pristininae, Rhyacodrilinae, Rhyacodriloidinae, Phallodrilinae, Telmatodrilinae, Limnodriloidinae, this based on the priority rule of zoological nomenclature (Erséus et al. 2008)

[c] The family Enchytraeidae is a large family of over 600 species that mainly occur in terrestrial habitats

proposal was broadly accepted by oligochaete taxonomists (*e.g.* Jamieson 2006; Martin et al. 2008). The proposal was also supported by previous evidence of phylogenetic relationships based on morphological characters (Erséus 1990) and molecular data, including the contribution of Christensen and Theisen (1998) using nuclear 23S rRNA and mitochondrial COI (Cytochrome c Oxydase 1) sequences. Based on molecular data, Christensen and Theisen concluded that naidids were a subordinate group within the family Tubificidae, and suggested a sister relationship between naidids and some rhyacodriline tubificids. This relationship has been further supported by molecular analyses using 18S rDNA (Erséus et al. 2000, 2002), and 12S, 16S and 18S rDNA sequences (Envall et al. 2006). However, the application of the priority rule of zoological nomenclature, based on the names first appearance, requires that the family Tubificidae be named Naididae (Erséus et al. 2008). Thus, the former family Tubificidae is now formally proposed as synonym of the family Naididae. The growing evidence from molecular data of an increasing number of oligochaete species (to date, about 200 of 1,700 aquatic oligochaete species have been reported in the Genbank, NCBI) suggests that this interpretation may change again in the near future. Therefore, in order to avoid nomenclatural confusion for readers of this book, we use the old term "tubificid" to refer to the species comprised in the old family Tubificidae and "naidid" for members of the old family Naididae. Wherever possible, the still accepted subfamily names are used (see Table 2.1).

Several oligochaete families (Table 2.1) contain a small number of species and have either a reduced or scattered geographic distribution and rare occurrence. For example, the families Tiguassidae, Narapidae, Biwadrilidae and Lutodrilidae are monospecific, and Capilloventridae, Dorydrilidae, Opistocystidae, Parvidrilidae, and Propappidae have less than seven species each, with geographical distributions that are often local and/or very scattered (Martin et al. 2008).

The family Haplotaxidae contains about 20 species and has a cosmopolitan distribution, although the scarcity of the species and their habitats (*e.g.* caves for the genus *Delaya* or remote sites for other genera) has made the study of this family difficult (Brinkhurst 1988). For these reasons, members of these ten families are rarely reported in field community studies or water quality assessments and have not been used in laboratory bioassays. Most field aquatic pollution studies only require knowledge of the most common tubificid and naidid oligochaete species, together with the less frequently occurring lumbriculids and enchytraeids.

2.3 Species Identification and Some Taxonomic Observations

Most aquatic oligochaete species can only be identified when sexually mature, except for naidines and pristinines and a few other characteristic species, such as *Branchiura sowerbyi* (with caudal gills) or *Psammoryctides barbatus* (with fanwise dorsal chaetae). Some morphological characters of the most common species used in toxicology and bioaccumulation studies are shown in Fig. 2.3.

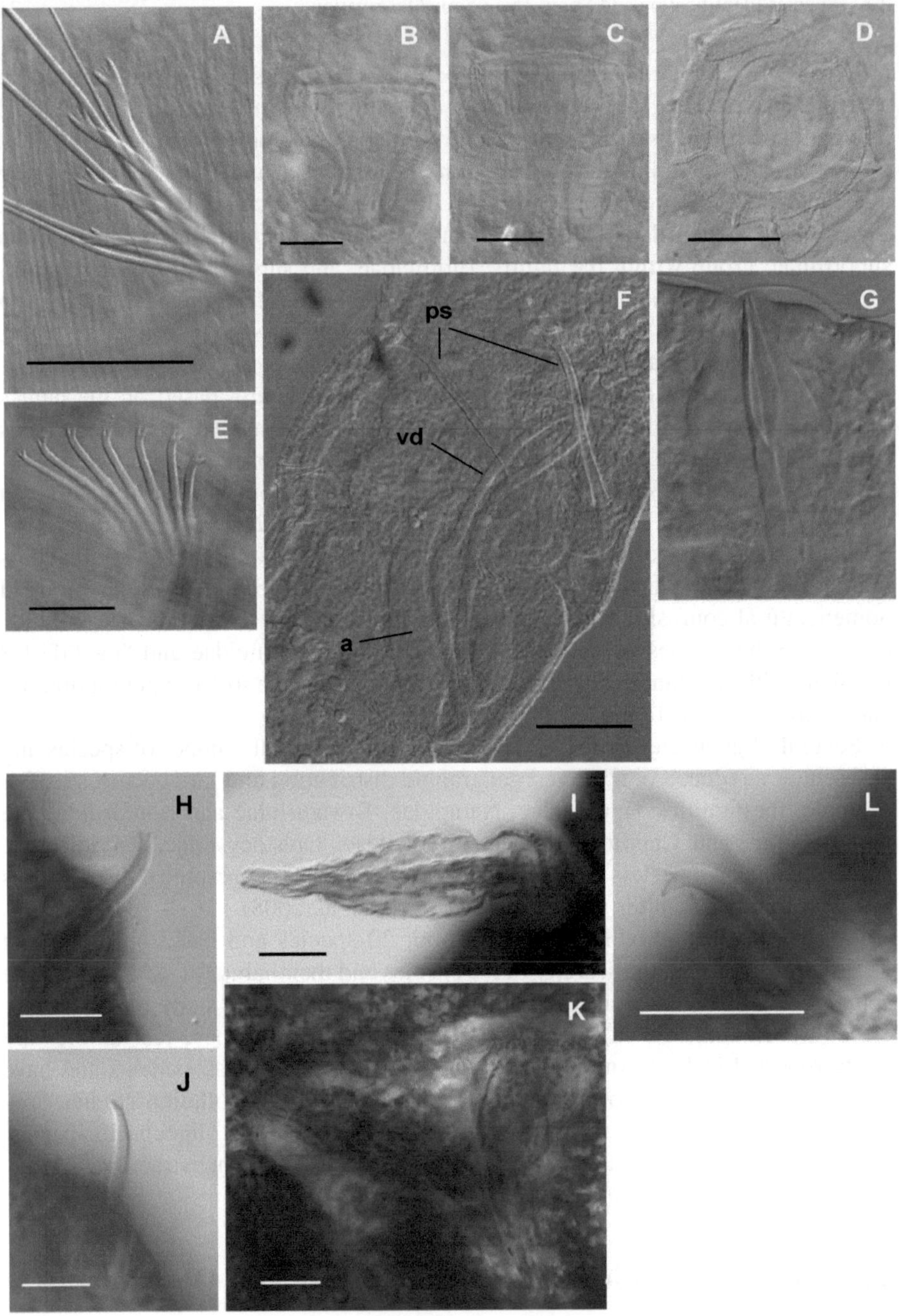

Fig. 2.3 Morphological characters of the oligochaete species *Tubifex tubifex*: (**a**) dorsal chaetae, (**b**, **c**) penial cuticle in lateral view, (**d**) penial cuticle in dorso-ventral view, showing a characteristic granulation. *Limnodrilus hoffmeisteri*: (**e**) chaetae, (**f**): clitellar segments showing cuticular penis sheath (*ps*), vas deferens (*vd*) and atrium (*a*). *Potamothrix hammoniensis*: (**g**) spermathecal chaeta. *Stylodrilus heringianus*: (**h** and **j**) bifid and simple-pointed chaetae; (**i**) penis in lateral view, (**k**) a pair of penes in ventral view. *Lumbriculus variegatus*: (**l**) chaeta (scale in the figures represents 50 μm length)

A practical solution to the difficulties of species identification when field samples include a high percentage of immature individuals is to sample several times a year, increasing the likelihood of finding the mature individuals required for the identification of most species. An alternative (but impractical) approach developed through the seventies to the nineties involved using the protein taxonomy of aquatic oligochaetes as a tool for distinguishing species among immature individuals as well as for species that reproduce primarily asexually, or for identifying sibling-species. Milbrink and Nyman (1973) first analysed non-specific esterases in 18 species of different oligochaete families using gel electrophoresis with promising results. Protein taxonomy, using an isoelectric focusing (IEF) technique, has also been used with enchytraeids (Westheide and Brockmeyer 1992; Schmelz 1996) showing constant patterns for *Enchytraeus* spp. and *Fridericia* spp. In the last 10 years, most molecular tools for species identification have focused on the study of mitochondrial DNA markers (mainly, cytochrome c oxydase subunit 1 -COI- gene sequence) and have successfully developed molecular barcodes for species identification (Hebert et al. 2003; Ratnasingham and Hebert 2007), a method that examines molecular differences at the species level. Studies on earthworms (Huang et al. 2007) and marine tubificids (Erséus and Kvist 2007) have provided evidence that a mitochondrial COI based identification system may be an efficient tool for oligochaete taxonomy in the future. For congeneric species, Erséus and Kvist (2007) have reported genetic divergence values of 19–25% in COI sequence, in contrast with intraspecific distances of usually less than 1%, thus allowing a 100% of tested species to be correctly identified (*i.e.* in agreement with the morphology-based identification). These results are promising and DNA barcoding may be a practical tool particularly for identifying species used in toxicological studies.

2.3.1 Naidids and Tubificids

The recently re-defined family Naididae (which includes the old family Tubificidae) includes more than 1,000 species, classified into eight subfamilies (Erséus et al. 2008; Martin et al. 2010), with about the 50% of the species being members of the Naidinae and the Tubificinae (Table 2.1). Members of these two subfamilies are the oligochaete worms with the greatest variation in chaetal form. Most freshwater species have hair-like and bifid chaetae with pectinations (thin teeth) between the two main prongs in the dorsal chaetal bundles and bifid chaetae in ventral bundles, although marine species generally have only bifid chaetae.

Many naidid species differ from other oligochaetes in that they inhabit the water column and graze on aquatic vegetation, rather than burrowing in the sediment, and a few are also predatory. They have adapted to this life style by developing eyes and long chaetae and have developed the ability to swim. All naidids (subfamilies Naidinae and Pristininae) are characterised by the position of the male pores in the anterior segments (V, VI or VIII) compared to segment XI in most tubificids.

The reproductive pattern in naidids alternates between sexual and asexual reproduction (mostly by paratomy or budding-off of zooids, but some by fragmentation). A few naidid species of the genus *Chaetogaster* are predaceous and commonly feed on other naidid species. While the free-living aquatic behaviour is typical for naidids, there are tropical species (*Pristina* spp.) that have adapted to living in soil (*e.g.* Collado and Schmelz 2001, 2002).

The subfamilies Phallodrilinae (reviewed by Erséus 1992) and Limnodriloidinae (reviewed by Erséus 1982) are mainly marine and contain numerous species, however, little work has been done with regard to water pollution. Some Phallodrilinae occur in subterranean freshwater habitats (*e.g.* Giani et al. 2001; Pinder et al. 2006), where it is likely that there are still many undiscovered species.

The small subfamily Telmatodrilinae has a cosmopolitan but scattered distribution (Alaska, California, Northern Europe, Siberia, Tasmania). The species inhabit a range of different freshwater habitats (rivers, marshes and lakes). The subfamily was reviewed by Holmquist (1974), although her proposals were not universally accepted (see Brinkhurst and Wetzel 1984).

Representatives of the subfamily Rhyacodrilinae are common in rivers, and in the marine environment. Baker and Brinkhurst (1981) and Finogenova (1982) reviewed the genus *Monopylephorus*, a common marine and estuarine genus that contains the species *M. rubroniveus* and *M. cuticulatus* and these have been used in ecotoxicity and bioaccumulation bioassays. The rhyacodriline *Branchiura sowerbyi* has been occasionally used in toxicity assessment of freshwater sediments and bioaccumulation studies, and is potentially a useful laboratory species relevant to tropical countries where the species is common in freshwater habitats. A new subfamily, the Rhyacodriloidinae, has recently been established for a small group of species (Martin et al. 2010).

Species of the subfamily Tubificinae are sediment-dwellers and are common in freshwater and estuarine benthic communities, with many being used as pollution bioindicators. Since 1971, several common genera of the subfamily Tubificinae have undergone taxonomic revision (see Brinkhurst and Wetzel 1984). The status of the old subfamily Aulodrilinae was revised by Giani et al. (1984) who proposed its inclusion into the Tubificinae. Two cosmopolitan tubificine species *Tubifex tubifex* and *Limnodrilus hoffmeisteri* are probably the most widely used tubificids in ecotoxicity and bioaccumulation studies (Fig. 2.3a–f). The taxonomic characters of *T. tubifex* were studied in detail by Holmquist (1983, 1985). More recently, several papers have been published on the study of molecular markers in different populations of *T. tubifex*. Sturmbauer et al. (1999) and Beauchamp et al. (2001) analysed mitochondrial-rDNA markers in different populations of this species in Northern Europe and North America, finding differences between lineages of 5–13% that the authors interpreted as evidence for the existence of cryptic species (*i.e.* species that have recently diverged into separate species that can only be identified for the present on molecular data: Bickford et al. 2007). Nevertheless, Bely and Wray (2004) have also reported great distances in the COI sequence (9–11%) between disjunct populations of a few common and widely distributed naidine taxa (*Nais variabilis*, *Paranais frici*, *P. litoralis* and *Stylaria lacustris*). Studies on other cosmopolitan oligochaete species have revealed similar results in the percent of divergence between intraspecific lineages (*e.g. Eisenia fetida*: Huang et al. 2007). Large amounts of molecular data are still unavailable and to date no standard levels of genetic divergence between species or even families have been established for distinguishing taxa (Will and Rubinoff 2004). Therefore, at present, one can only say that species such as *T. tubifex* show more within-species genetic diversity than is apparently normal, but how different this needs to be before species with this degree of diversity are considered more than a single species has yet to be determined. New gene or methodological approaches need to be developed to correctly evaluate intraspecific divergences in ubiquitous or peregrine species (Pop et al. 2007). Single-gene assays are promising tools for identifying individuals to species in support of identifications based on morphological characters, or for revealing inconsistencies between molecularly- and morphologically-defined species, but it should not be a primary criterion for recognizing species boundaries (Moritz and Cicero 2004). Such a decision would require a re-definition of the species-concept. Before molecular data can be used as a biological tool in species identification, more species need to be analysed to develop an understanding of the degree of variation that is normal within species of local as compared with cosmopolitan distribution. Further, questions on how polyploidy, parthenogenesis, latitude (*e.g.* influence of glaciations), or genetic drift may have affected molecular divergence in oligochaetes need to be considered.

2.3.2 Lumbriculids

All species of the family Lumbriculidae occur in freshwater habitats and their geographical distribution is restricted to the Holarctic region, except for *Lumbriculus variegatus* and *Stylodrilus heringianus* which have been recorded in the Southern hemisphere, likely through introduction. Both these species have been used for ecotoxicity and bioaccumulation studies, and *L. variegatus* has been included in several protocols. While *S. heringianus* is a species that is relatively easy to identify, from the presence of long penis hanging backwards, at an angle or parallel to the axis of the body (Fig. 2.3), this is not the case for *L. variegatus,* which is widely used in toxicology studies. *Lumbriculus variegatus* has a complex taxonomy because of the different patterns observed in the composition and location of reproductive organs. It comprises several subspecies distributed across the Holarctic region (Cook 1971) each defined by a relative frequency in the position of male pores and number of gonads in different segments. However, the high degree of intraspecific variation in this species and the fact that adult individuals in the same population of *L. variegatus* can show different patterns in composition and location of the sexual organs (gonads, and male and female ducts, number and position of spermathecae) (Mrázek 1906; Cook 1971; Timm 1979) makes the taxonomy of this species even more complicated and almost certainly the use of molecular labelling to identify the species will be required. The taxonomy of the species is further complicated by the fact that it reproduces primarily by asexual fragmentation (autotomy), thus identification is usually done on immature worms and is mainly based on the form of the chaetae. This is problematic as the chaetae are alike in most *Lumbriculus* species. In the laboratory, the species does not reproduce sexually (P. Egeler, T. Timm, and R. Collado, pers. comm.), but in the field mature individuals are occasionally found. Most recently, Gustafsson et al. (2009) have examined mitochondrial 16S, the COI and the nuclear ITS region in *L. variegatus* from 20 different sites in Europe, North America, Japan and Australia, including several particular culture stocks. Analyses concluded the existence of two clades (proposed as species, but unnamed) due to a mean genetic distance of 14.6% for 16S and the COI sequences (combined) or 17% for the COI gene. Therefore, until the taxonomy of the species used in toxicity tests is resolved, it would seem to be essential that researchers know the genetic strain and history of the specimens that they use or alternatively use organisms from a common strain. Further work should be undertaken to resolve the taxonomy of this species.

In the last 20 years, 12 new lumbriculid genera have been described, 7 of them in North America, *i.e. Phagodrilus* McKey-Fender and Fender 1988; *Tenagodrilus* Eckroth and Brinkhurst 1996; *Secumbelmis* Fend and Gustafson 2001; *Eremidrilus* Fend and Rodriguez 2003; *Pilaridrilus* Fend and Lenat 2007; *Martinidrilus* Fend and Lenat 2007 and *Altmanella* Fend 2009. The description of so many new genera illustrates the fact that knowledge of the family in North America in particular is still low and significant progress in taxonomy can still be anticipated. Several recent papers have contributed to the identification of species of the genera *Trichodrilus* (Rodriguez and Giani 1994), *Eclypidrilus* (Fend 2005), and *Rhynchelmis* (Fend and Brinkhurst 2000). Some of these genera are associated with river habitats where water quality is good and have been suggested as useful indicator taxa (Fend and Lenat 2007). There are two known predaceous genera of lumbriculids, *Agriodrilus* and *Phagodrilus* (see Brinkhurst and McKey-Fender 1991).

2.3.3 *Phreodrilids*

The family Phreodrilidae contains approximately 50 valid species, occurring mainly in Australia, where the family has been revised (Pinder and Brinkhurst 1997a). Its geographical distribution is largely limited to Southern Hemisphere, although three species have been described from collections in North Africa (Giani et al. 1995), the Arabian Peninsula (Martinez-Ansemil et al. 2002a) and Europe (Gunn et al. 2003). Most phreodrilids are sediment dwellers, although some species have been encountered amongst saturated moss and weeds (Pinder and Brinkhurst 1997a). Two species (*Schizodrilus*) are terrestrial. Phreodrilids are mostly associated with streams and wetland areas where very little pollution other than some nutrient enrichment or salinization occurs, and there are no species that are dominant in disturbed situations (Adrian Pinder, pers. comm.).

2.3.4 Enchytraeids

Of all the clitellate families, the family Enchytraeidae is found in the greatest range of habitats, and occurs in soil as well as in a wide variety of limnic, brackish and marine habitats (Erséus 2005). The revision of Brinkhurst and Jamieson (1971) only included a short chapter on the taxonomy of the family Enchytraeidae, and for a comprehensive taxonomy of this family in Europe one should still use the 1959 revision by Nielsen and Christensen. The exclusion of enchytraeids from many treatises and aquatic oligochaetes keys cannot be justified as approximately one third of the total enchytraeid species (Martin et al. 2008) inhabit aquatic habitats. Enchytraeid worms are not only an important part of most aquatic communities but they can also be dominant in subterranean (Giani et al. 2001) or extremely cold habitats (Bauer 2002). Many species (*e.g. Enchytraeus, Marionina*, or *Cernosvitoviella* species) are very thin, small worms (usually less than 1 cm long) that probably inhabit interstitial water in both aquatic and terrestrial habitats. There have been major contributions to the taxonomy of enchytraeids in the last 50 years (see Table 2.1), notably a recent revision of *Fridericia* by Schmelz (2003), a common genus in European rivers. The publication series *Newsletter on Enchytraeidae* has contributed to updating the taxonomy of this family since 1985.

2.3.5 Lumbricine Worms

Representatives of different lumbricine families are frequently found in some aquatic environments. They consist of microdrile (families Alluroididae and Tiguassidae) and megadrile species. The latter are commonly members of the crassoclitellate families Lumbricidae, Sparganophilidae, and Almidae. In European rivers it is not uncommon to find *Eiseniella tetraedra* Savigny in the benthic communities of all kinds of clean water habitats (rivers, lakes, caves, springs, both in high mountains and lowlands). Among the members of the family Sparganophilidae, *Sparganophilus tamensis* has been reported from aquatic habitats in Europe and North America, generally in lakes (Spencer 1980; Zicsi and Vaucher 1987), and less frequently in rivers (Omodeo 1984). *Sparganophilus pearsei* has been used in aquatic toxicology and bioaccumulation studies (Vidal and Horne 2003a, b). *Criodrilus lacuum* has been reported in aquatic habitats in Europe, the Near East (Moubayed et al. 1987) and North Africa (Boumaiza et al. 1986), and according to Brinkhurst and Jamieson (1971) is associated with both freshwater and brackish waters. Data on species diversity and the geographic distribution of families of megadrile oligochaetes present in aquatic habitats are provided by Martin et al. (2008).

In conclusion, the identification of the species in any applied branch of biology, including ecotoxicology, is not only necessary but a crucial step in guaranteeing the quality of work. Erroneous identification or the mixture of several species in a single culture can invalidate the results of a whole research project. Good taxonomy in biological studies is as critical as good laboratory practice in chemical laboratories, therefore, confirmation of the identity of the species by a specialized taxonomist, or by molecular analysis of COI gene sequence and verification in the Genbank are recommended practices. Researchers are also concerned about the nomenclature and various synonyms used in naming species. The taxonomic nomenclature of the annelids has undergone changes over the years particularly following modern phylogenetic analyses that have revised both genera and families. For the various synonymies, we strongly rely on the large revision of *Aquatic Oligochaeta of the World* by Brinkhurst and Jamieson (1971). This publication is out of press, but is still an indispensable reference work for the history of the synonymies of many aquatic oligochaete species (except enchytraeids). However, in the last 30 years many taxa have been further revised and new ones added, particularly marine species (see Brinkhurst and Wetzel 1984; Bisby et al. 2011). The correct identification of the common species used in toxicology and bioaccumulation studies is not usually problematic and can be achieved with some training. The identification of the many species present in field communities requires specific training and this is now possible using regional keys, such as those published for North America (Milligan 1997; Kathman and Brinkhurst 1998), South America (Brinkhurst and Marchese 1989), Northern and Central Europe (Timm and Veldhuijzen van Zanten 2002; Timm 2008), Lake Baikal (Semernoy 2004) and Southeastern Asia (Pinder and Ohtaka 2004). A checklist of the North American freshwater oligochaetes is also available from Wetzel et al. (2008). The taxonomy of marine species of microdrile oligochaetes has largely been resolved by Christer Erséus, who has lead the development of the systematics of marine oligochaetes over the last 35 years. The World Register of Marine Species (Appeltans et al. 2011) is a reference work where recent synonymies of marine taxa can be verified. Zoological systematics is nowadays subject to active revision of the classification of higher taxa down to the species level, as a result of the use of new molecular tools. Therefore, in the next few years we will probably see a process that will revise and re-configure the classification of oligochaetes. This is an expected result following the introduction of new methods although along with this process confusion and high nomenclatorial instability must be avoided.

References

Adiyodi KG (1988) Annelida. In: Adiyodi KG, Adiyodi RG (eds) Reproductive biology of invertebrates, Chapter 9. Wiley, Chichester, pp 189–250

Appeltans W, Bouchet P, Boxshall GA, Fauchald K, Gordon DP, Hoeksema BW, Poore GCB, van Soest RWM, Stöhr S, Walter TC, Costello MJ (eds) (2011) World Register of Marine Species. http://www.marinespecies.org

Avel M (1959) Classe des Annelides Oligochètes. In: Grassé PP (ed) Traité de Zoologie. V. Masson, Paris, pp 224–470

Baker HR, Brinkhurst RO (1981) A revision of the genus *Monopylephorus* and redefinition of the subfamilies Rhyacodrilinae and Branchiurinae (Tubificidae: Oligochaeta). Can J Zool 59:939–965

Bauer R (2002) Survival of frost and drought conditions in the soil by enchytraeids (Annelida; Oligochaeta) in Arctic, subalpine and temperate areas. Eur J Soil Biol 38:251–254

Beauchamp KA, Kathman RD, McDowell TS, Hedrick RP (2001) Molecular phylogeny of tubificid oligochaetes with special emphasis on *Tubifex tubifex* (Tubificidae). Mol Phylogenet Evol 19:216–224

Bely AE, Wray GA (2004) Molecular phylogeny of naidid worms (Annelida: Clitellata) based on cytochrome oxidase I. Mol Phylogenet Evol 30:50–63

Benham WB (1890) An attempt to classify earthworms. Q J Microsc Soc (NS) 31:201–315

Bickford D, Lohman DJ, Sodhi NS, Ng PKL, Meier R, Winker K, Ingram KK, Das I (2007) Cryptic species as a window on diversity and conservation. Trends Ecol Evol 22:148–155

Bisby FA, Roskov YR, Orrell TM, Nicolson D, Paglinawan LE, Bailly N, Kirk PM, Bourgoin T, Baillargeon G, Ouvrard D (eds) (2011) Species 2000 & ITIS Catalogue of Life. www.catalogueoflife.org. Species 2000: Reading, UK

Boumaiza M, Martinez-Ansemil E, Giani N (1986) Les Oligochètes et Aphanoneura des eaux courants de Tunisie, I – Données faunistiques. Annls Limnol 22:231–237

Brinkhurst RO (1981) A contribution to the taxonomy of the Tubificinae (Oligochaeta, Tubificidae). Proc Biol Soc Wash 94:1048–1067

Brinkhurst RO (1982) Evolution of the Annelida. Can J Zool 60:1043–1059

Brinkhurst RO (1988) A taxonomic analysis of the Haplotaxidae (Annelida, Oligochaeta). Can J Zool 66:2243–2252

Brinkhurst RO (1991) A phylogenetic analysis of the Phreodrilidae (Annelida, Oligochaeta), with a description of a new species. Can J Zool 69:2031–2040

Brinkhurst RO, Jamieson BGM (1971) The aquatic Oligochaeta of the world. Oliver & Boyd, Edinburgh, 860 pp

Brinkhurst RO, Marchese M (1989). Guía para la Identificación de Oligoquetos Acuáticos Continentales de Sud y Centro America. Colección Climax No.6, 2nd edn. Asociación Ciencias Naturales del Litoral (ACNL), Sto. Tomé, 179 pp

Brinkhurst RO, McKey-Fender D (1991) The anatomy of the pharynx of two predatory aquatic oligochaetes. Can J Zool 69:669–675

Brinkhurst RO, Wetzel MJ (1984) Aquatic Oligochaeta of the world: supplement. A catalogue of new freshwater species, descriptions and revisions. Canadian technical report of hydrography and ocean sciences 44, 101 pp

Christensen B, Theisen BF (1998) Phylogenetic status of the family Naididae (Oligochaeta, Annelida) as inferred from DNA analyses. J Zool Syst Evol Res 36:169–172

Coates KA (1986) Redescription of the oligochaete genus *Propappus* and diagnosis of the new family Propappidae (Annelida, Oligochaeta). Proc Biol Soc Wash 99:417–428

Coates KA, Locke JM, Healy B, Wetzel MJ (2006) Families Enchytraeidae and Propappidae. In: The aphanoneuran and clitellate Annelida occurring in the United States and Canada. www.inhs.uiuc.edu/~mjwetzel/EnchytraeidaeNA.html

Coler RA, Gunner HB, Zuckermann BM (1967) Selective feeding of tubificids on bacteria. Nature 216:1143–1144

Collado R, Schmelz RM (2001) Descriptions of three *Pristina* species (Naididae, Clitellata) from Amazonian forest soils, including *P. marcusi* sp. nov. Hydrobiologia 463:1–11

Collado R, Schmelz R (2002) *Pristina trifida* sp.nov., a new soil-dwelling microannelid (Oligochaeta: Naididae) from Amazonian forest soils, with comments on species recognition in the genus. Zootaxa 118:1–14

Cook DG (1971) Lumbriculidae. In: Brinkhurst RO, Jamieson BGM (eds) Aquatic Oligochaeta of the world. Oliver & Boyd, Edinburgh, pp 200–285

Diaz RJ, Erséus C, Boesch DF (1987) Distribution and ecology of Middle Atlantic Bight Oligochaeta. Hydrobiologia 155:215–225

Eckroth MC, Brinkhurst RO (1996) *Tenagodrilus musculus*, a new genus of Lumbriculidae (Clitellata) from a temporary pond in Alabama, USA. Hydrobiologia 334:1–9

Envall I, Källersjö M, Erséus C (2006) Molecular evidence for the non-monophyletic status of Naidinae (Annelida, Clitellata, Tubificidae). Mol Phylogenet Evol 40:570–584

Erséus C (1982) Taxonomic revision of the marine genus *Limnodriloides* (Oligochaeta, Tubificidae). Verh naturwiss Ver Hamburg 25:207–277

Erséus C (1990) Cladistic analysis of the subfamilies within the Tubificidae (Oligochaeta). Zool Scr 19:57–63

Erséus C (1992) A generic revision of the Phallodrilinae (Oligochaeta, Tubificidae). Zool Scr 21:5–48

Erséus C (1993) Taxonomy of *Capilloventer* (Capilloventridae), a little-known group of aquatic Oligochaeta. J Nat Hist 27:1029–1040

Erséus C (1999) *Parvidrilus strayeri*, a new genus and species, an enigmatic interstitial clitellate from underground waters in Alabama. Proc Biol Soc Wash 112:327–337

Erséus C (2005) Phylogeny of oligochaetous Clitellata. Hydrobiologia 535(536):357–372

Erséus C, Gustavsson L (2002) A proposal to regard the former family Naididae as a subfamily within Tubificidae (Annelida, Clitellata). Hydrobiologia 485:253–256

Erséus C, Källersjö M (2004) 18S rDNA phylogeny of Clitellata (Annelida). Zool Scr 33:187–196

Erséus C, Kvist S (2007) COI variation in Scandinavian marine species of *Tubificoides* (Annelida: Clitellata: Tubificidae). J Mar Biol Assoc UK 87:1121–1126

Erséus C, Prestegaard T, Källersjö M (2000) Phylogenetic analysis of Tubificidae (Annelida, Clitellata) based on 18S rDNA sequences. Mol Phylogenet Evol 15:381–389

Erséus C, Källersjö M, Ekman M, Hovmöller R (2002) 18S rDNA phylogeny of the Tubificidae (Clitellata) and its constituent taxa: dismissal of the Naididae. Mol Phylogenet Evol 22:414–422

Erséus C, Wetzel M, Gustavsson L (2008) ICZN rules – a farewell to Tubificidae (Annelida, Clitellata). Zootaxa 1744:66–68

Fend SV (2005) A review of the genus *Eclipidrilus* (Annelida:Clitellata: Lumbriculidae), with description of new species of western North America. Zootaxa 969:1–42

Fend SV (2009) An evaluation of the genus *Kincaidiana* Altman, 1936, with the designation of *Altmanella* n.gen. (Annelida, Clitellata, Lumbriculidae). Zootaxa 2077:1–30

Fend SV, Brinkhurst RO (2000) New species of *Rhynchelmis* (Clitellata, Lumbriculidae), with observations on the Nearctic species. Hydrobiologia 428:1–59

Fend SV, Gustafson DL (2001) A new lumbriculid genus and species from North America (Clitellata, Lumbriculidae). Hydrobiologia 463:13–22

Fend SV, Lenat DR (2007) Two new genera of Lumbriculidae (Annelida, Clitellata) from North Carolina, USA. Zootaxa 1666:1–22

Fend SV, Rodriguez P (2003) *Eremidrilus* n. gen. (Annelida, Clitellata, Lumbriculidae) and new species from California, U.S.A. Can J Zool 81:515–542

Ferraguti M, Erséus C (1999) Sperm types and their use for a phylogenetic analysis of aquatic clitellates. Hydrobiologia 402:225–237

Finogenova NP (1982) Revision of the genus *Monopylephorus* (Oligochaeta, Tubificidae). Zool Zh 61:5–17

Gelder S (1980) A review of the symbiotic Oligochaeta (Annelida). Zool Anz 204:69–82

Giani N, Martinez-Ansemil E, Brinkhurst RO (1984) Révision du status taxonomique des Aulodrilinae (Tubificidae, Oligochaeta). Bull Soc Hist Nat Toulouse 120:17–22

Giani N, Martin P, Juget J (1995) A new species of Phreodrilidae (Oligochaeta), *Astacopsidrilus naceri* sp.nov. from Morocco (North Africa), with notes on the biogeography of the family. Can J Zool 73:2375–2381

Giani N, Sambugar B, Rodriguez P, Martinez-Ansemil E (2001) Oligochaetes in southern European groundwater. Hydrobiologia 463:65–74

Glasby CJ, Timm T (2008) Global diversity of polychaetes (Polychaeta, Annelida) in freshwater. Hydrobiologia 595:107–115

Graefe U, Römbke J (1985) Checklist of new taxa 1960–1985. Newsl Enchytraeidae 1:1–20

Gunn RJM, Pinder AM, Walker BM (2003) Phreodrilidae (Annelida: Oligochaeta) a family new to Europe. Ir Nat J 27:315–317

Gustafsson DR, Proce DA, Erséus C (2009) Genetic variation in the popular lab worm *Lumbriculus variegatus* (Annelida: Clitellata: Lumbriculidae) reveals cryptic speciation. Mol Phylogenet Evol 51:182–189

Harman WJ, Loden MS (1978) A re-evaluation of the Opistocystidae (Oligochaeta) with descriptions of two new genera. Proc Biol Soc Wash 91:453–562

Harper RM, Fry JC, Learner MA (1981a) A bacteriological investigation to elucidate the feeding biology of *Nais variabilis* (Oligochaeta, Naididae). Freshwater Biol 11:227–237

Harper RM, Fry JC, Learner MA (1981b) Digestion of bacteria by *Nais variabilis* (Oligochaeta) as established by autoradiography. Oikos 36:211–218

Hebert PDN, Cywinska A, Ball SL, Dewaard JR (2003) Biological identification through DNA barcodes. Proc R Soc Lond B 270:313–321

Holmquist C (1974) On *Alexandrovia onegensis* Hrabě from Alaska, with a revision of the Telmatodrilinae (Oligochaeta, Tubificidae). Zool Jb Syst 101:249–268

Holmquist C (1983) What is *Tubifex tubifex* (O.F. Müller) (Oligochaeta, Tubificidae)? Zool Scr 12:187–201

Holmquist C (1985) A revision of the genera *Tubifex* Lamark, *Ilyodrilus* Eisen, and *Potamothrix* Vejdovsky & Mrázek (Oligochaeta, Tubificidae), with extensions to some connected genera. Zool Jb Syst 112:311–366

Huang J, Xu Q, Sun ZJ, Tang GL, Su ZY (2007) Identifying earthworms through DNA barcodes. Pedobiologia 51:301–309

Jamieson BGM (1981) The ultrastructure of the Oligochaeta. Academic, London, 462 pp

Jamieson BGM (1988) On the phylogeny and higher classification of the Oligochaeta. Cladistics 4:367–410

Jamieson BGM (2006) Non-leech Clitellata (with contributions by M Ferraguti). In: Rouse G, Pleijel F (eds) Reproductive biology and phylogeny of Annelida, vol 4, Jamieson BGM (Series ed). Science Publishers, Enfield, pp 235–392

Jamieson BGM, Tillier S, Tillier A, Justine J-L, Ling E, James S, McDonald K, Hugall AF (2002) Phylogeny of the Megascolecidae and Crassiclitellata (Annelida, Oligochaeta): combined versus partitioned analysis using nuclear (28S) and mitochondrial (12S, 16S) rDNA. Zoosystema 24:707–734

Kathman D, Brinkhurst RO (1998) Guide to the freshwater oligochaetes of North America. Aquatic Resources Center, Tennessee, 264 pp

Martin P (2001) On the origin of the Hirudinea and the demise of the Oligochaeta. Proc R Soc Lond B 268:1089–1098

Martin P, Kaygorodova I, Sherbakov D, Verheyen E (2000) Rapidly evolving lineages impede the resolution of phylogenetic relationships among Clitellata (Annelida). Mol Phylogenet Evol 15:355–368

Martin P, Martinez-Ansemil E, Pinder A, Timm T, Wetzel MJ (2008) Global diversity of oligochaetous clitellates ("Oligochaeta"; Clitellata) in freshwater. Hydrobiologia 595: 117–127

Martin P, Martinez-Ansemil E, Sambugar B (2010) The Baikalian genus *Rhyacodriloides* in Europe: phylogenetic assessment of Rhyacodriloidinae subfam. n. within the Naididae (Annelida). Zool Scr 39:462–482

Martinez-Ansemil E, Giani N, Sambugar B (2002a) Oligochaetes from underground waters of Oman with descriptions of two new species of Phreodrilidae (Oligochaeta): *Antarctodrilus arabicus* n.sp. and *Phreodrilus stocki* n.sp. Contrib Zool 71:147–158

Martinez-Ansemil E, Sambugar B, Giani N (2002b) First record of Parvidrilidae (Annelida, Oligochaeta) with a description of a new species (*Parvidrilus spelaeus* sp.nov.) and comments on the family and its phyletic relationships. J Zool Lond 256:495–503

McKey-Fender D, Fender WM (1988) *Phagodrilus* gen. nov. (Lumbriculidae) systematics and biology of a predaceous oligochaete from western North America. Can J Zool 66:2304–2311

Milbrink G, Nyman L (1973) On the protein taxonomy of aquatic Oligochaeta. Zoon 1:29–35

Milligan M (1997) Identification manual for the aquatic Oligochaeta of Florida, vol 1, Freshwater oligochaetes. vol 2, Estuarine and nearshore marine oligochaetes. South Florida Water

Management District in cooperation with Florida Department of Environmental Protection and Riverwoods Field Laboratory
Moritz C, Cicero C (2004) DNA barcoding: promise and pitfalls. PLoS Biol 2:28–39
Moubayed Z, Giani N, Martinez-Ansemil E (1987) Distribution of aquatic Oligochaeta and Aphanoneura in the Near East. In: Krupp F, Schneider W, Kinzelbach R (eds) Proceedings of the symposium on the fauna and zoogeography of the Middle East, Mainz, 1985, pp 78–89
Mrázek A (1906) Die Geschlechtsverhältnisse und die Geschleschtsorgane von *Lumbriculus variegatus* Gr. Zool Jahrb (Anat) 23:381–462
Nielsen CO, Christensen B (1959) The Enchytraeidae, critical revision and taxonomy of European species. Natura Jutl 8–9:1–160
Nielsen CO, Christensen B (1961) The Enchytraeidae, critical revision and taxonomy of European species. Natura Jutl 10(Suppl 1):1–19
Nielsen CO, Christensen B (1963) The Enchytraeidae, critical revision and taxonomy of European species. Natura Jutl 10(Suppl 2):1–23
Omodeo P (1984) On aquatic Oligochaeta Lumbricomorpha in Europe. Hydrobiologia 155:187–190
Omodeo P (1987) Some new species of Haplotaxidae (Oligochaeta) from Guinea and remarks on the history of the family. Hydrobiologia 155:1–13
Omodeo P (1996) *Kathrynella*, a new oligochaete genus from Guyana. Hydrobiologia 334:11–15
Omodeo P, Coates KA (2001) New alluroidids (Annelida, Clitellata) from Guyana. Hydrobiologia 463:39–47
Pinder AM, Brinkhurst RO (1997a) Review of the Phreodrilidae (Annelida, Oligochaeta, Tubificida) of Australia. Invertebr Taxonomy 11:443–523
Pinder AM, Brinkhurst RO (1997b) The family Capilloventridae (Annelida, Clitellata) in Australia with description of two new species of *Capilloventer*. Zool Scr 26:255–265
Pinder A, Ohtaka A (2004) Annelida: Clitellata, Oligochaeta. In: Yule CM (ed) Freshwater invertebrates of the Malaysian Region. Academy of Sciences of Malaysia, Kuala Lumpur, pp 162–174
Pinder AM, Eberhard SM, Humphreys WF (2006) New phallodrilines (Annelida: Clitellata: Tubificidae) from western Australian groundwater. Zootaxa 1304:31–48
Pop AA, Cech G, Wink M, Csuzdi C, Pop VV (2007) Application of 16S, 18S rDNA and COI sequences in the molecular systematics of the earthworm family Lumbricidae (Annelida, Oligochaeta). Eur J Soil Biol 43:543–552
Purschke G, Westheide W, Rohde D, Brinkhurst RO (1993) Morphological reinvestigation and phylogenetic relationship of *Acanthobdella peledina* (Annelida, Clitellata). Zoomorphology 113:91–101
Ratnasingham S, Hebert PDN (2007) BOLD: the Barcode of Life Data System (www.barcodinglife.org). Molecular Ecology Notes 7: 355–364
Righi G, Varela ME (1983) *Narapa bonettoi* gen. nov., sp. nov. (Oligochaeta, Narapidae, Fam. Nov.) de agua doce da Argentina. Rev Asoc Cienc Nat Litoral 14:7–15
Righi G, Ayres I, Bittencourt ECR (1978) Oligochaeta (Annelida) do Instituto Nacional de Pesquisas da Amazonia. Acta Amazonica 8(Suppl 1):1–49
Rodriguez P, Giani N (1994) A preliminary review of the taxonomic characters used for the systematics of the genus *Trichodrilus* Claparède (Oligochaeta, Lumbriculidae). Hydrobiologia 278:35–51
Römbke J (1992) New taxa since 1985. Newsl Enchytraeidae 3:14–15
Römbke J, Dozsa-Farkas K (1996) New taxa since 1992. Newsl Enchytraeidae 5:5–9
Rouse GW, Fauchald K (1995) The articulation of annelids. Zool Scr 24:269–301
Särkkä J (1996) Meiofauna of the profundal zone of the northern part of Lake Ladoga as an indicator of pollution. Hydrobiologia 322:29–38
Schmelz R (1996) Species separation and identification in the Enchytraeidae (Oligochaeta, Annelida): combining morphology with general protein data. Hydrobiologia 334:31–36
Schmelz R (2003) Taxonomy of Fridericia (Oligochaeta, Enchytraeidae): revision of species with morphological and biochemical methods. Goecke & Evers, Keltern-Weiler, p 415 (Germany)

Schmelz R, Römbke J (1999) New enchytraeid taxa since 1996. Newsl Enchytraeidae 6:9–10
Schoch-Bösken JS, Römbke J (1993) Bibliography of the Enchytraeidae (1950–1991). Acta Biologica Benrodis, Supplementband 1, 76 pp
Semernoy VP (2004) Oligochaeta of Lake Baikal. In: Timm T, Mazapova GF, Timoshkin OA (eds) Russian Academy of Sciences, Siberian Division, Novosibirsk "Nauka", 527 pp
Siddall ME, Apakupakul K, Burreson EM, Coates KA, Erséus C, Gelder SE, Källersjö M, Trapido-Rosenthal H (2001) Validating Livanow: molecular data agree that leeches, branchiobdellidans and *Acanthobdela peledina* form a monophyletic group of oligochaetes. Mol Phylogenet Evol 21:346–351
Spencer DR (1980) The aquatic Oligochaeta of St. Lawrence Great Lakes region. In: Brinkhurst RO, Cook DG (eds) Aquatic oligochaete biology. Plenum Press, New York, pp 115–164
Stephenson J (1930) The Oligochaeta. Clarendon Press, Oxford, 978 pp
Struck TH, Purschke G (2005) The sister group relationship of Aeolosomatidae and Potamodrilidae (Annelida: "Polychaeta" – a molecular phylogenetic approach based on 18S rDNA and cytochrome oxidase I. Zool Anz 243:281–293
Struck TH, Hessling R, Purschke G (2002) The phylogenetic position of the Aeolosomatidae and Parergodrilidae, two enigmatic oligochaete-like taxa of the 'Polychaeta', based on molecular data from 18S rDNA sequences. J Zool Syst Evol Res 40:155–163
Sturmbauer C, Opadiya GB, Niedersträtter H, Riedmann A, Dallinger R (1999) Mitochondrial DNA reveals cryptic oligochaete species differing in cadmium resistance. Mol Biol Evol 16:967–974
Timm T (1979) The Estonian Lumbriculidae. Eesti NSV Teaduste Akad Toimetised (Biol) 28:158–171
Timm T (1981) On the origin and evolution of aquatic Oligochaeta. Eesti NSV Teaduste Akad Toimetised (Biol) 30:174–181
Timm T (2008) A guide for the freshwater Oligochaeta and Polychaeta of Northern and Central Europe. Int J Faunistics Floristics Eur Inland Waters 66:1–235 (+102 plates)
Timm T, Veldhuijzen van Zanten HH (2002) Freshwater Oligochaeta of North-West Europe. World biodiversity database. CD-ROM Series. Biodiversity Center of ETI, Multimedia Interactive Software, version 1.0
Vidal DE, Horne AJ (2003a) Mercury toxicity in the aquatic Oligochaete *Sparganophilus pearsei*: I. Variation of resistance among populations. Arch Environ Contam Toxicol 45:184–189
Vidal DE, Horne AJ (2003b) Mercury toxicity in the aquatic oligochaete *Sparganophilus pearsei* II: autotomy as a novel form of protection. Arch Environ Contam Toxicol 45:462–467
Wavre H, Brinkhurst RO (1971) Interactions between some tubificid oligochaete and bacteria found in the sediments of Toronto Harbour, Ontario. J Fish Res Bd Can 28:335–341
Westheide W (1971) Intertitial Polychaeta (excluding Archiannelida). Smithson Contrib Zool 76:57–70
Westheide W (1997) The direction of evolution within the Polychaeta. J Nat Hist 31:1–15
Westheide W, Brockmeyer V (1992) Suggestions for an index of enchytraeid species (Oligochaeta) based on general protein patterns. Zeits f zool Syst Evolutionfors 30:89–99
Wetzel MJ, Kathman D, Fend S, Coates KA (2008) Classification and checklist of the freshwater oligochaetes occurring in North America and North of Mexico. www.inhs.uiuc.edu/~mjwetzel/FWOligoNAChklst.html
Will KW, Rubinoff D (2004) Myth of the molecule: DNA barcodes for species cannot replace morphology for identification and classification. Cladistics 20:47–55
Zicsi A, Vaucher C (1987) On the occurrence of *Sparganophilus tamiensis* Benham in the lake Genève. Rev Suisse Zool 94:861–864

Chapter 3
Ecology and Field Studies

Abstract In this chapter we have attempted to provide the reader with an overview of the ecological response of aquatic oligochaetes to anthropogenic disturbance. There is a long history of the use of oligochaete community structure in assessment studies which has demonstrated that distribution patterns exist that are associated with pollution. There is a considerable range in species response both to nutrient contamination and metals and organic chemicals. Some species assemblages have been correlated with stressors, but species assemblage composition can change regionally. Recent multivariate statistics are particularly useful in identifying those responses to disturbance and synthesizing them in an easily presented and understood way. The traditional view of oligochaetes as a group associated with polysaprobic conditions needs reconsideration, as most oligochaete species are associated with oligo- and β-mesosaprobic conditions and the species are distributed along the entire range of ecological classes of water quality. The use of tolerance values to weight relative abundance data fails to distinguish between stress factors, and the values correspond primarily to organic enrichment tolerance. Worm species do not respond in the same way to all types of contaminant which is implied by the general use of the tolerance value approach. To identify specific causative agents and causality requires tailored site specific studies and will often require an experimental approach. We do not see focussing on oligochaetes alone in field community based approaches to be a useful strategy, as information on the entire community can just as easily be obtained and provides richer information. However, neither do we recommend ignoring oligochaetes or treating all the species as a single group because of the great response range exhibited within the group, and the fact that in many habitats they comprise the major part of the community. It is necessary to improve the training, at present largely neglected, of benthic zoologists to include knowledge on the most common oligochaete species.

The true test of any assessment method lies in the extent of the loss of valuable information while attempting to render results convincing to non-biologists

R.O. Brinkhurst

P. Rodriguez and T.B. Reynoldson, *The Pollution Biology of Aquatic Oligochaetes*,
DOI 10.1007/978-94-007-1718-3_3, © Springer Science+Business Media B.V. 2011

3.1 Introduction

In this chapter we have tried to provide the reader with an overview of the ecological responses of aquatic oligochaetes to anthropogenic disturbance by reviewing studies, largely at the population and community level, over the past 100 years. Arguably, the association of aquatic oligochaetes with aquatic pollution was observed more than 2000 years ago by Aristotle (384–322 BC) who identified the presence of red threads (probably tubificid worms) dominating in organically polluted situations. However, there was little ecological literature on oligochaetes published prior to 1960, other than their inclusion in the saprobic system, developed at the beginning of the twentieth century. The first quantitative pollution surveys that included aquatic oligochaetes were published in the 1960s (Goodnight and Whitley 1960; Brinkhurst 1965a, 1966a, b; Brinkhurst and Kennedy 1965), and over the next 40 years there has been a gradual evolution in the approaches used to understand the response of aquatic ecosystems in general to anthropogenic stress. These include new approaches to population and community field studies, new data analysis and interpretation methods, the use of biomarkers, and bioaccumulation studies. We have approached our synthesis of the oligochaete literature on population and community responses to anthropogenic disturbance by examining different types of stresses. The large majority of studies have investigated the response to organic enrichment and eutrophication (Table 3.1). We have examined both the overall utility of both indicator species and community assemblages as well as some stress specific responses.

3.2 Freshwater Community Studies

One of the earliest modern community level studies of the response of oligochaetes to gross pollution was conducted by Brinkhurst in a series of quantitative surveys, beginning in 1958. The River Derwent in Derbyshire, England, was a grossly polluted river and subjected to so many abuses that the river itself failed to meet the local legislative standard for sewage effluent. With the installation and operation of

Table 3.1 Summary of ecological studies addressing the response of aquatic oligochaetes to various types of stress

Stress	Number of publications reviewed
General	21
Eutrophication and organic enrichment	71
Metals	14
Organic chemicals	9
Sediments	8
Thermal stress	2
Salinity	2
Miscellaneous	3

Table 3.2 Number of tubificid species recorded from the River Derwent in 1958 and in 1959–1962 (From Brinkhurst 1966b)

River Derwent	*Sites*	*Tt*	*Lh*	*It*	*Li*	*Lu*	*Lc*	*Ph*	*Pm*	*Rc*	*Bv*	*Pb*	*Ap*	S	S max per survey
Before	*4*	+	+	–	–	–	–	–	–	–	–	–	–	2	2
improvement	*6*	+	+	–	–	–	–	–	–	–	–	–	–	2	2
(2 surveys)	*7*	+	+	–	–	–	–	–	–	–	–	–	–	2	2
	13	+	+	–	–	–	–	–	–	–	–	–	–	2	2
After	*4*	+	+	–	–	–	–	+	+	+	–	+	–	6	5
improvement	*6*	+	+	–	+	+	+	+	–	+	+	+	–	9	7
(7 surveys)	*7*	+	+	+	–	–	+	+	–	+	+	+	–	8	5
	13	+	+	–	–	+	+	+	–	–	+	+	+	8	5

Abbreviations: *Ap: A. pluriseta, Bv: B. vejdovskyanum, It: I. templetoni, Li: L. ignota, Lc: L. claparedianus, Lh: L. hoffmeisteri, Lu: L. udekemianus, Pb: P. barbatus, Ph: P. hammoniensis, Pm: P. moldaviensis, Rc: R. coccineus, Tt: T. tubifex, S:* number of species

new sewage treatment facilities, following a successful prosecution of the many parties involved, the river improved very rapidly to the point where it became a "normal degraded" stream. Engineering predictions had suggested a slow return to this condition, based on the deep layers of accumulated sludge on the river bed. These, however, were quite quickly flushed downstream, indicating their transitory nature. In 1958 there were only tubificids and some Orthocladinae chironomids below the major effluent, and at subsequent stations there were also some naidid and enchytraeid worms. There was a dramatic change in the tubificid species recorded from 1958, prior to the improvement, and in 1959 and subsequent years (Table 3.2). The fauna of the River Derwent after recovery included members of the genera *Potamothrix* (reported as *Euilyodrilus), Psammoryctides* (as *Psammoryctes*) and *Aulodrilus*, but also genera later considered typical of more clean water assemblages, such as *Rhyacodrilus* and *Bothrioneurum.* This study was published in detail (Brinkhurst 1965a), and was referred to in a number of early summaries on the ecology of aquatic oligochaetes and prompted early schemes to investigate the response of aquatic oligochaetes to anthropogenic stress (Brinkhurst 1966a, b; Brinkhurst and Cook 1974). A number of life history studies were also undertaken at that time to determine if oligochaete species could be expected to occur throughout the year or if sampling should take place at specific times. It was quickly recognized that worms could be categorized based on their maturity (immature, mature, mature with sperm in the spermathecae or mated worms with or without eggs). In some species, cohorts could be recognized with some difficulty but worms generally seemed to breed once a year, although some species have more than one reproductive period. Work with cultures indicated that there might be several generations a year which would be undetectable from field samples taken at monthly intervals. It was apparent that tubificids remained in benthic communities year round (Brinkhurst 1966a) and tolerant and adaptable species such as *Limnodrilus hoffmeisteri* did not have fixed seasonal reproductive patterns, but each population could be breeding at any time of the year (Kennedy 1966a, b) if conditions were suitable.

The oligochaetes (especially the tubificids) are important as primary consumers, primary decomposers, modifiers of the substrate, and as food for predators (Schwank 1982). They can constitute the greatest proportion of the macrofauna present in natural freshwater systems and they are particularly useful as biological indicators of pollution because of their high tolerance to a wide range of variation in environmental quality, and additionally have a life cycle that is completed in the aquatic system and are, therefore, only exposed to contaminants from aquatic sources, providing good site specificity. Aquatic oligochaete species cover a wide spectrum of distribution and tolerance in the aquatic environment compared with other taxa whose distribution is much restricted.

At an early meeting on water quality criteria held in Cincinnati in 1962, Brinkhurst first outlined the biology of tubificid oligochaetes in relation to water pollution (Brinkhurst 1965b). He outlined three basic approaches for the identification of pollution using aquatic oligochaetes: (1) establishing tolerance limits for individual species; (2) searching for indicator species; and (3) analysing the community structure. The three approaches suggested by Brinkhurst (1965b) are still appropriate to the understanding of much of the information that follows in this chapter, since most contributions fit one of those three categories. Largely based on his early work in the UK, Brinkhurst summarised the knowledge at that time on the sensitivity of various species. He stated that there was no universal indicator species of aquatic oligochaete, the presence of which could be associated with a particular degree of pollution. However, he summarized a few consistent features of species assemblages associated with organic pollution: (1) a *Tubifex tubifex – Limnodrilus hoffmeisteri* species complex which are the species most resistant to organic and inert mineral pollution; (2) *Potamothrix* (as *Euilyodrilus*) *hammoniensis, Psammoryctides barbatus,* and *Limnodrilus udekemianus* which appeared to occur in organically enriched areas, but were apparently less tolerant than the *Tubifex – Limnodrilus* complex; (3) *Rhyacodrilus coccineus* which appeared to be associated with sandy sediment, and tolerant of slightly enriched areas, and; (4) other species occurring in only clean sites. This grouping of species was based strictly on empirical data available at that time. We have re-analyzed those data presented by Brinkhurst from three studies presented in his 1962 paper and using more quantitative analysis, ordination (MDS in PRIMER 6) methods, and there is little evidence of those Brinkhurst groupings.

The association among the oligochaete species (Fig. 3.1) shows the two pollution species associations reported by Brinkhurst (1 and 2) to be in the centre of the ordination and the species occurring at clean sites (4) encompass Brinkhurt's polluted associations. Furthermore, *Spirosperma ferox* which is considered a more sensitive species is closely associated with *L. hoffmeisteri*. ANOSIM shows no significance to the species associations proposed by Brinkhurst. This provides little support for the concept of indicator species-associations among the oligochaetes, as reported at that time.

The concept of indicator species has extended from their use as presence/absence of certain species to other more biologically meaningful concepts (*e.g.* reduction in abundance which can be interpreted as reproductive impairment, or species

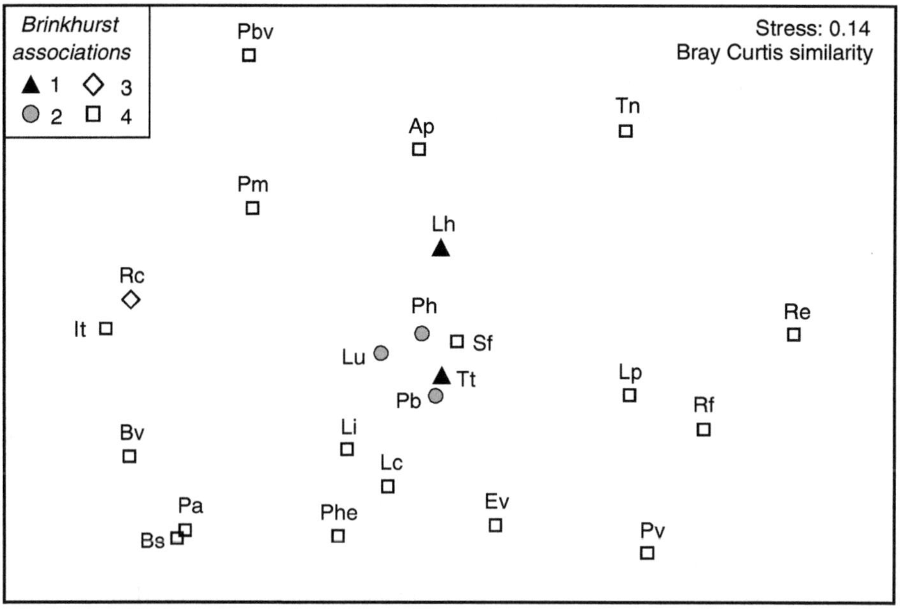

Fig. 3.1 Re-analysis of Brinkhurst (1965b) data from the United Kingdom, ordination showing four groups identified by Brinkhurst. Abbreviations: *Aulodrilus pluriseta* (Ap), *Bothrioneurum vejdovskyanum* (Bv), *Branchiura sowerbyi* (Bs), *Embolocephalus velutinus* (Ev), *Ilyodrilus templetoni* (It), *Limnodrilus hoffmeisteri* (Lh), *L. claparedianus* (Lc), *L. profundicola* (Lp), *L. udekemianus* (Lu), *Lophochaeta ignota* (Li), *Potamothrix hammonienis* (Ph), *P. heuscheri* (Phe), *P. moldaviensis* (Pm), *P. bavaricus* (Pbv), *P. vejdovskyi* (Pv), *Psammoryctides albicola (Pa), P. barbatus* (Pb), *Rhyacodrilus coccineus* (Rc), *R. ekmani* (Re), *R. falciformis* (Rf), *Spirosperma ferox* (Sf), *Tubifex newaensis* (Tn), *T. tubifex* (Tt)

assemblages). Brinkhurst and Kennedy (1965) investigated the biology of a small stream filled with fine particles of coal and the effluent from a pig farm. These studies all revealed that *T. tubifex* and *L. hoffmeisteri* were the most abundant tubificids in grossly polluted sites, and that other *Limnodrilus* species and *Ilyodrilus* (as *Tubifex*) *templetoni* were the first species to be added to the list in slightly less extreme situations.

The impression gained from examining species lists from lakes at that time was that worm species did not seem to be associated with different degrees of eutrophication. Nevertheless, if a quantitative approach was adopted, it became apparent that some species, like *Potamothrix hammoniensis, Spirosperma ferox* and *Embolocephalus velutinus* might be useful in this regard. A survey of the lakes in English Lake District found that the tubificid fauna of a lake could not be established by sampling at only one location, and that there was no strong correlation between species presence and the marked trophic gradient in these or other European lakes (Brinkhurst 1964a). Reynoldson (1990) revisited the English Lake District surveyed rather superficially 25 years earlier. The resulting data was analyzed by factor analysis and classification and showed a strong correlation between tubificid abundance and primary productivity, and individual species could be related to trophic state. However, distinct communities were not recognized, nor were any strong species associations identified.

Aston (1973) reviewed some of the literature on tubificids and water quality of that period, citing a great deal of the early North American work. He concluded that the most polluted situations contained the usual species group *T. tubifex* and *L. hoffmeisteri*, and that other species were "eliminated in a rough sequence". He also suggested respiratory physiology and life history adaptations as the basis for this observation.

Studies of North American habitats using data on identified worm taxa began in the late 1960s. Prior to this, the absence of species level keys prevented any detailed work and the percentage contribution of oligochaetes to the community (Goodnight and Whitley 1961) was the only method of incorporating oligochaetes in ecosystem assessment. In extreme situations, such as the outward expansion of areas in western Lake Erie dominated by tubificids from 1930 to the mid 1960s adjacent to the major inflows from the Detroit, Maumee, and Raisin rivers, this approach was sufficient (Carr and Hiltunen 1965). However, it is somewhat depressing that this very simple expression is still used as a "metric" in some assessment methods when far more information can be gained from the examination of the composition of the oligochaete community. Following the development of taxonomic keys to most of the families (Brinkhurst 1964b, 1965c; Brinkhurst and Cook 1966), the distribution patterns of species in the St. Lawrence Great Lakes area in particular could be established (Table 3.3). Hiltunen (1969) also reported the specific patterns present in those contaminated regions such as western Lake Erie. Work on the St. Lawrence Great Lakes with species level identification began with representative samples drawn from extensive surveys executed by the Bureau of Commercial Fisheries, Ann Arbor, and the Great Lakes Institute of the University of Toronto. The Ann Arbor material was used to establish the taxonomy and distribution of species in the Great Lakes. The Toronto samples had been collected with a Franklin dredge and many had been sorted by a rough elutriation technique, and so they were not strictly quantitative (R.O. Brinkhurst, pers. commun.). Despite problems of sample collection and study design, general trends in the distribution of taxa in relation to a trophic gradient were discerned (Brinkhurst et al. 1968; Brinkhurst 1969). Brinkhurst (1967) also plotted the distribution of oligochaete species across Saginaw Bay, Lake Huron, and found that the classic water quality variables (depth, dissolved oxygen, *etc.*) and the worm distributions were poorly correlated, except for *S. ferox* (negatively correlated with organic matter and positively with depth). This reflects the fact that distribution of the worms relates more to conditions in the sediment rather than those in the water column. The patterns were not as clear as those seen in Lake Erie, with the exception of *L. hoffmeisteri* which appeared to be distributed according to the major flow patterns of the river around the perimeter of Saginaw Bay. In a study on the effects of sewage and the iron and steel industry on Hamilton Harbour, Lake Ontario, Johnson and Matheson (1968) found no macroinvertebrates in sediments containing more than 25% Fe_2O_3. Oligochaete worms increased in abundance

Table 3.3 Classification of some aquatic oligochaetes from the St. Lawrence Great lakes by trophic level (Data from several Brinkhurst's publications)

Species	Oligotrophic	Mesotrophic	Eutrophic	Hypereutrophic
Tubifex tubifex	+	–	–	+
Lophochaeta ignota	?	–	–	–
Tasserkidrilus superiorensis	+	–	–	–
Tasserkidrilus kessleri	+	–	–	–
Psammoryctides californianus	+	–	–	–
Potamothrix moldaviensis	–	–	+	–
Potamothrix vejdovskyi	–	–	+	–
Potamothrix bedoti	–	–	+	–
Potamothrix bavaricus	–	–	+	–
Limnodrilus udekemianus	–	–	+	–
Limnodrilus hoffmeisteri	–	–	–	+
Limnodrilus profundicola	+	–	–	–
Limnodrilus claparedianus	–	–	+	–
Limnodrilus angustipenis	–	+	–	–
Limnodrilus cervix	–	–	+	–
Limnodrilus maumeensis	–	–	+	–
Ilyodrilus templetoni	–	–	+	–
Spirosperma ferox	–	+	–	–
Spirosperma nikolskyi	+	–	–	–
Quistadrilus multisetosus	–	–	–	+
Aulodrilus limnobius	–	+	–	–
Aulodrilus pluriseta	–	+	–	–
Aulodrilus pigueti	–	+	–	–
Rhyacodrilus coccineus	+	–	–	–
Rhyacodrilus sodalis	?	–	–	–
Rhyacodrilus montana	+	–	–	–
Bothrioneurum vejdovskyanum	–	?		
Branchiura sowerbyi	–	–	–	+
Phallodrilus hallae	+	–	–	–
Eclipidrilus lacustris	+	–	–	–
Lumbriculus variegatus	–	–	–	+
Stylodrilus heringianus	+	–	–	–

with depth, a fact that authors related to a higher proportion of silt/clay or organic matter in sediments. *Limnodrilus hoffmeisteri*, *L. cervix* and *T. tu*bifex comprised 92% of total macroinvertebrates communities in the study area.

Howmiller and Beeton (1971) compared the fauna of Green Bay, Lake Michigan, from 1952 to 1969 samples, but were still using numbers of unidentified oligochaetes. Their comment that *"The Chironomidae includes many species adapted to a wide range of environmental conditions. However, as a group, the midges display pollution tolerance second only to the Oligochaeta"* indicates the then prevailing view that all oligochaetes were very tolerant simply because a few species survive in very large numbers in the most polluted situations. However, in another paper,

Howmiller and Beeton (1970) had reported that *Stylodrilus heringianus* and *Tasserkidrilus* (as *Tubifex) kessleri* were found at the oligotrophic end of the bay, while at the innermost eutrophic stations *L. hoffmeisteri*, other *Limnodrilus* species, and sometimes *Dero digitata* dominated. In between, and by implication under mesotrophic conditions, members of the genera *Aulodrilus, Spirosperma, Quistadrilus* and *Potamothrix* were found. In studies on the open water of Lake Michigan and the smaller water bodies of Wisconsin, Howmiller (1974a, b) pointed to the apparent discrepancy between the early findings on European lakes described by Brinkhurst (1964a) and the situation that both authors were establishing for the lakes of North America. In fact, later European work where the sampling was more detailed and quantitative showed a better agreement between the two continents. Howmiller found a low diversity of tubificids in his study lakes. *Tubifex tubifex* was found at the two extremes of the trophic gradient, in grossly polluted sites and oligotrophic sites, *L. hoffmeisteri* was ubiquitous but was most abundant where organic pollution was greatest, while *I. templetoni* was fairly abundant and identified as tolerant, but less so than *L. hoffmeisteri*. The diversity of the tubificid fauna of the St. Lawrence Great Lakes was found to be clearly higher than that of smaller lakes which implies that worms are more valuable as indicators of water quality in the larger systems than in the smaller ones. Perhaps a reflection of the dispersal mechanisms available to worms compared to insects.

The large quantity of data on the benthos of the Great Lakes collected during this period was summarized both by Cook and Johnson (1974) and by Mozley and Howmiller (1977). They provided some general remarks about the relationship between American and Eurasian data, but presented very few references on the latter. Cook and Johnson (1974) noted the existence of similar oligochaete assemblages for eutrophic and mesotrophic lakes on both sides of the Atlantic, and a less frequent oligotrophic community dominated by *Stylodrilus heringianus*. The collections described in these reviews were based on offshore sampling from large vessels. Nearshore work began later with surveys by authors such as Stimpson et al. (1975), Nalepa and Thomas (1976), Krieger (1984), and especially Barton, whose early work is referred to in his 1986 study of Lake Ontario. These authors identified the extent of pollution in the harbours, with corresponding gradients of improving conditions from the harbour mouths to the open lake. *Stylodrilus heringianus* was the dominant organism in much of the clean water and sandy habitats of the open lakes, but the usual series of tubificids were found along the trophic gradient. These shoreline studies often included information on the biology of naidine species such as *Chaetogaster, Uncinais, Vejdovskyella, Nais, Piguetiella, Ophidonais*, and *Arcteonais*. Stimpson et al. (1975) recorded *Piguetiella michiganensis, Uncinais uncinata* and *Potamothrix moldaviensis* in the shallow, wave-washed sites. In the sandy, deeper sites the lumbriculid *S. heringianus* was dominant, and was associated with the tubificine *Potamothrix* species and *L. hoffmeisteri*. Silty sediments were dominated by the latter, associated with *Aulodrilus* species, other *Limnodrilus* species, as well as *Ilyodrilus templetoni, Quistadrilus multisetosus* (as *Peloscolex*) and *Potamothrix vejdovskyi*. Barton (1986) used ordination techniques to identify organic enrichment, depth and susceptibility to upwelling as the primary factors determining the composition of benthic communities. Surprisingly, exposure to wave action and substratum type had little effect, while these had been important in Lake Huron where there was no organic enrichment.

In a study in Toronto Harbour, Brinkhurst's focus shifted to trophic gradients in nearshore areas of the Great Lakes. The study led to detailed work on the interspecific interaction between *T. tubifex, L. hoffmeisteri*, and *Q. multisetosus*, perhaps the most tolerant trio of tubificine species (Brinkhurst 1970). This study showed that worms were under the least ecological stress when they were in a mixed species "flock", and that the production of worms was twice as high in flocks as compared to the same species in pure culture (Brinkhurst et al. 1972; Brinkhurst and Austin 1979). This research was directed by the interest at that time in productivity (through the International Biological Program) and generated the hypothesis that worms in mixed culture would prove to be more tolerant to pollution stress in the laboratory, as was demonstrated later (see Chapman et al. 1982a, Sect. 4.2.1). The work on the Bay of Quinte, Lake Ontario, developed into a productivity model for benthic communities along a trophic gradient (Johnson and Brinkhurst 1971a, b). A few oligochaete species, in particular, *L. hoffmeisteri* at four sites and *T. tubifex* at one site, together with some chironomids and crustaceans accounted for most of the secondary production. *Stylodrilus heringianus* was found in the deep waters of Lake Ontario where macroinvertebrate production was lower.

An exhaustive review of the taxonomy, zoogeography and ecology of Great Lakes species was reported by Spencer (1980) at the first of a series of international symposia dedicated to aquatic oligochaete studies. The author suggested the need for future surveys especially in the northern unpolluted lakes to provide of baseline data for evaluating degradation, and additional surveys in small lakes and rivers of the area to provide a better understanding of the geographic distribution of oligochaete species. The open water oligochaete fauna and its relationship with pollution was re-examined by Lauritsen et al. (1985). They took samples from 286 sites in Lake Michigan. *Stylodrilus heringianus* was most common, its abundance being inversely related to the organic content of the sediments according to the abstract, but this relationship was not statistically significant. Problems with high variances made statistical testing impossible, and these are probably related to the use of the Shipek sampler. Several methods of estimating water quality from ecological data on worms were used, and results agreed with the earlier ideas about species distribution, worm abundance and pollution. This account classified worm assemblages at each site assigning the oligochaete community to one of the four trophic groups proposed by Mozley and Howmiller (1977) (Table 3.4), and the numerically dominant species determined the trophic classification of each site.

In North America, there have been relatively few studies of contaminated areas outside the St. Lawrence Great Lakes where worms have been identified to species. Much of the work has been done by consultants or government agencies and the results are not published in the open literature. An example of this type of publication is that by Watt et al. (1973), which concluded that the population density of oligochaetes in the St. John River Basin (New Brunswick, Canada) was closely related to water pollution levels, and that the correlation was improved when the

Table 3.4 A classification of oligochaete species in the St. Lawrence Great Lakes and Scandinavian lakes, based on the degree of enrichment (From Howmiller and Scott 1977, Table II, with permission of Water Environment Federation, and Milbrink 1983a, Table 1, with permission of Springer Publ., © conveyed by Copyright Clearance Centre, Inc.)

Howmiller and Scott (1977)	Milbrink (1983a)
Group 0: Species largely restricted to oligotrophic situations *Stylodrilus heringianus* (O) *Embolocephalus variegatus* *Spirosperma nikolskyi* (O) *Tubifex superiorensis* (M) *Limnodrilus profundicola* (O) *Tasserkidrilus kessleri* (O) *Rhyacodrilus coccineus* (O) *Rhyacodrilus montana* (O)	Group 0: Species characteristically found in oligotrophic situations or the initial stages of enrichment *Stylodrilus heringianus* *Rhynchelmis limosella* *Spirosperma ferox* *Tubifex tubifex* *Lampodrilus isoporus* *Eiseniella tetraedra*
Group 1: Species of mesotrophic areas or only slightly enriched areas *Spirosperma ferox* (M) *Isochaetides freyi* (M) *Ilyodrilus templetoni* (M) *Potamothrix moldaviensis* (M) *Potamothrix vejdovskyi* (M) *Aulodrilus* spp. (M) *Arcteonais lomondi* *Dero digitata* *Nais elinguis* *Slavina appendiculata* *Uncinais uncinata*	Group 1: Species associated with mesotrophic areas or permanently slightly enriched areas *Psammoryctides barbatus* *Limnodrilus profundicola* *Rhyacodrilus coccineus* *Aulodrilus pigueti* *Bothrioneurum vejdovskyanum* *Aulodrilus limnobius* *Psammoryctides albicola*
Group 2: Species tolerating extreme enrichment or organic pollution *Limnodrilus angustipenis* (SP) *Limnodrilus hoffmeisteri* (SP) *Limnodrilus udekemianus* (SP) *Limnodrilus cervix* (SB) *Limnodrilus claparedianus* (SB) *Limnodrilus maumeensis* (SB) *Quistadrilus multisetosus* (SB) *Tubifex tubifex* (SB)	Group 2: Species tolerating considerable organic pollution or else eutrophic conditions *Ilyodrilus templetoni* *Lophochaeta ignota* *Limnodrilus claparedianus* *Limnodrilus cervix* *Potamothrix hammoniensis* *Potamothrix heuscheri* *Potamothrix moldaviensis* *Potamothrix vejdovskyi* *Potamothrix bedoti* *Aulodrilus pluriseta*
	Group 3: Species found in great abundance in situations of extreme organic enrichment *Limnodrilus hoffmeisteri* *Tubifex tubifex* *Eiseniella tetraedra*

The parentheses indicate the classification of Lauritsen et al. (1985) for St Lawrence Great Lakes species (O = oligotrophic, M = mesotrophic, SP = saprophilic, SB = saprobiont) species were identified. They recommended that oligochaete monitoring be included in future water resource management plans because of the value of the worm data in making distinctions at "lower levels" of pollution.

In Europe, Timm has been publishing on the taxonomy and ecology of aquatic oligochaetes since 1958, and the appearance of his summary of the Estonian Oligochaeta in 1970 was the first to appear in English (Timm 1970). His contribution has mainly focussed on the study of macrozoobenthic communities in lakes and in particular the oligochaete species and assemblages in Estonian and Russian lakes (Timm 1994; Timm et al. 1996a; Kangur et al. 1998). Grigelis (1980) summarized Russian ecological studies up to that time and these mainly provided an extensive bibliography of the population size and biomass of dominant species across all regions of the former Soviet Union. From 1967 to 1987, Soviet biologists held a series of meetings on aquatic oligochaete biology, the proceedings have been published (Belyaev et al. 1972; Kurashvili 1983; Kachalova and Parele 1987), and two are available as English translations. In these proceedings, there are important contributions to the study of life cycles of naidines and tubificines, population ecology and role of oligochaetes as prey for fish. Poddubnaya and Bakanov (1983) conducted a detailed study of the spatial microdistribution of species complexes and their seasonal dynamics in Lake Pleshcheevo. In the littoral, a 7-m transect from the shore to the centre was sampled (each sample area 50 cm^2, at 20 cm depth). The authors showed that each species was characterised by its own specific aggregation recurrence frequency. *Limnodrilus udekemianus* and *L. hoffmeisteri* showed a uniform distribution with a low value of patchiness compared to *P. barbatus* with the highest patchiness value. Patches were organized with adult individuals occupying the central part of the patch while the immature worms surrounded them. Some interspecific significant correlations between tubificines were found, the highest being for *P. hammoniensis* with *P. barbatus* ($r=0.66$). No antagonistic interspecific interactions were detected.

Milbrink (1973) began a series of studies of the tubificid and lumbriculid fauna along a pollution gradient in larger Swedish lakes. As in North America, the primary focus was the nutrient status of the systems. He was able to arrange the species from Swedish lakes into a eutrophication series, but noticed that competition and food supply were important in addition to oxygen concentration, the variable through which most of the eutrophication processes affect benthos. He noted that *T. tubifex* appeared to be affected by competition with other species, that *L. hoffmeisteri* could be an indicator of pollution, and that *P. hammoniensis* was common in eutrophic lowland lakes as it is all over Europe. This species has never been definitively identified in North America (Mark Wetzel, pers. comm.) despite some early records and the presence of other *Potamothrix* species with more limited European distributions. The genus is thought to have a Ponto-Caspian origin (Milbrink and Timm 2001) and was possibly introduced to North America through ship ballast water. In Sweden, *Spirosperma ferox* was found with *Stylodrilus heringianus* in oligotrophic water bodies, whereas in North America *S. ferox* was considered to be a mesotrophic species between *S. nikolskyi* and *Q. multisetosus*. Milbrink (1973) also reported *P. barbatus* as a somewhat tolerant species (limited to the estuary of the St. Lawrence

River, in North America), and the scarcity of *I. templetoni* was noted, in contrast to its classification as a common and tolerant species in North America. Finally, *Limnodrilus profundicola* and the lumbriculid *Rhynchelmis limosella* were regarded as cold stenotherms in Swedish lakes, while *Branchiura sowerbyi* was found in warm water effluents supposedly unrelated to the trophic scale. This latter species seems to have originated in the tropics, although there is a continuous distribution from south-eastern Asia to temperate European waters, with it being more abundant where there is considerable organic matter input or warm waters (Aston 1968).

Milbrink (1973) proposed an indicator community approach, as opposed to indicator species, using the biology of *T. tubifex,* with its occurrence in both very enriched and oligotrophic habitats, to illustrate the problem of indicator species. His publication provides a good review of this type of approach. In 1978, Milbrink provided a summary of the overlapping distributions of species along the trophic gradient (Table 3.4), and in 1983 he proposed an improved index (see below), which followed the trend of moving away from simple descriptions of species occurrence and association with trophic condition to a more formalized quantitative description, including a measure of species tolerance and the distribution and abundance of the organisms. Other studies used a similar community approach in Scandinavia (*e.g.* in the Baltic sea by Leppäkoski and Lindstrom 1978, and in freshwater by Särkkä 1987). For the six most common species Särkkä and Aho (1980) calculated Levins' niche breadth (B_i) and the ecological overlap or probability of species co-occurrence, concluding that there are low interactions between the species, this, in contrast to the Toronto Harbour findings of positive species interactions (Brinkhurst et al. 1972). The largest B_i values were for *P. hammoniensis* in heavily polluted eutrophic lakes and *S. ferox* in less polluted habitats. Särkkä (1989, 1994, 1996) examined meiobenthic oligochaete species as indicators of lake eutrophication and pollution. *Amphichaeta leidigii*, *Specaria josinae* and *Vejdovskyella comata* preferred eutrophic conditions, and Särkkä pointed out that species level identification was desirable if using oligochaetes for biomonitoring purposes. A more recent contribution on the indicator value of oligochaete assemblages of small Swedish lakes (Milbrink and Timm 2002) concluded that *S. ferox*, *S. heringianus* and *T. tubifex* were characteristic of oligotrophic conditions where abundance is low (<300 ind. m^{-2}), whereas *L. hoffmeisteri* and/or *P. hammoniensis* largely dominated eutrophic small lakes with high densities (>5,000 ind. m^{-2}). These authors concluded that lake classification based upon total-P concentration and oligochaete community composition generally agreed.

Studies of larger alpine lakes (Adreani et al. 1981; Probst 1987) followed the same pattern, initially using descriptive analysis of relative abundances of species. Experimental work (Adreani et al. 1984) with cohort cultures of lacustrine species contributed to the understanding of population biology of tubificines in lakes. These authors described that the eutrophic-oligotrophic species sequence of *T. tubifex – L. hoffmeisteri – P. barbatus – S. ferox* corresponded with an increase in generation time, and often a decrease in fecundity. Their results also indicated density regulation mechanisms operating through the proportion of mature animals and the fecundity of the species.

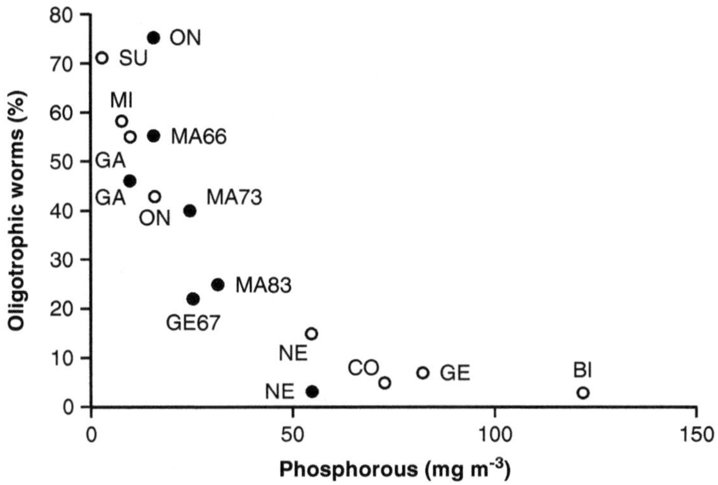

Fig. 3.2 Mean relative abundance (%) of oligotrophic worm species plotted against mean concentrations of phosphorus in Lakes Superior (*S*), Michigan (*MI*), Gardia (*GA*), Maggiore (*MA*), Ontario (*ON*), Geneva (*GE*), Neuchatel (*NE*), Constance (*CO*) and Bienne (*BI*). In some lakes, the year in which samples were taken is indicated by the last two numbers. Solid circles indicate samples taken at the deepest points in the lake (From Lang 1990, Fig. 1, modified and reproduced with permission of John Wiley & Sons Ltd., © conveyed by Copyright Clearance Centre, Inc.)

Early work done by Juget (1958) on the oligochaete communities of Lake Leman was expanded through collaboration with Lafont (Lafont and Juget 1985; Lafont et al. 1991) on smaller subalpine French lakes. These studies described the distribution of species in relation to water depth, oxygen and ammonium dissolved in water. The biological status of lakes was described by biotic indices which integrated the richness and abundance of oligochaete species (see below).

Lang was one of the earlier users of multivariate statistical methods to describe community change and response (Lang 1978; Lang and Lang-Dobler 1979) in oligochaetes. In his studies on the indicator value of oligochaete communities for the assessment of eutrophic conditions in Swiss lakes, he identified *Potamothrix hammoniensis (*with *P. heuscheri* and *T. tubifex* in some water bodies*)* as the species numerically dominant in eutrophic lakes; *P. vejdovskyi* in mesotrophic lakes; and *S. heringianus* in oligotrophic lakes (Lang 1985, 1986). The relative abundance of the three groups of species indicated lake trophic state and sedimentation patterns associated with primary production. However, in his later papers, Lang focussed on the assessment of eutrophy in lakes by the measurement of the relative abundance of oligotrophic species (mostly *S. heringianus* and *E. velutinus*) (Lang 1989a, 1991, 1992; Lang and Reymond 1995a, b, c). Using data from three large lakes in North America and eight large lakes in western Europe, Lang (1990) computed a linear regression model of the relationship between the mean relative abundance of oligotrophic species (OS) and the mean total-P concentration (TP, mg m^{-3}): $OS = 80.29–8.35\ TP^{0.5}$, $r^2 = 0.81$ (Fig. 3.2). The model, modified for the data from the deepest areas only, predicts the disappearance of oligotrophic species at a total-P value of 55.5 mg m^{-3}, which corresponds to the value recorded in Lake Geneva 5 years before the oligotrophic species were no longer found in the profundal zone (35–59 mg TP m^{-3}). Lang (1993, 1999; Lang and Reymond 1995a) used this model to predict the change in the relative abundance of oligotrophic species following a reduction in total-P. At a small scale, oligochaete populations in Lake Geneva appeared in patches, where oligotrophic species are more abundant in furrows or troughs in the bottom sediments, and mesotrophic and eutrophic species in the ridges, a distribution that was dependent on the organic sedimentation (higher on the ridges) (Lang 1989b). The prediction of the relative abundance of oligotrophic species may be modified by the influence of temperature, where increased water temperature may reduce the abundance of oligotrophic species and favour other species (Lang and Reymond 1995c) independent of the total-P content in water. This modifying factor may play an important role in the future, within the context of climate change.

In southern European rivers, a number of authors have described the response of oligochaete species to nutrient enrichment and pollution. Using data on oligochaete species assemblages more quantitative determinations were made of the relationships and responses of the entire community to disturbance, again associated primarily with nutrients. In a polluted mountain stream, Lafont (1977) found a community dominated by *T. tubifex*, *Nais elinguis*, and *S. heringianus*, and not by *Limnodrilus* species. Lafont stated that local conditions could modify the community, for example, in shallow, fast flowing waters where oxygen levels were high, heavy organic inputs supported more species than in areas of low flow where oxygen became limiting. In a review on the value of oligochaete species as indicators of pollution, Lafont (1984) added the naidines *Amphichaeta leydigii, Vejdovskyella intermedia, Dero digitata* and *Specaria josinae* as well as the enchytraeid *Marionina riparia* to the list of known tolerant species. He also recommended conducting studies, such as those of Chapman et al. (1982b, c), on the relative tolerance of species to a variety of pollutants and environmental factors in laboratory controlled conditions, to reveal further insight into the tolerance values of the species.

A few studies have focused on water pollution by metals instead of nutrients. In the River Mort, Giani (1984) found that simple parameters such as density, richness, and dominance were useful for classifying heavily polluted sites. Giani emphasized the convenience of oligochaete assemblages for assessing highly polluted sites as they allowed the discrimination of communities where other taxa (*e.g.* arthropods) are absent.

In 1981, Verdonschot published the first of a series of contributions on oligochaete assemblages in ditches and brackish waters in the Netherlands. Nijboer et al. (2004) reviewed and evaluated 20 years of monitoring data for oligochaetes and concluded that width and depth of the water body (even if the maximum depth was only 6 m in the study sites) were the most important factors explaining the distribution of the species. These authors examined the relationship between species distributions in the Netherlands and environmental variables (habitat characteristics, pH, phosphorous and chloride concentration). In this study, *Aulodrilus pluriseta* and *Slavina appendiculata* appeared to be associated with small water bodies, with low pH, chloride, and total P-concentration, while *L. udekemianus*, *L. claparedianus*, *L. profundicola*,

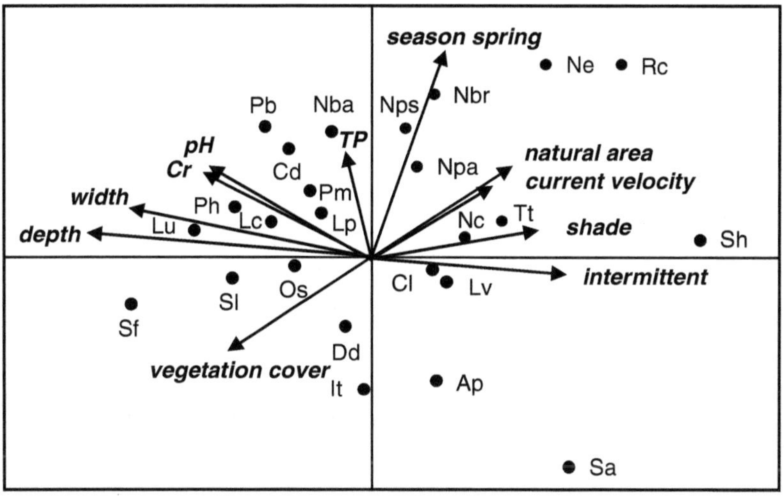

Fig. 3.3 Ordination diagram of the CCA showing the projection of oligochaete species in the Netherlands, along environmental gradients of 11 variables contributing significantly ($p<0.05$). *Arrows* show the direction in which a variable explains most of the variation of the species composition (From Nijboer et al. 2004, Fig. 3, redrawn and reproduced with permission of Springer Publ. © conveyed by Copyright Clearance Centre, Inc.). Chromium (Cr), Total Phosphorous (TP), Oligochaete species: *Aulodrilus pluriseta* (Ap), *Chaetogaster diaphanus* (Cd), *C. limnaei* (Cl), *Dero digitata* (Dd), *Ilyodrilus templetoni* (It), *Limnodrilus hoffmeisteri* (Lh), *L. profundicola* (Lp), *L. udekemianus* (Lu), *Lumbriculus variegatus* (Lv), *Nais barbata* (Nba), *N. bretscheri* (Nbr), *N. communis* (Nc), *N. elinguis* (Ne), *N. pardalis* (Npa), *N. pseudobtusa* (Nps), *N. variabilis* (Nv), *Ophidonais serpentina* (Os), *Potamothrix hammonienis* (Ph), *P. moldaviensis* (Pm), *Psammoryctides barbatus* (Pb), *Rhyacodrilus coccineus* (Rc), *Slavina appendiculata* (Sa), *Spirosperma ferox* (Sf), *Stylaria lacustris* (Sl), *Stylodrilus heringianus* (Sh), *Tubifex tubifex* (Tt)

P. hammoniensis, *P. moldaviensis*, *P. barbatus*, *S. ferox*, *Nais barbata*, and *Chaetogaster diaphanus* tended to occur in larger water bodies with high pH and chloride concentration (Fig. 3.3). However, the analysis included only a few variables associated with pollution (total-P and NO_3^- concentration, pH, NH_4^+, conductivity), the inclusion of other variables such as organic content could have provided further explanation of the species distribution patterns. Probably the most comprehensive examination of the indicator value of oligochaetes has also been undertaken by Verdonschot (2006), who used data from 772 samples from eight European countries. His results (Fig. 3.4) indicate that oligochaete taxa are present across a wide range of water velocities and are not restricted to stagnant water (limnophil in Fig. 3.4). Taxa are also evenly distributed over eight microhabitat types and ten different river zones. His general conclusion was that when higher taxonomic levels (Class and Family) are used in water quality assessment, results become biased.

Factorial analysis was used by Martinez-Ansemil (1990) to determine communities of worms from four families in the River Tambre (Spain), and he reviewed a large amount of earlier literature on species distribution and associations, particularly European work. He suggested three site/species assemblages: first, sites with

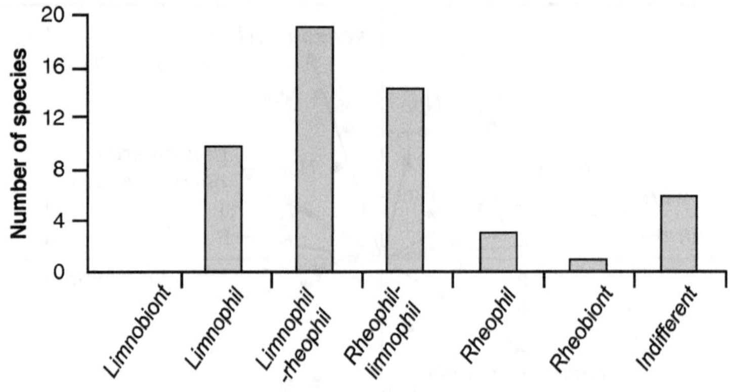

Fig. 3.4 Number of oligochaete taxa in different current velocity classes of European water bodies, as present in AQEM database (From Verdonschot 2006, Fig. 4, redrawn and reproduced with permission of Springer Publ., © conveyed by Copyright Clearance Centre, Inc.)

stony riverbeds (with or without sand or gravel) with a moderate to high flow velocity were characterised primarily by *Stylodrilus heringianus* and *S. parvus* with *E. velutinus*; second, sites with sandy or muddy sediments, with low or null flow velocity, were characterised by a group of species including *N. elinguis, U. uncinata, A. leydigii, L. ignota, T. tubifex, L. hoffmeisteri* and *B. vejdovskyanum;* and finally, a group of sites with macrophytes or plant debris, was characterised by the naidines *Stylaria lacustris, Vejdovskyella comata,* and *Nais alpina.*

Studies in reservoirs are much less frequent, although oligochaetes can be the major component of the benthic community. From two field surveys of the profundal benthos in 1973/75 and 1987/88 Real et al. (1993) described factors influencing the abundance and distribution of oligochaetes in Spanish reservoirs. Oligochaetes represented 91.7% and 89.2% of the macrozoobentos, respectively, and the increase in chironomids accompanied by a decrease in oligochaete relative abundances in 1987/88 when compared to the previous study was attributed to an improvement in oxygen conditions. Oligochaete densities increased with depth and were higher in summer and lower in winter. The improvement in quality in the reservoirs was also accompanied by a change in species composition, with an increase in abundance and frequency of oligo- and mesotrophic species (*e.g. S. ferox, E. velutinus* and *A. pluriseta*), while *L. hoffmeisteri* was absent in the more recent survey. Rieradevall and Real (1994) made similar observations in a study on a Spanish lake where there was a high relative abundance of oligochaetes in the macrobenthic community (up to 64%), with increasing density with depth and in finer sediments. In some areas of the lake, where a chemocline persisted for up to 4 months and was associated with anoxic conditions, only *Potamothrix heuscheri*[1] occurred.

[1] This species can be confused with *P. hammoniensis* if the identification is based on only the shape of spermathecal chaetae, since specimens of *P. heuscheri* with spermathecal chaetae similar to *P. hammoniensis* have been reported in the literature (*e.g.* Timm et al. 1996b).

Work on other continents is generally more recent. In South America, Marchese (1987) in the Paraná River, Argentina, calculated simple and multiple correlations between environmental variables and the density of several oligochaete species. The author showed the value of separately analysing data from the banks and the middle of the channel, and multiple correlation coefficients explained 52% of the variation of *Narapa bonettoi* in the bank areas and 35% variation of *L. hoffmeisteri* in the middle of the channel. In Ivinhema River and Patos Lake, Brazil, the association of the most abundant oligochaete species was examined using Detrended Correspondence Analysis (Takeda 1999; Montanholi-Martins and Takeda 2001). Density and spatial patterns were mainly associated with grain size and the organic content of sediments. A few studies in China have used diversity and other simple metrics (*e.g.* Qi 1987, on the Lower Pearl River China). Interestingly, some of the taxa in the Chinese work are common to the Pacific Rim of North America and the responses of the North American communities may be applicable to pollution ecology studies in China.

The identification of stress specific responses to environmental factors other than organic enrichment began in field surveys. However, tolerance ranges of oligochaete species to stress factors other than organic enrichment have largely been established through laboratory testing (see Sects. 4.2 and 4.3) and this work has demonstrated that all worms are not equally tolerant to contaminants. Although such laboratory work is valuable, the approach does not replace well designed field studies in the estimation of effects of contaminants outside the controlled conditions of the laboratory. There are a number of studies that have identified stress factors for aquatic oligochaetes under field conditions. For example, acid tolerance reached as low as pH 4 in six lakes in Northern Ontario (Roff and Kwiatkowski 1977) for *Stylodrilus heringianus* and *Rhyacodrilus montana*, but *L. hoffmeisteri* was missing from lakes with pH lower than 5.5–6.6.

Brinkhurst (1965b) first pointed out the intolerance of aquatic oligochaetes to metals. However, the interpretation of field data is not always straightforward, and responses to stress are also affected by other biotic or abiotic factors. Thus, according to Wentsel et al. (1977), metal levels in sediments of up to 969 ppm Cd, 14032 ppm Zn and 2106 ppm were tolerated by *Limnodrilus* sp. in Palestine Lake. However, the population size was reduced at lower levels of metal contamination as competition from other taxa, especially *Chironomus tentans* and *Chaoborus punctipennis*, increased. Aston (1973) reviewed the effects of metals and pesticides on tubificids in field studies and reported on a few references of the sensitivity of tubificids to metals. Thus, 0.12–1.20 ppm Cu in rivers eliminated tubificids from the River Churnet, while worms were tolerant to pesticides (such as hexachlorobenzene).

Branchiura sowerbyi appears to be a thermally tolerant species since it has been reported in Europe and North America in areas receiving thermal cooling effluents from power stations (Aston 1973). Attempts to relate the distribution and abundance of 24 oligochaete species to the effects of a thermal effluent (Nichols 1981) on the Keowee Reservoir, South Carolina, was considered to be compromised by infrequent sampling, as the only observed effect was a stimulation of population growth in the immediate vicinity of the power plant.

There are few studies that have addressed the tolerance range values of oligochaetes based on field work because of the difficulty in controlling the many environmental variables that can confound results. Future research with mesocosms (see Sect. 3.5.2) could be an approach that would provide the required data.

In summary, there is a general pattern common to most of the work on oligochaete response to disturbance. Initial studies began after the development of taxonomic keys, this is perhaps clearest in North America where prior to Brinkhurst's preparation of keys to the major families oligochaetes were treated as a single taxon. With the availability of keys, the initial studies were descriptive and simply noted associations between the degree of disturbance and the species found. More formal approaches were then developed that used indices that incorporated tolerance values and presence/absence or relative abundance to generate an overall score of some type that was usually range standardised. This approach has the problems associated with all systems that ascribe tolerance values; primarily, that some species are tolerant of a range of conditions, but are poor competitors (*e.g. T. tubifex*) and can be typically found at both ends of a stress gradient; and second, that tolerance values and indices have to be developed for each type of stress, and most systems generally ascribe tolerance values to organic enrichment only. The interpretation of the absence or the population size of some indicator species is not straightforward, since many different factors can cause a similar response. These difficulties have been circumvented to some degree by the development of multi-metrics, but these apply to the entire community and not just oligochaetes. Parallel with the development of whole community multi-metric indices has been the use of multivariate statistical methods to describe worm assemblages and their relationship with the environment. These methods have resulted in the development of the predictive modelling approach used in various Reference Condition Approaches which form the basis of freshwater bioassessment programmes in the UK and other European countries, Australia, Canada, and the USA. Again, as with the multi-metric approach, the trend is to use the entire macroinvertebrate community, not only oligochaete species assemblages.

3.3 Estuarine and Marine Studies

Extraordinarily high numbers of oligochaetes can be found in estuarine environments occurring under stones, on algae and in other habitats. The importance of tidal and sub-tidal marine oligochaetes in the benthos is not well represented in the literature as there have been few studies on the distribution and tolerance of brackish and euryhaline oligochaete species. In the past, a lack of knowledge on the taxonomy of the estuarine and marine taxa was one of the main obstacles to research on their ecology, however, over the last 3 decades the publications of Christer Erséus, and other researchers, have largely addressed these deficiencies. For marine and estuarine oligochaete ecology, the reader is referred to the extensive reviews by Giere (1980, 2006) and Giere and Pfannkuche (1982), where production, life cycles, zonation data, as well as the impacts of pollution are among the topics covered.

Further studies at a regional scale have addressed the habitat preference of marine oligochaetes mainly focussing on sediment particle size heterogeneity and stability and the oxygen regime at the sediment interface (Diaz et al. 1987; Diaz and Erséus 1994).

Although at that time there was little knowledge of brackish water species, Brinkhurst and Simmons (1968) were able to relate an increase in the relative abundance of oligochaetes to the degree of pollution in San Francisco Bay, and to describe the tolerance of *L. hoffmeisteri* to low levels of salinity. The biological index developed by Sanders (1960) was used by Siegfried et al. (1980) in the San Francisco Bay. The five most abundant organisms at each station were rated on an abundance scale (from 1 to 5) and the rankings for each taxon were summed to give the *Biological Index of Dominance* for that species. In this study, oligochaetes were the dominant taxon at all sites but one where the amphipod *Corophium stimpson* was most abundant.

As with other benthic invertebrates, the distribution of oligochaete species in estuaries is highly dependent on their salinity tolerance (Fig. 3.5). The transition of species across a salinity gradient was examined in the Fraser River (Chapman and Brinkhurst 1981). In the Fraser Estuary, it appears that while the salinity in the water column close to the sediments changes with the tidal cycle the pore water salinity, in mud, is more stable. However, sediment does have an annual cycle of salinity change, based on alternation between increased runoff after snow melt in the mountainous headwater area and greater salt water penetration when runoff is low in the winter. The benthic freshwater communities seem to respond quite rapidly downstream in pace with the seaward shift of freshwater, but to return more slowly upstream. The implication here is that monitoring at a specific field stations might appear to show a decline in diversity as the more estuarine communities occupy a site for part of the year. Sampling at fixed dates could also produce anomalies because of inter-annual variation in the timing of the spring melt. Therefore, quality assessment in estuaries using benthic communities becomes more complex due to the spatial change in the salinity gradient in response to both tidal and seasonal cycles. Habitat characteristics also have a major influence on species composition, thus the expected increase in species richness from the oligohaline to freshwater zone in the James River Estuary was not observed until the upstream lotic freshwater areas with greater habitat diversity (Diaz 1989).

The salinity gradient and the amount of organic matter (TOC) are both factors that structure macrofaunal communities in polluted estuaries. Henderson (1983) described long-term changes in species composition, faunal density, and dominance patterns of the macrobenthic invertebrate communities in the upper Clyde Estuary (UK) affected by organic pollution between 1974 and 1980. Communities were dominated by both oligochaetes and polychaetes, with *Monopylephorus rubroniveus*, *Tubificoides pseudogaster*, *T. benedii*, (as *T. benedeni*), *Heterochaeta costata* (as *Tubifex costatus*), *Lumbricillus lineatus*, *Nais elinguis*, and *Paranais litoralis* being characteristic of polluted brackish areas. During high river flows *T. tubifex* and *L. hoffmeisteri* were also recorded. The estuarine species *T. benedii*, *T. pseudogaster* and *H. costata* reached peak densities of 75,000, 20,000 and >100,000 ind. m^{-2}, respectively. *Paranais litoralis* always occupied a wide range of locations in the

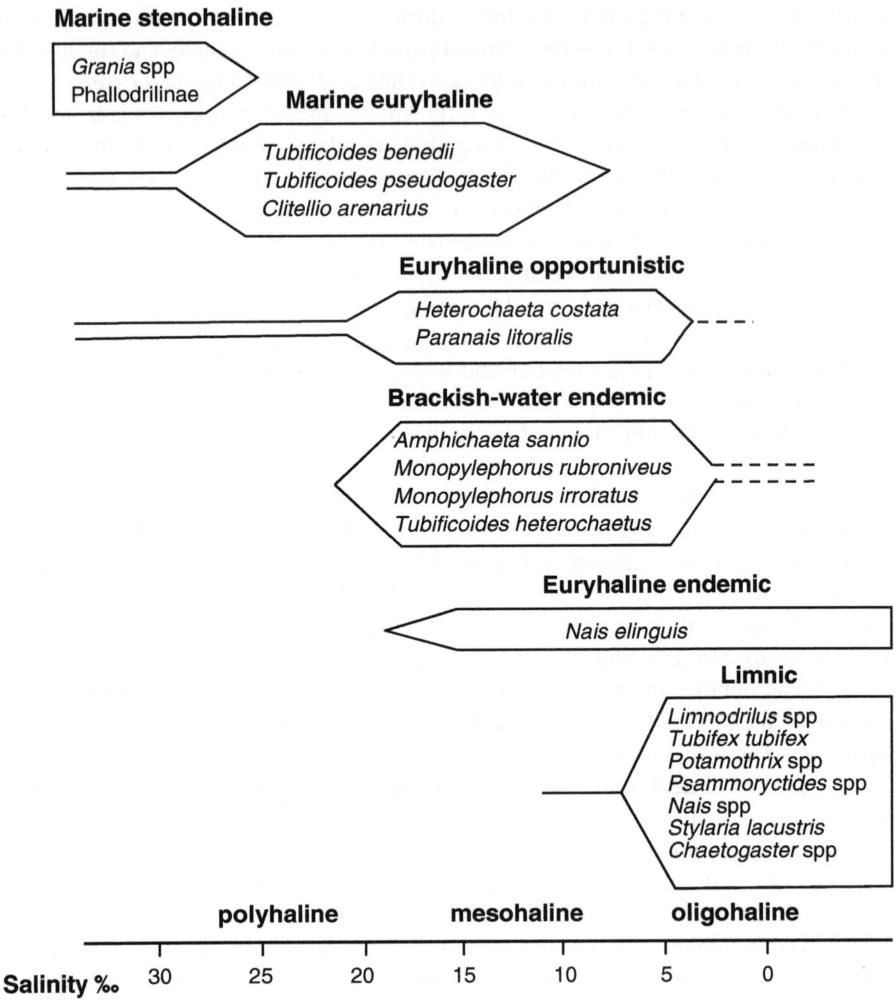

Fig. 3.5 Association of common eulittoral and sublittoral European oligochaetes along a hypothetical salinity gradient (Redrawn from Giere and Pfannkuche 1982)

Clyde Estuary with maximum densities of over 500,000 ind. m^{-2}. This species was dominant in the estuary comprising 50–60% of the population in 1975, and more recently 80–90% of the population, with a density response to fluctuating dissolved oxygen in the summer.

In the Thames Estuary, Hunter (1981) studied the survival of eight tubificid species which formed a succession, similar to others reported from other European estuaries. The freshwater species of the oligochaete community included *L. hoffmeisteri*, *L. udekemianus*, *L. cervix*, *P. barbatus* and *T. tubifex* and these extended seaward to

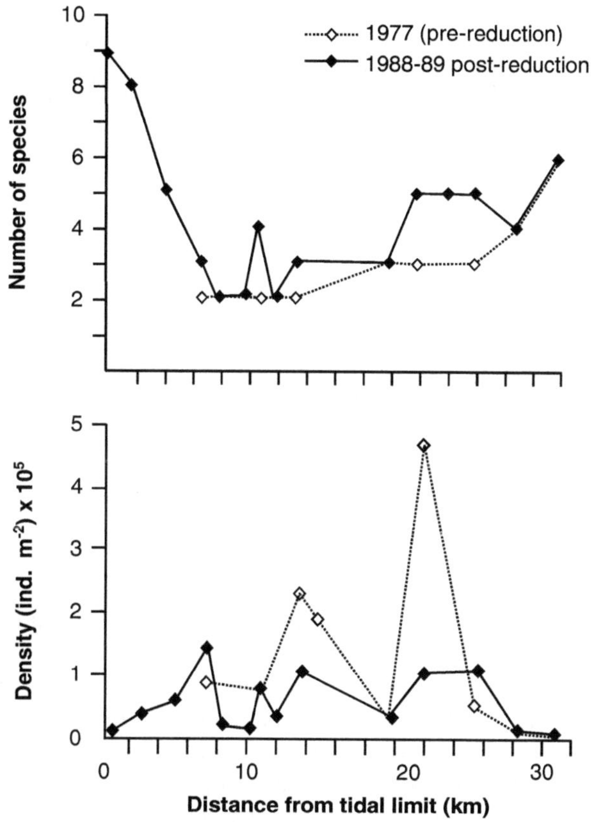

Fig. 3.6 The number and abundance (ind. m^{-2}) of oligochaete species in intertidal section of the upper Forth Estuary in 1977 and 1988/89, pre- and post-reduction of organic inflow (From McLusky et al. 1993, Figs. 5, 6, redrawn and reproduced with permission of Springer Publ., © conveyed by Copyright Clearance Centre, Inc.)

a salinity of 5‰. The euryhaline species *H. costata* and *M. rubroniveus* occurred in areas with salinities of 2–20‰. Further seaward, *T. benedii* was reported at salinities over 26‰. A similar succession in increasing salinity of *T. tubifex* and *L. hoffmeisteri,* followed by *H. costata* and furthest seaward *T. benedii* was described in the Forth Estuary (McLusky et al. 1980, 1993) which was polluted by sewage and industrial effluents. After reduction of organic discharges to the estuary, similar or slightly higher species richness, but much lower oligochaete density was observed at most sites (Fig. 3.6). *Tubificoides benedii* was dominant in the lower part of the estuary with populations attaining a maximum biomass of 13.0 g dw m^{-2} at some sites, where it represented approximately 57% of the total worm mean biomass (polychaetes and oligochaetes), this area was recognised as an important habitat for feeding birds. In the upper part of the Forth Estuary, *H. costata* was dominant in the intertidal oligochaete community and reached a maximum biomass of 78.04 g dw m^{-2}. Birtwell and Arthur (1980) showed that *T. benedii* in the Thames Estuary had an exceptional ability to tolerate anoxia (LT_{50}=58.8 h at 20°C, 26.6 h at 25°C, and

Table 3.5 Oligochaete species characteristic of different tidal zones, derived from cluster analysis of tidal benthic community in James River Estuary (Virginia, USA) (From Diaz 1989, Table 3, with permission of Springer Publ., © conveyed by Copyright Clearance Centre, Inc.)

Species	Upper tidal freshwater zone	Lower tidal freshwater zone	Oligohaline zone	Mesohalyne zone
Limnodrilus udekemianus	+	–	–	–
Branchiura sowerbyi	+	–	–	–
Quistadrilus multisetosus	+	–	–	–
Dero digitata	+	–	–	–
Limnodrilus spp.	+	+	–	–
Limnodrilus hoffmeisteri	+	+	–	–
Ilyodrilus templetoni	+	+	–	–
Nais spp.	–	+	–	–
Tubificoides heterochaetus	–	–	+ [a]	–
Tubificoides brownae	–	–	–	+ [b]

[a]Highest constancy in oligohaline zone
[b]More frequent and abundant in mesohaline zone

17.8 h at 30°C) as measured in experiments with worms acclimated to 20°C, although the species appears to prefer relatively high dissolved oxygen (DO) levels (critical DO = 5–6%). *Limnodrilus hoffmeisteri* and *H. costata* showed a seasonal pattern in their vertical distribution which followed movement of the black, reduced sediment layer, such that most worms remained above this layer. As the Thames Estuary recovered in response to pollution abatement activities, Andrews (1984) described an incremental increase in food-web interconnectivity, and the oligochaete worms that had occupied a central position were displaced by the polychaete *Nereis* sp. The formerly very abundant tubificid *L. hoffmeisteri* was replaced by *H. costata* a few kilometres from the vicinity of outfalls, a change that had been predicted from laboratory studies performed by Birtwell and Arthur (1980).

In a study of the James River Estuary (Virginia, USA) polluted by industrial and municipal wastes, Diaz (1989) characterized the tidal freshwater benthic community which included cosmopolitan species of *Limnodrilus, Nais, Branchiura sowerbyi*, and *Dero digitata*, and also the American species *Quistadrilus multisetosus* (Table 3.5). The benthic community in the oligohaline zone included the estuarine oligochaete *Tubificoides heterochaetus,* and in the mesohaline zone *Tubificoides brownae* was reported as the most abundant species. The author related pollution variables to the whole benthic community using multivariate methods, but the effects of pollution were difficult to evaluate except at sites near outfalls, where species such as *T. heterochaetus* and *T. brownae* occurred.

The benthic macrofaunal species of coastal habitats are largely euryoecious, opportunistic species, well adapted to extreme ecological conditions and can recolonise quickly. Gamenick et al. (1996) in two field experiments studied the colonisation of sediments in the Baltic sea depopulated through anoxia and sulphide accumulation. The presence of *P. litoralis* in shore communities exposed to frequent high sulphide events is most likely associated with its ability to escape adverse oxygen/sulphide sediment conditions by drifting into the overlying water. Moreover, mass mortality can be quickly compensated for through asexual reproduction by paratomy. In contrast, recolonisation by *H. costata* only occurred through the sediments. Both *H. costata* and *T. benedii* occurred in deeper layers where sulphide is present in high concentrations. Tolerance experiments with both these endobenthic species underlined their adaptation to life under hypoxia and extreme sulphide conditions. In field experiments, the abundance and vertical distribution of *H. costata* in sediment was affected after 1 week of anoxia, and the species tolerated sulphide increases to 1,443 μmol l^{-1} for up to 7 days.

The dominance of the oligochaetes in nutrient enriched tidal areas was described by Gray (1971), Leppäkoski (1975), and Birtwell and Arthur (1980). Total abundance of oligochaete worms has been used as an indicator of organic pollution in sediments in the Forth Estuary (McLusky et al. 1993), where abundance was reduced from a maximum of 476,933 ind. m^{-2} down to 107,350 ind. m^{-2} over an 11–12 year period. These high densities are in fact similar to those observed in grossly polluted freshwater systems. These authors also considered changes in species diversity, including non-oligochaete species in a sediment pollution assessment. Increases in oligochaete abundance were also reported in another study on biodeposits in mussel beds, where oligochaetes were the dominant macrofaunal group, and reached up to *ca.* 37,000 ind. m^{-2} along a transect following the biodeposition plume (compared with a maximum value of 276 ind. m^{-2} in a control site) (Kröncke 1996).

In a long-term experiment in Great Sippewissett Marsh (MA, USA) Sardá et al. (1996) used the ratio of polychaetes to oligochaetes to assess the response of benthos to nutrient enrichment in muddy tidal creeks. This ratio was lower in fertilized creeks than in control creeks, both in terms of abundance (0.5 vs. 1.2) and biomass (1.4 vs. 4.0). The two common species *Paranais litoralis* and *Monopylephorus evertus* peaked at different times of the year, the former dominating during the spring and the latter in autumn. Half of the secondary production in fertilized creeks was due to oligochaetes compared to only 18% in the controls. The contribution of *M. evertus* and *P. litoralis* to the mean annual production was 3.7 g dw m^{-2} y^{-1} and 0.5 g dw m^{-2} y^{-1}, respectively. Other oligochaete species, as M*onopylephorus irroratus* and *Lumbricillus* sp. were absent in control sites and contributed 0.7 g dw m^{-2} y^{-1} to the secondary production in fertilized creeks. The authors also discussed the possible interaction of polychaetes and oligochaetes, and both inhibition of larval settlement and direct competition for food were considered as plausible explanations for the shift in dominance from polychaetes to oligochaetes.

In the tidal creeks of Charleston Harbor, in the southeastern USA, benthic communities are dominated by annelids. Weinstein and Sanger (2003) compared the tolerance of the oligochaete *Monopylephorus rubroniveus* and polychaete species to fluoranthene, and tolerance to the toxicant appears to be related to their tolerance to hypoxic conditions, with their tolerance being consistent with their relative abundance in the field.

In a coastal area of British Columbia (Canada) polluted by paper mill effluents, *Lumbricillus lineatus* was the dominant species up to 1.5 km from the mill outfall, and then replaced by an association of other enchytraeid species of the genus *Marionina*, *Enchytraeus* and *Lumbricillus* (Coates and Ellis 1980). In this study, the percentage of *L. lineatus* to the total adult enchytraeids was proposed as a simple index for the assessment of pulp mill effluents in intertidal areas. *Lumbricillus lineatus* and other *Lumbricillus* species are able to feed on oiled organic debris (Giere and Hauschildt 1979; Coates 1995), therefore, high densities of individuals may be important in the breakdown of oil after spills. In fact, several studies on the impact of crude oil pollution on marine fauna reveal the presence of some tolerant oligochaete species. The study by Giere (1979) was the first on the impact of an oil spill on the intertidal meiofauna. Six weeks after the large spill from the "Urquiola" in north west Spain in 1976, the littoral meiobenthic fauna was dominated by *Turbellaria* species and the few oligochaetes were mainly enchytraeid species, such as *Marionina subterranea* and a small number of individuals of *M. preclitellochaeta*, collected from surface samples (0–5 cm depth). Even in areas affected by moderate oil pollution, the meiofauna was greatly reduced, and a few oligochaete species were present (mainly *Marionina achaeta*). One year later, the increase in both faunal diversity and abundance in the vertical distribution of the meiofauna was interpreted as a sign of habitat recovery. Oligochaete species found in this study were *Marionina subterranea*, *M. preclitellochaeta*, *M. achaeta*, *M. southerni*, *Phallodrilus monospermathecus*, and other unidentified species of the genera *Lumbricillus*, *Enchytraeus*, *Marionina*, and *Phallodrilus*.

The papers described above used species level identification of oligochaetes in estuarine surveys on water pollution. However, most studies have either ignored the oligochaetes or only use family level classification. An assessment study of estuarine sites in the Gulf of Mexico (Engle et al. 1994) used stepwise discriminant analysis and identified three metrics which contributed most to the overall model determination coefficient (r^2). A linear combination of the proportion of the expected Shannon's diversity index (adjusted to remove the effects of salinity), the proportion of tubificid oligochaetes, and the proportion of bivalve molluscs to total benthic fauna in the model produced a score, used as an index for estuarine benthic integrity. The score was normalised to range from 0 to 10, and benthic integrity was considered degraded when the score was >4.1.

Peterson et al. (1996) reviewed the marine pollution literature on benthic responses associated with complex gradients of organic pollution and toxicants. Gradients included oil spills, natural petroleum seeps, sewage discharges and industrial discharges. The review included both mesocosms and field studies and authors assumed that covariance in multiple environmental variables is typical of most anthropogenic discharges into marine environments. They characterize oligochaetes together with polychaetes and nematodes as "not especially sensitive to toxicants", including opportunistic non-selective deposit feeding species, typically showing substantial increases when exposed to organic pollution. However, in their Table 1, summarizing the relevant literature on soft-sediment benthic community responses, there are only two references to oligochaetes, both related to natural oil seeps, as one of the benthic groups with increased abundance, as reported by Spies et al. (1980), and the other as a group with greatly reduced abundance, as reported by

Steichen et al. (1996). This clearly shows the value of discriminating species or species groups as opposed to using the Class level identification for oligochaetes.

In summary, there is a great deal of evidence from field biology that the tolerance levels to various stress factors by worms vary markedly from one species to another. There is also some indication that toxicology work (see Chap. 4) supports some of the broader generalizations about the respective field tolerance range of species, but there has been far too little work done, especially on the supposedly less tolerant taxa. The combined effects of multiple stressors on taxa distribution in the field requires further research and this will contribute to understanding the intertidal and marine habitat interactions of oligochaetes, polychaetes, and other macrofaunal species, as well as contribute to the interpretation of changes in community structure and composition. Some species are extending their geographical ranges quite rapidly through anthropogenic transport. The presence of some species may alter the trophic status of others through competition for food resources or habitat, so that zoogeography has to be accounted for in field studies using species composition of aquatic communities.

3.4 Analytical Methods

Methods used to analyze data describing oligochaete assemblages and their responses to pollution are the same as those used in bioassessment using the whole larger benthic community. One can define three developmental stages in the interpretation of community level data. The first was simple description of the species present and very simple metrics based on the total count of oligochaetes. The second phase was the development of various indices and metrics. These were attempts to integrate the taxonomic composition, and by default the ecological preferences, of the benthic community into simpler quantitative values, thus providing a measure of the effects of disturbance based on an indicator or measure of tolerance. The final phase was the use of multivariate methods that are designed for dealing with multiple variables and data matrices, but only became widely available with the development of desktop computers and associated software (Fig. 3.7).

It is essential that knowledge of the taxonomy of oligochaetes inhabiting water bodies and the species-level keys be available before ecological surveys are undertaken, and these are still not available for some regions of the world. The use of descriptive approaches in oligochaete field studies was in part due to the fact that those involved in the studies were more taxonomically oriented and methods for data analysis, for the purposes of environmental assessment, were secondary. In particular, this was the case in Eastern Europe and Russia where there is a strong history of taxonomy and biogeographic description. While there is a significant volume of literature on the ecology of aquatic oligochaetes in these countries, much of it consists of listing the abundance of the commonest species and interspecies aggregations in different water bodies. This type of work and the retention of this knowledge and skills is absolutely essential before further and more sophisticated

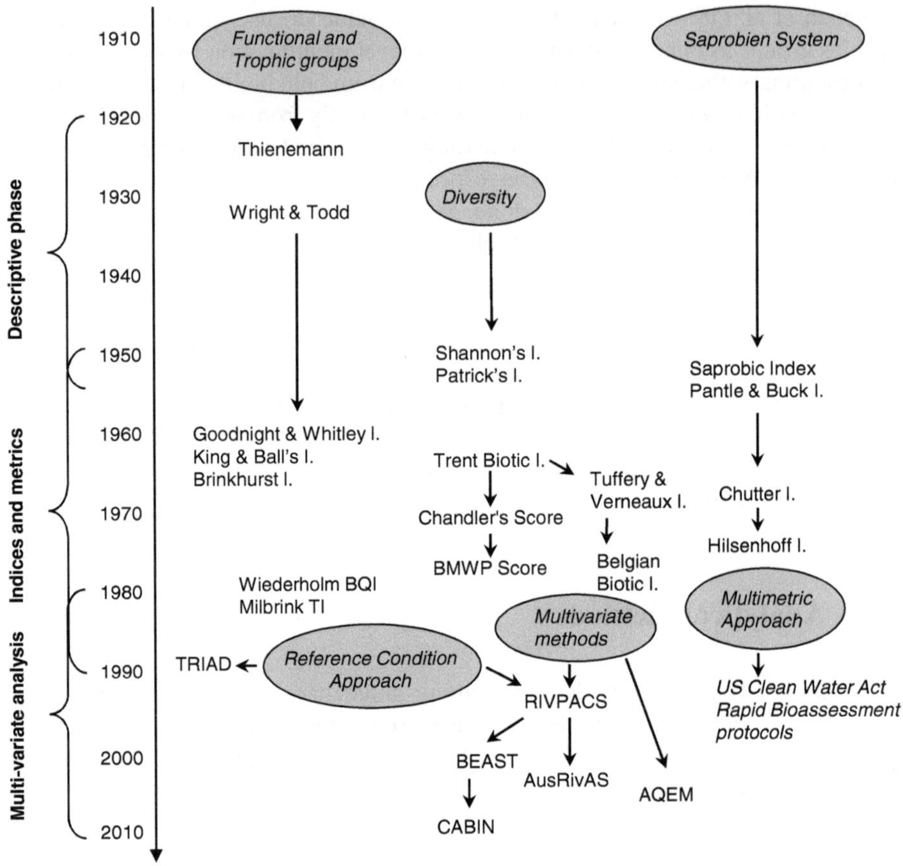

Fig. 3.7 Summary of the history of some of the biotic indices and metrics that have most influenced the water quality assessment, and in particular those that used oligochaetes

analytical methods can be applied, but new statistical tools can allow the testing of appropriate hypothesis related to pollution assessment that cannot be done with descriptive approaches. It is perhaps ironic that in western countries, and in North America in particular, as analytical methods have become more sophisticated resources and funding for taxonomic work is either in decline or is extremely limited. There is little investment in training new, young scientists in taxonomy, a field that is the basis of most branches of biological science.

The starting point for most analytical methods is a data matrix of samples or sites and an estimation of the abundance of taxa (either biomass but usually counts). These data are estimates of the actual population and are, therefore, biased to some degree, regardless of the semi-quantitative or quantitative nature of the sampling method. The bias depends on many factors, including the relationship between the substratum and the sampler used, the mesh size used (if any) for processing the sample and the magnification used in sorting (see Sect. 6.2). However, so long as the sampler behaves consistently over the range of substrata sampled, and all other variables are controlled, relative abundance values will be comparable. Sampling sites should be selected to minimize variation in microhabitat, or separate sampling protocols should be used for different microhabitats. Many of these sources of error are ignored in reports and publications, but should be considered or reported before applying the methods described below. The data matrix will begin as raw counts and may then be transformed in some way, the most extreme form being presence/absence data. The type of transformation used will depend on the type and assumptions of the analysis method used. Some statistical methods, typically univariate parametric methods, assume normality of distribution and therefore a transformation is almost always required to meet this assumption. Most multivariate and all univariate non-parametric methods make no assumptions about the distribution of the data and in those cases transformations are used to upweight or downweight the contribution of taxa in the analysis.

There is a very large literature on indices and statistical methods that have been applied to the benthos in general without specific reference to the oligochaetes. This literature is too extensive to be reviewed in detail here, but three reviews (Burd et al. 1990; Kathman and Brinkhurst 1991; Norris and Georges 1993) will provide an adequate introduction to this subject. A few additional sources are Ghetti and Bonazzi (1977), Herricks and Cairns (1982), Furse et al. (1984), Wright et al. (1984), and Johnson and Wiederholm (1989). Herricks and Cairns (1982) made many points that still have validity. For instance, the assumption that abiotic factors are the major controlling factors in organism distribution and community structure, and that biomonitoring tends to focus on descriptions of physical and chemical habitat factors, although the methods used to measure these variables are often not relevant to the experience of the animals themselves. These authors also mentioned the fact that biological factors, *e.g.* competition for space and food, are usually ignored due in part the limited knowledge of species autecology. Geographic and temporal separation of sampling stations are important but rarely mentioned factors related to benthic community structure (Burd et al. 1990) as well as the importance of understanding the scale and purpose of replication at site and regional levels (Bailey et al. 2004).

3.4.1 Indices and Metrics

The early approach using benthic community data for pollution detection and assessment involved expert interpretation of the data matrix by examination of trends in species distribution and abundance. These faunistic studies evolved into the subjective recognition of certain “indicator” taxa and their abundance relative to other forms, but in fact no universal indicator oligochaete species have been found. Quantitative data reduction methods, most notably diversity indices, had been proposed for stream quality assessment by Patrick as early as the late 1940s (Patrick 1949) and became increasingly popular through the 1970s. However, the diversity concept was never popular in interpreting the field response of oligochaetes to disturbance since similar or even lower diversity values were obtained at both ends of trophic gradients, that is both in unpolluted headwaters and in polluted reaches (*e.g.* Shannon's diversity index in the River Nervion ranged from 0.99 in unpolluted headwaters to 2.19 in a site partially recovered from pollution, Rodriguez 1984). Instead, a number of very simple indices involving oligochaete data were proposed (Table 3.6).

The first indices to be developed were very simple, one of the earliest being that of Wright and Tidd (1933) used in western Lake Erie, which was simply the number of oligochaetes per square metre. Total density was divided into categories representing the levels of pollution proposed by Wright and Tidd of: <1,000 ind. m^{-2} for natural, 1,000–5,000 ind. m^{-2} for moderately enriched, and >5,000 ind. m^{-2} for strongly enriched water bodies. These values were used in a number of studies in the Great Lakes (*e.g.* Nalepa and Thomas 1976). In a later version of this index, the number indicating severe pollution was raised to 10,000 ind. m^{-2}, over several replicates. A major problem with using abundance is that numbers are highly dependent on time of year, the sampling device and the sieve mesh size. In the early 1960s, Goodnight and Whitley (1961) and King and Ball (1964) developed simple indices that were based on the proportion of the community (abundance and biomass, respectively) represented by the oligochaetes. Goodnight and Whitley (1961) decided that >80% oligochaetes represented "highly polluted", and <60% "good" condition. The use of this classification in the Great Lakes was later limited to tubificids, excluding *Stylodrilus heringianus,* an early tacit acknowledgment of the fact that some oligochaetes were not associated with organic enrichment and the inclusion of the tolerance concept. Similarly, Brinkhurst (1966b) developed an index that incorporated the number of species of tubificids present and the proportion of *Limnodrilus hoffmeisteri* to total oligochaetes. This is, in fact, again one of the earliest applications of the relative tolerance of different species to enrichment assessment. The distribution of *S. heringianus* in the Great Lakes proved useful in that it was missing at shallow-water sites, near urban centres known to be polluted. However, the species was present in large numbers where large organic inputs were mediated (presumably) by an adequate oxygen supply.

One of the first applications of what are essentially tolerance values to the development of a quantitative index for oligochaetes was that of Howmiller and Scott (1977), who adapted an index originally developed for chironomids. Howmiller and Scott divided 26 oligochaete species, including tubificids, naidids and the lumbriculid *Stylodrilus heringianus* into three groups (Table 3.4). Seven species were classified as oligotrophic (Group 0), 11 species as mesotrophic (Group 1), and eight species as tolerant to organic pollution or extreme enrichment (Group 2). They used the Trophic Condition Index (*TCI*) for the oligochaete assemblage and created a simulated data set representing a hypothetical distribution of oligochaetes along a pollution gradient at 12 stations. They compared the results of using species diversity and the proposed index (*TCI*) on this data set. The diversity values were low at both ends of the gradient, but the *TCI* varied steadily from 2.0 at the heavily polluted site to 0.0 at the clean station 12. The authors then examined a real data set for

Table 3.6 Environmental metrics for pollution assessment based on aquatic oligochaete community data

Index	Source	Metric
Total density of oligochaetes	Wright and Tidd (1933)	Number of oligochaetes per square metre
% Oligochaetes	Goodnight and Whitley (1960)	Relative abundance of oligochaete (or tubificids) to the benthic community: < 60% Good, 60–80% Doubtful, ≥ 80% Highly polluted (either organic or industrial)
Oligochaete biomass	King and Ball (1964)	Wet weight ratio of aquatic insects to tubificids. In extreme conditions the ratio tends to zero
% Limnodrilus hoffmeisteri	Brinkhurst (1966a, b)	Abundance percentage of *L. hoffmeisteri* to total tubificids
Trophic Condition Index	Howmiller and Scott (1977)	$\text{TCI}=\frac{\sum n_0+2\sum n_2}{\sum n_0+\sum n_1+\sum n_2}$ where n_0, n_1, and n_2 are the total number of the oligochaetes belonging to species in Groups 0, 1 and 2 respectively (Table 3.4)
Wiederholm Lake Index	Wiederholm (1980)	No. oligochaetes/(No. oligochaetes + No. chironomids)
Benthic Quality Index	Wiederholm (1980)	$BQI=\sum_{j=0}^{5}\frac{h_j y_{ij}}{\sum_{j=0}^{5} y_{ij}}$, where y_{ij} = no. individuals of each indicator group j in site i; $\sum y_{ij}$ = total no. individuals of all indicator groups j in site i; h_j = score (0–5) according the indicator values of different taxa
Trophic Index	Milbrink (1983a)	$\text{TI}=C\frac{0.5\sum n_0+\sum n_1+2\sum n_2+3\sum n_3}{\sum n_0+\sum n_1+\sum n_2+\sum n_3}$ C = 1 where oligochaete abundance >3,600 ind. m^{-2} (using a sieve opening of 0.6 mm), C = 0.75 for 1,200–3,600 ind. m^{-2}, 0.5 for 400–1,200, C = 0.25 for 130–400 ind. m^{-2}, and C = 0 when abundances fall below 130 ind. m^{-2}; n_0, to n_3 as in Trophic Condition index
% Tubificids	Slepukhina (1984)	D = number Tubificidae/Total number of oligochaetes. Range of values 0–1
Index of biological quality	Lafont (1984)	$Io = 10S\ T^{-1}$ where S = total number of oligochaete species, T = relative abundance of tubificids without hair setae to the total oligochete worms

(continued)

Table 3.6 (continued)

Index	Source	Metric
Composite index of biological quality	Lafont (1984)	Eo, is codified by letters representing classes of relative abundance. Subindex is a code of oligochaete species richness
Oligochaete index of sediment bioindication	Rosso et al. (1994)	$IOBS = 10S\ T^{-1}$ where S = total number of oligochaete species, T = percentage dominant tubificid group (either with or without hair chaetae) to the total oligochete worms
Environmental Quality Class (EQC) value	Verdonschot (2006)	EQC value and tolerance values at family, genus and species level

55 stations in three areas of Green Bay, Michigan, defined by physical and chemical parameters. Five indices, the new Trophic Condition Index, Shannon‘s species diversity, total abundance of oligochaetes,% oligochaetes and % *L. hoffmeisteri* among tubificids (Table 3.6) were calculated for these data, and only the *TCI* discriminated the three areas, at a statistically significant level (using the non-parametric Mann-Whitney *U* test). This supported the view that incorporating species-level information did refine the interpretation of the data and improved the sensitivity of the analysis.

The *TCI* of Howmiller and Scott (1977) was modified by Milbrink (1983a) who proposed several changes. He added a fourth group of species that consisted of *T. tubifex* when it occurs with *L. hoffmeisteri* alone and recognized the fact that *T. tubifex* can exist at both ends of the trophic scale which is one of the problems with the approach of assigning a tolerance value to species. He added both an n_0 group to the numerator, multiplied by 0.5, and also a new n_3 group, multiplied by 3 (see Table 3.6). Both *TCI* approaches upweight the count for eutrophic species and downweight the contribution of oligotrophic species, so that the value of the index increases as systems become more eutrophic. The inclusion of total abundance in Milbrink's version of the *TCI* simply weights the index further based on overall productivity, so that it ranges from 0 to 3, compared to the range of 0–2 for the original Howmiller and Scott index. Again, an artificial data set representing taxa common in Swedish lakes was used and showed the Trophic Index (*TI*) performed well. However, using simple total abundance produced a very similar result. As before, diversity indices of two types failed to discriminate between stations well, but there were only 8 taxa in the data set. Milbrink also calculated *TI*s for 16 lake basins in Sweden. The results again showed good agreement between total worm abundance and the *TI*, and between both descriptors and a factor composed of total phosphorus divided by mean depth.

In Russia, Slepukhina (1984) compared some methods for assessing water quality. She found the percentage of oligochaetes to be useful in detecting pollution in rivers at three levels (clean, doubtful, polluted), although in Lake Ladoga where oligochaetes can represent 50–90% of the fauna in the central part of the lake, Slepukhina seemed to imply that the area should be classified as clean, despite stating that there were local discharges of organic pollution into the lake. This demonstrated that using proportion of oligochaetes alone has severe limitations, as some systems, particularly the deep profundal in lakes where natural anoxia can occur, are naturally dominated by oligochaetes. She also applied a formula relating the number of tubificids to the total number of oligochaetes, essentially the Goodnight and Whitley method, without success in Lake Ladoga, since in the mesosaprobic zone this ratio can encompass the entire range of possible values (0–1).

The Saprobien system is an empirical approach to water quality assessment that uses the tolerance values of the invertebrate assemblages to organic pollution. It was applied to benthic communities by Kolkwitz and Marsson (1908) and has been refined through numerous contributions, some of which applied to species of aquatic oligochaete. Currently, it has been incorporated in the implementation of the European Water Framework directive for the evaluation of the river Danube and in several central European countries. In Bulgarian rivers, Uzunov applied the Saprobien system using aquatic oligochaetes. The saprobic valence (s_i) was determined for each species as the score corresponding to the maximum on the normal distribution curve for an environmental variable, and the saprobic index for each site was expressed by the Zelinka and Marvan formula. Uzunov (1982a) evaluated the significance of the type of substratum and the saprobity as abiotic factors conditioning the oligochaete distribution in rivers. Using six abundant species, the author calculated the action power (η_i^2) of the species dispersion caused by any abiotic factor and determined a mean Saprobity η^2 value of 20.5%, and a mean Substrate η^2 value of 4.8%. He concluded that saprobity (in association with the oxygen regime) has a leading role in the distribution of aquatic oligochaetes, and the influence of the nature of substratum is significant but secondary. In another paper, Uzunov (1982b) developed theoretical distribution curves of abundance relative to dissolved oxygen and organic (saprobic) gradients for six naidid species and four tubificids. He estimated oxygen concentration to be responsible for 35% of the distribution factors, and the organic matter (saprobity) of the water and the substrate were less significant, but still important. This work on Bulgarian rivers enabled the author to estimate saprobic values for 113 oligochaete species (Uzunov et al. 1988). The values ranged between 0.1 (for *Trichodrilus leruthi*) and 3.8 (for *Limnodrilus udekemianus*), with a mean value of 2.0, which corresponds to a β-mesosaprobic condition. Verdonschot (2006) also used a modified saprobic valence to classify oligochaete species from an extensive European database of 101 taxa from about 400 streams. In the classification, the sum of oligo and β-mesosaprobic species is higher than the α-mesosaprobic and polysaprobic species (Fig. 3.8), which is contrary to the general view that considered oligochaetes as mainly restricted to the most organically enriched conditions.

Analogous to the Saprobien system of classifying aquatic macroinvertebrates was the Hilsenhoff Index developed in North America. Originally developed in 1977 by Dr. William Hilsenhoff from the University of Wisconsin to assess the effects of low dissolved oxygen caused by organic loading in streams, it has been

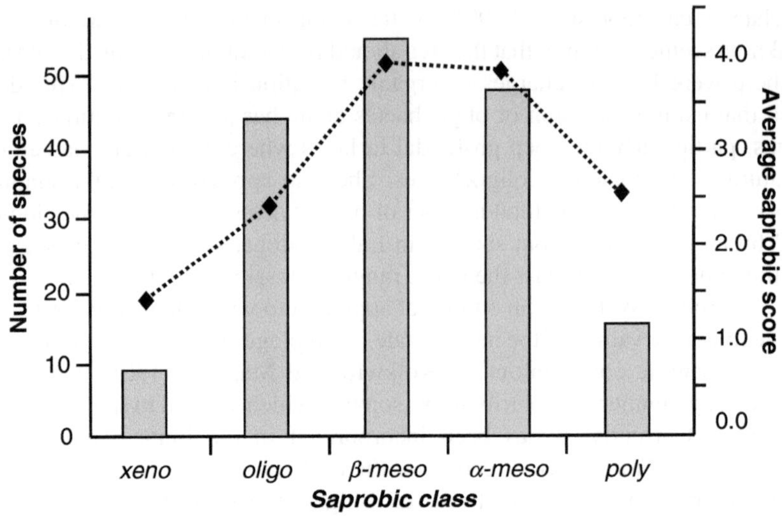

Fig. 3.8 Oligochaete diversity (*bars* – number of species) and average Saprobic indices (*line*), according to Zelinka and Marvan, for different European water quality classes as reported in the AQEM database (From Verdonschot 2006, Fig. 3, redrawn and reproduced with permission of Springer Publ., © conveyed by Copyright Clearance Centre, Inc.)

refined and values have been developed for oligochaete species (Table 3.7). This index is still widely used and is often a component of the multimetric indices that contribute to the Index of Biological Integrity (*IBI*) used widely in the USA for stream and river bioassessment. Lenat (1993) and Bode et al. (1996, 2002), as well as Barbour et al. (1999), have published tolerance values at the species level for oligochaete taxa for use in a modified Hilsenhoff Index ranging from 0 (sensitive) to 10 (tolerant). The values for oligochaete taxa vary from 2.8 to 10, but very few species from unpolluted sites have been included. For instance, single but different tolerances values are reported by Lenat (1993) and Bode et al. (1996) for the whole family Lumbriculidae (Table 3.7) and values have not been developed for the many lumbriculid species that are associated with pristine headwaters (S. Fend, pers. comm.).

Looking for simple indices, Gross (1976) found the following criteria to be useful in detecting mild organic pollution in running waters in France: the specific diversity of the oligochaetes, the percentage of oligochaetes in the benthos, and the importance of the association between *T. tubifex*, *L. hoffmeisteri*, and *L. udekemianus*. Using some of these criteria, Lafont (1984) and his co-workers proposed several indices (Table 3.6). These evolved from simple calculations such as the total number of oligochaete species in the sediment or the relative abundance of tubificid worms without hair chaetae (in fact a slight modification of the index proposed by Brinkhurst based on the % of *L. hoffmeisteri* as the species without chaetae found were mostly *Limnodrilus* spp.), to other indices such as the *Io* which include both

Table 3.7 Tolerance levels for oligochaete species used in the Modified Hilsenhoff Index calculations as reported by Bode et al. (1996, 2002) and Lenat (1993, reproduced with permission of the North American Benthological Society)

Taxon	Bode et al. (1996, 2002)	Lenat (1993)
Tubificinae		
Aulodrilus americanus	7	–
Aulodrilus limnobius	7	5.2
Aulodrilus pigueti	7	4.7
Aulodrilus pluriseta	7	–
Bothrioneurium vejdovskyanum	7	–
Haber speciosus		2.8
Ilyodrilus templetoni	10	9.4
Isochaetides freyi	8	7.6
Isochaetides curvisetosus	10	7.2
Limnodrilus cervix	10	10.0
Limnodrilus claparedianus	10	–
Limnodrilus hoffmeisteri	10	9.8
Limnodrilus profundicola	10	–
Limnodrilus udekemianus	10	9.7
Limnodrilus spp.	–	4.9
Potamothrix moldaviensis	8	–
Potamothrix vejdovskyi	8	–
Quistadrilus multisetosus	10	6
Spirosperma ferox	6	–
Spirosperma nikolskyi	–	7.7
Spirosperma (Peloscolex) spp.	10	8.8
Tubifex tubifex	10	10.0
Tubificidae with hair chaetae	10	–
Tubificidae without hair chaetae	10	–
Rhyacodrilinae		
Branchiura sowerbyi	6	8.4
Rhyacodrilus spp.	10	–
Naidinae		
Amphichaeta americana?	6	–
Arcteonais lomondi	6	–
Chaetogaster diaphanus	7	–
Chaetogaster diastrophus	7	–
Chaetogaster limnaei	7	–
Chaetogaster setosus	7	–
Dero digitata	10	–
Dero furcata	10	–
Dero nivea	10	–
Dero obtusa	10	–
Dero sp.	10	10.0
Haemonais waldvogeli	8	–
Nais barbata	8	–
Nais behningi	6	–

(continued)

Table 3.7 (continued)

Taxon	Bode et al. (1996, 2002)	Lenat (1993)
Nais bretscheri	6	–
Nais communis	8	–
Nais elinguis	10	–
Nais pardalis	8	–
Nais simplex	6	–
Nais variabilis	10	–
Nais sp.	8	9.1
Ophidonais serpentina	6	–
Paranais frici	10	–
Piguetiella michiganensis	6	–
Pristina aequiseta	8	–
Pristina breviseta	8	–
Pristina leidyi	8	–
Pristina synclites (menoni?)	8	–
Pristina sp.	8	9.9
Pristinella jenkinae	8	–
Pristinella osborni	8	–
Pristina spp.	8	–
Ripistes parasita	8	–
Slavina appendiculata	6	7.1
Specaria josinae	6	–
Stylaria lacustris	6	8.5
Vejdovskyella comata	6	–
Vejdovskyella intermedia	8	–
Vejdovskyella sp.	6	–
Lumbriculidae	5	7.3
Eclipidrilus sp.	5	–
Stylodrilus heringianus	5	–
Enchytraeidae	10	10

Values range from 0 (very intolerant of organic waste) to 10 (very tolerant of organic waste)

richness and abundance of tubificid worms without hair chaetae. The values of the *Io* index of Biological Quality ranged between 0.1 and 0.9 for the pollution stage before azoic sediments (1–8 oligochaete species) to 5.1–8 for slightly polluted sediments (21–28 oligochaete species). This index was later modified (Rosso et al. 1994) to a dominance index called the Oligochaete Index of Sediment Bioindication (*IOBS*). Both indices are unusual as they reduce the complexity of information by rolling up information to a higher taxonomic level, (*viz.* those animals with or without hair chaetae), when identification was first performed to species level. The sampling procedure was apparently quantitative (a Surber sampler of 100 cm^2 or a core of 25 cm^2 sampling area), however the incorporation of number of individuals with or without hair chaetae is effectively lost by using relative abundances. Lafont (1984) also proposed an alphanumeric code system which integrates oligochaete richness and abundance information (Composite Index of Biological Quality, *Eo*). In this system, the numeric subscript indicates the richness class (1: 1–2 species, 2: 2–3 species, and so on) and the letter the relative abundance of tubificids without hair chaetae. The index could range from A_1 where the site would be represented by a community with one or two species with over 91% of the tubificids without hair chaetea (*i.e.* largely *Limnodrilus* spp.) to F_{11} where the community would have 21 or 22 species with ≤15% tubificids without hair chaetae. In fact, one of the assumptions of this system, that there are differences in the indicator value of tubificids with and without hair chaetae, is not supported by the results of a large European study (Verdonschot 2006).

In another French study, Giani (1984) examined the performance of various indices in determining different levels of heavy pollution. Forty six oligochaete taxa were recorded from seven sites along a small river and the number of oligochaetes increased progressively from upstream to downstream, except for a dramatic reduction at one site and a near total absence of worms at a site below a zinc foundry. Fourteen potential indices were evaluated and Giani, using Milbrink's Trophic Index (*TI*), suggested that two species (*L. udekemianus* and *L. profundicola*) be assigned to the next higher group of species than proposed by Milbrink (1983a). Comparing the original (*TI*) with his modification (TI_{mod}) of the index, Giani suggested that the indices do appear to respond differently.

Re-examination of the data reported by Giani using MDS ordination shows that measures that incorporate richness tend to characterise one type of response (*Io* and richness), and those based on relative abundance of tubificids another (Gp 3a in Fig. 3.9). According to Giani, the Trophic Index (Milbrink 1983a) was the least responsive and the modification suggested by Giani had little effect. Perhaps not surprising as it is an index based on response to the trophic gradient (as are most tolerance based indices) and not metal contamination. This is a good illustration of the fact that tolerance based systems need calibration for the type of stress they are being used to assess. The trophic condition indices had a similar response range as the three diversity indices (Shannon-Wiener, Simpson and Pielou's eveness). The M_3 index of Bournaud Keck has a very different behaviour to all the others, and is particularly responsive to dominance. In the study by Giani (1984), where contamination was severe, the simplest indices, such as density and richness, the proportion of the families and species dominance, were adequate to characterise system response. In an earlier paper on this study, Giani (1983) concluded that, when analyzing the total benthic community, identification of the oligochaete species added little to the analysis. Giani also found that the total study of the fauna and the use of various simple indices gave very similar results. However, the study was conducted in a system with severe effects and most methods should be effective, and therefore the conclusions from this study should not be extrapolated too far.

A more quantitative assessment of seven indices was undertaken by Verdonschot (1989). He examined data from regulated streams in the Netherlands. Twenty four sites were sampled along a pollution gradient with ten replicates taken at each season, and at each site the habitat was sampled proportionally according to its occurrence. Based on the whole macroinvertebrate fauna Verdonschot observed an

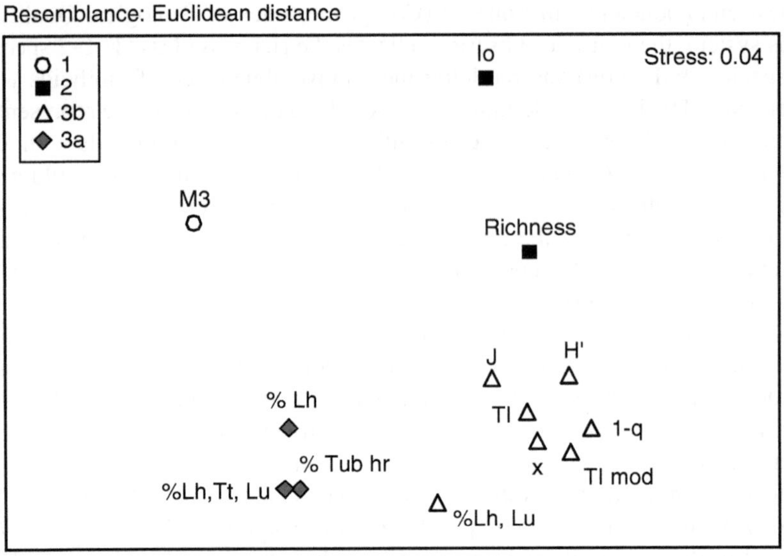

Fig. 3.9 Comparison of 14 indices of oligochaete communities with nMDS ordination from the River Mort (Giani 1984) using standardised data. Shown are four associations among the indices from cluster analysis (*Lh: Limnodrilus hoffmeisteri, Lu: L. udekemianus, Tt: Tubifex tubifex, Tub hr*: Tubificids with hair chaetae, *H'*: Shannon-Wiener diversity index, *M3* and *x*: Bournaud Keck indices, I_o: Lafont Index, *J*: Pielou Index, *TI*: Milbrink'sTrophic Index, see Table 3.6)

environmental gradient from reference to polluted by sewage (two sites formed single clusters and were not considered part of the organic gradient). Verdonschot then examined the oligochaete communities to see if they could discriminate the same four clusters from unpolluted to heavily polluted. Neither total oligochaete abundance nor number of oligochaete taxa showed a clear pattern. Verdonschot also compared seven of the oligochaete indices. Examination of how well the indices discriminated the four clusters showed the % tubificids to the total oligochaetes, the Benthic Quality Index, Trophic Condition Index and Milbrink's Index to be insensitive, while the average values for three of the indices, % oligochaetes, % *L. hoffmeisteri* and the Lafont's Index *Io* followed the pollution gradient, but did not discriminate among the clusters, and Verdonschot concluded that none of the indices were sensitive enough to be useful for assessing these streams.

Probably the most recent development in the tolerance value approach for oligochaetes is the work of Verdonschot (2006) on the development of a macroinvertebrate assessment system for European streams. This was done as part of the requirements of the European Water Framework Directive (WFD). Almost 900 samples were taken from approximately 400 streams covering 29 stream types in eight countries. Verdonschot analysed the oligochaete data from this European database. Almost half a million specimens of oligochaetes were collected in 772 samples. Eight families, 41 genera and 69 species were recorded, and 189 environmental

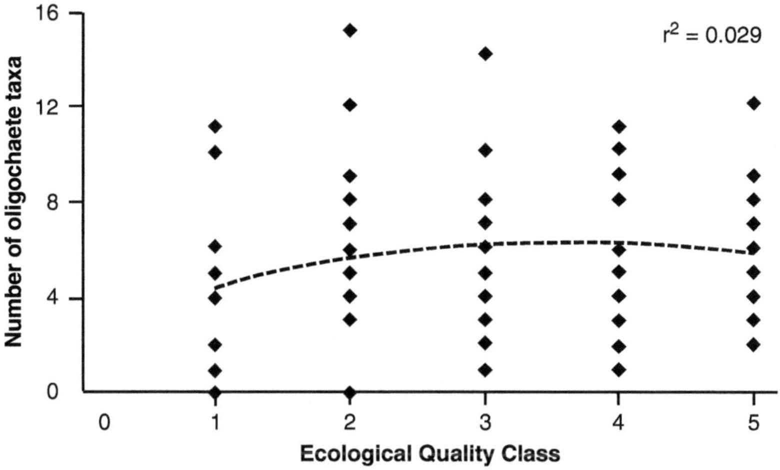

Fig. 3.10 Number of oligochaete taxa at each site and the assigned quality class according to the AQEM assessment system in European water bodies. Ecological classes: *1* – bad quality to *5* = high quality (From Verdonschot 2006, Fig. 5, redrawn and reproduced with permission of Springer Publ., © conveyed by Copyright Clearance Centre, Inc.)

variables were included in the analysis. Streams were assigned to one of 28 categories using the Water Framework Directive "System A" descriptors which are categorical at the landscape scale and include: ecoregion, catchment size, geology, and altitude. The multimetric AQEM assessment system (Hering et al. 2004) was used to classify each site into an ecological quality class and oligochaete richness had no relationship with the classification of sites into Ecological Quality Classes (EQC), although oligochaete relative abundances shows a slight ($r^2 = 0.168$) inverse relationship with stream quality. It was clear that the number of oligochaete species occurring across the ecological quality classes (EQC) was the same (Fig. 3.10), but there was a slight trend to increase relative abundance of oligochaetes in poorer quality waters. Verdonschot also showed that different species followed distributions according to ecological quality class, and that oligochaetes at the species level can discriminate the range of ecological quality described across Europe. Weighted averaging was used to evaluate the relationship between oligochaete distribution and the ecological quality class, and also showed two major variables explained much of the observed distribution patterns: first, altitude, which represented current gradients, slope and temperature, and second, total phosphorus concentration, which represents both organic pollution and eutrophication gradients. From these data Verdonschot (2006) was able to generate both optima and tolerances for the EQC, altitude and total phosphorus for 69 oligochaete species (Table 9 in Verdonschot 2006). For instance, the EQC optimum for the whole oligochaete fauna was 2.24, in a scale of 1 = bad and 5 = high quality, where the lowest values were for the so-called "Tubificidae without hair chaetae" (EQC = 1.78) and maximum for Propappidae (EQC = 4.11).

Interestingly, among Tubificinae species are some of the highest EQC values (*e.g. Spirosperma ferox*, EQC = 4.47), as well as the lowest (*e.g. Potamothrix bavaricus*, EQC = 1.0). The lumbriculids *Stylodrilus parvus* and *S. brachystilus* showed the highest EQC values (4.43 and 4.54, respectively). So, in fact, the oligochaetes encompass a very wide range of environmental quality in streams. It is also noteworthy that the values of EQC do not show any correlation with saprobity indices (Table 3.8).

Although these later studies were more quantitatively rigorous, these are the exception in the index based approach. Most studies that have compared indices either qualitatively or quantitatively have found that the indices developed elsewhere do not work that well in studies in other locations, suggesting that they have little capacity for extrapolation, and that simple indices work well, primarily when effects on benthic community are so severe that any method will identify disturbance. This leads us to conclude that indices only work locally which makes their application somewhat circular and of limited utility. They were developed primarily as a method for integrating larger amounts of information and other more rigorous and quantitative approaches are now available to do precisely that and are described below.

3.4.2 Multivariate Approaches

The development of the metrics and indices described above was largely driven by the need to reduce the information on multiple species assemblages into a single description of the community that captured the interactions and influences of environment on that assemblage. There is, in fact, a whole branch of statistics specifically suited to the task of recognizing patterns with large numbers of input variables and for determining relationships among matrices, that is, multivariate statistics. The first steps in developing multivariate statistical methods were taken in the 1870s with the paper on singular value decomposition by Eugenio Beltrami, together with work on linear and multiple regression by Sir Francis Galton and his collaborator Karl Pearson, who is sometimes considered as being the originator of Principal Component Analysis. However, until the advent of mainframe and personal computers multivariate analysis, although theoretically of value, was of no practical application to the environmental scientist because of the number and intensity of calculations involved. In the last 10 years, both the accessibility of relatively inexpensive personal computers and specialized multivariate statistical software (*e.g.* CANOCO, PATN, PCORD, PRIMER) or the inclusion of multivariate options in more popular software packages (*e.g.* MINITAB, SAS, SPSS, SAS, STATISTICA) have made these methods available to the *non-cognoscenti*.

The particular attributes of multivariate statistical methods that make them especially suitable for use with field collected community or species assemblage data are that they extract information from data sets with multiple variables by either grouping sites or variables into units that are more similar to each other than to other

Table 3.8 Trophic ranks of aquatic oligochaetes from various sources (Modified after Verdonschot 1989), and Ecological Quality Class (EQC) values from DCCA ordination (From Verdonschot 2006, Table 9, with permission of Springer Publ, © conveyed by Copyright Clearance Centre, Inc.)

	Citation[a]					
Taxon	MH	HS	M	W	S	EQC Optimum
Aulodrilus pluriseta	2	2	3	–	3	2.57
Dero digitata	–	2	–	–	3	2.67
Ilyodrilus templetoni	2	2	3	–	3	2.02
Limnodrilus claparedianus	4	3	3	–	2–3	2.18
Limnodrilus hoffmeisteri	3	3	4	4	3	1.57
Limnodrilus profundicola	1	1	2	–	2–3	1.08
Limnodrilus udekemianus	3	3	3	–	2–3	1.71
Lumbriculus variegatus	–	–	–	–	2	2.83
Nais elinguis	–	2	–	–	–	1.85
Nais pseudobtusa	–	–	–	–	–	1.0
Ophidonais serpentina	–	–	–	–	2	2.07
Potamothrix hammoniensis	–	–	3	3	3	1.66
Psammoryctides barbatus	–	–	2	–	1	2.19
Rhyacodrilus coccineus	1	1	2	–	1	3.30
Stylaria lacustris	–	–	–	–	i	1.94
Tubifex tubifex	4	3	1–4	–	1–3	1.42

MH Mozley and Howmiller (1977), *HS* Howmiller and Scott (1977), *M* Milbrink (1973), *W* Wiederholm (1980), *S* Särkkä (1987)

Codes: *1* oligotrophic, *2* mesotrophic, *3* eutrophic, *4* hypereutrophic, *i* indifferent. EQC optimum is the mode for each species along the environmental gradient under study, and it ranges from *1* (= low) to *5* (= high) quality scores

[a]Authority reference from Verdonschot (1989)

units (classification methods) or by reducing the number of variables required to explain the variation within a data matrix (ordination methods). As it relates to analysis of community assemblages, classification and ordination methods make use of information on the taxa present as well as their diversity and relative or absolute abundance.

There have been two main proponents of the application of multivariate approaches to the analysis and interpretation of oligochaete field communities in response to environmental disturbance. In Switzerland, Claude Lang has published many accounts of aquatic oligochaetes in relation to the trophic status of lakes and, in the Netherlands, Piet Verdonschot has looked at the response of oligochaete communities in small ditches and streams.

In an early paper, Lang (1978) proposed a multivariate approach and using Factorial Correspondence Analysis with a community data set for the tubificids and lumbriculids from 31 stations sampled in the profundal zone of Lake Geneva (Switzerland). From the analysis Lang concluded that local spatial and seasonal variation were small, based on paired samples and ten sites sampled in both spring and fall. This conclusion was not entirely supported by the data of nine paired

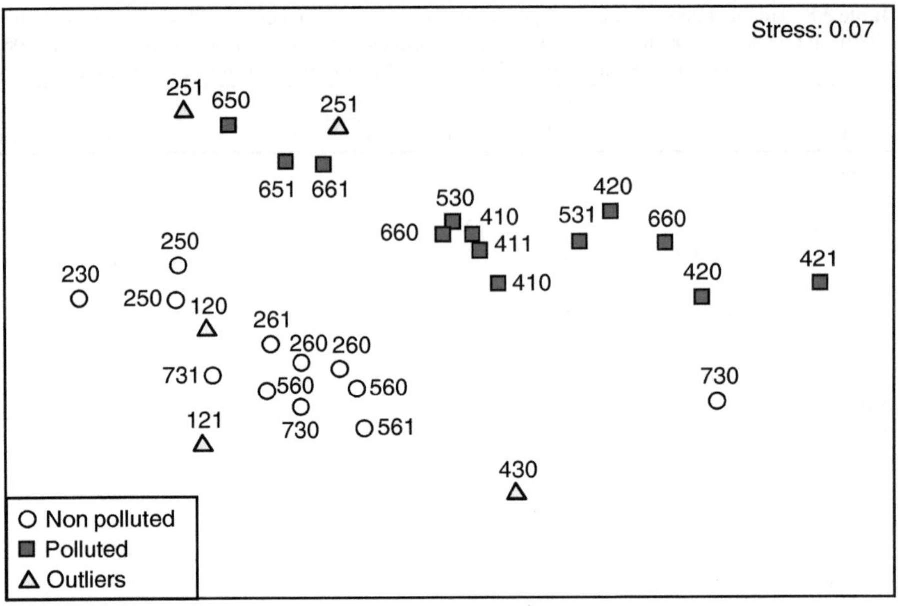

Fig. 3.11 Ordination plot derived from MDS analysis, after re-analysing Lang's data (1978) on oligochaete communities from the profundal zone of Lake Geneva (Switzerland). Sites which were outliers (Δ) in Lang's analysis are clearly in either the polluted (•) or unpolluted (○) group

samples, and at least one (St 730) or possibly two (St 410) showed considerable variation and a more quantitative analysis comparing the paired samples with randomized pairs of stations would have determined if the pairs were more similar. Lang also used data from a previous study to identify stations as either polluted or non-polluted, based on a sewage treatment plant impact study that characterized the sites based on worm species abundance, and interstitial water and sediment chemistry using Discriminant Analysis. The main points of Lang's thesis are: small scale distribution is not important, the oligochaete assemblages discriminate polluted from non-polluted sites, and there is little seasonal variation. Using more modern and appropriate methods developed over the last 30 years, we can reanalyze these data to determine whether these patterns can be confirmed or more rigorously quantified. The 31 sites were analyzed using nMDS in PRIMER 6, with the similarity matrix being calculated using the Bray-Curtis dissimilarity index and data were not transformed. From the ordination plot (Fig. 3.11) there are several points to note. The stress value is very low (0.07), indicating this is an excellent description of the relationship between ordination and the similarity matrix, and therefore 2 axes are quite sufficient to describe the variation, compared to the three axes used by Lang. There is a clear separation of the sites predefined as polluted and non polluted, furthermore, the outlying sites from the Lang analysis are clearly contained within either the polluted (120, 121, 430) or non-polluted (251) groups, and there is only one site that is equivocal (Station 730). There is clearly no seasonal pattern as spring and fall samples were not discriminated in ordination space. Using Analysis of Similarity (ANOSIM) it is possible to quantify both the seasonal and pollution source effects. Season was not significant ($p = 0.428$), however, exposure to pollution was significant ($p = 0.005$). The occurrence of worm species at sampling locations was used as an indicator of their sensitivity to pollution. This confirmed the oligotrophic status of *Stylodrilus heringianus*. Lang also considered *Spirosperma ferox* as a mesotrophic species (unlike Milbrink or Särkkä who considered it to be oligotrophic, see above) and a transitional series of species from oligotrophic to eutrophic within the genus *Potamothrix*. Other species are discussed in relation to their use as indicators. Lang concluded that pollution *per se* is not a scientific concept, and should be regarded as a comparative notion. Hence, the establishment of reference sites is of value, and this study demonstrated the progressive drift towards eutrophy in the lake at that time.

In 1979, Lang and Lang-Dobler tried to relate individual oligochaete species to specific sediment chemical variables. The mean concentrations of several pollutants from samples containing each species were used to define the species chemical environment. Six groups of species were identified using Factorial Correspondence Analysis. Each group was associated with a single major chemical factor. Pollution level in the sediment decreased from Group 1 to Group 6, as did the pooled coefficient of variation of the ten chemical variables. Lang later adopted an approach based on classifying species into trophic groups and quality assessments were based on the relative abundance of the oligotrophic group (see Sect. 3.2). There are some alterations to the species lists from publication to publication, but none of the habitats involved pollution levels to the degree that the communities were restricted to a hypereutrophic assemblage. Grouping the species in this manner overcomes many problems: zero records are eliminated, immature specimens can often be attributed to the groups, seasonal variations are damped, and zoogeographical problems are reduced.

Verdonschot (1987, 1989) worked on drainage ditches in the Netherlands, a very different habitat, using complex multivariate methods. In a first study, 97 sampling sites contained 22 identified oligochaete species and 33 environmental variables were also measured. Cluster analysis, ordination (by Detrended Correspondence: DCA and Detrended Canonical Correspondence Analysis: DCCA), and loglinear regressions were used. The combination of the three methods reveals all of the information in the data matrix, in contrast to simple indices that only capture a small part of the richness in the data. Oligochaetes were the third most abundant group among the benthos but had relatively low species richness. From ordination, the first four axes accounted for only 50% of the variation in the weighted average of the oligochaete species in relation to the environmental variables, and the first axis only 20%, suggestive of weak relationships between species distributions and the measured variables. Station clusters were characterized principally by abundant species and secondarily by frequent species. The site clusters, species, and environmental variables are presented in one complex figure from which a great deal of information can be derived.

In the second paper, Verdonschot (1989) addressed some of the problems related to using such data in water management. A pollution gradient of 24 sites was sampled to study the utility of various indices. Nineteen oligochaete taxa were recorded at replicated sites. It was found that a single sample, taken in December, was sufficient to collect 73% of the taxa, and that three replicate samples would obtain most of the species present over the study period. A similar result was obtained in April, but in July and September 6– 8 replicates would be required to collect all the common taxa. However, only a single sample would be sufficient to collect all those commonly present in the summer months. Species abundances were more useful when transformed on a lognormal scale. Cluster analysis was used to identify four groups of stations, and two sites were separate outliers from these groups. Various indices were used on the mean data from each cluster. It was anticipated that the indices would not discriminate the rather similar Groups 1 and 2, but that the rest should indicate a gradient of pollution impact. These data, with the environmental variables, were then analyzed by DCCA and Verdonschot illustrated the tolerance or saprobic ranking of the species according to different sources (Table 3.8). Verdonschot concluded that monitoring should include a Reference Condition scheme based on the total benthos that identifies clusters of stations, that this be used as background data in an ordination that includes biotic and abiotic data, and that each new site be assessed by projecting it into the reference scheme. This approach is in fact analogous to the RIVPACS and Reference Condition approaches used in the UK, Canada and Australia.

In a later attempt to develop an index specifically for lakes based on multivariate analyses, Lafont et al. (1991) used Principal Components Analysis (PCA) on 12 physico-chemical variables, and projected species richness and abundance of oligochaetes as supplementary data. Through stepwise linear multiple regression, the authors computed an index (*EOLA*) which equals the mean number of oligochaete species plus three times the mean log of their abundance (mean number of individuals per 0.1 m^2). In a study of 12 lakes, the index ranged between 3 and 20.

In North America, the Triad approach was initially proposed by Peter M. Chapman and co-workers, for particular use in sediment assessment. Originally used to convey the idea of comparing three data sets (bulk chemistry, toxicity testing, and community structure), the idea became associated with the development of numerical biological criteria. Reynoldson and Zarull (1993) discussed this approach in the context of sediment contamination in the St. Lawrence Great Lakes. They suggested that structural components of the benthic community cannot demonstrate causality which requires functional (toxicological) testing, a feature of the Triad approach. Some of the previous studies indicated the possibility of producing meaningful correlations between physical-chemical data and community structure through statistical evaluations. While correlation is not proof of cause and laboratory toxicology can easily miss effects on field species, when responses in laboratory tests are observed, provide direct evidence of sediment contaminants being the causal agent. The authors described a step-wise strategy of increasing the intensity of investigation based on failure to meet set criteria at each step. This approach was developed over a 10-year period and summarised by Grapentine et al. (2002) where information from four lines of evidence – surficial sediment chemistry, laboratory toxicity, invertebrate community structure and invertebrate tissue biomagnification – was integrated within each line to produce a pass ('–') or fail ('+') conclusion, and then combined across lines resulting in one of 16 outcome scenarios. For each scenario, the current status of the site, interpretation, and management recommendations were given. Management recommendation(s) ranged from "no action" to "risk management required" (the later in 9 of the 16 scenarios). Within each line of evidence, the strength of each response can also be ranked (*e.g.* score of 1 to 4), providing managers with more information to aid decision options.

In summary, there is a clear trend in all these studies from simple descriptions to the development of indices and scoring systems that try increasingly to capture the rich complexity of species assemblage data and incorporate ecological understanding. Multivariate techniques are the most recent development in this trend and at the present time are probably the most appropriate tool for analysing complex data sets. However, these methods simply identify patterns in the data. Their use should never preclude actual examination of the trends in species distributions and developing an understanding of what is responsible for these patterns by application of the knowledge available on the sensitivities and biological requirements (habitat, food, etc.) of the species.

3.5 Other Field Approaches

3.5.1 Size Spectrum Analysis

Among recent developments, size spectrum analysis is a method that uses variables that avoid the expertise required and the time taken to identify taxa. Real et al. (1992) used data on the width of the eighth segment together with physical and chemical data from 100 reservoirs in Spain in order to determine those factors of most significance in determining oligochaete distributions, and those reservoirs subjected to stress. Multivariate statistical methods and cluster analysis were used. Oligochaetes represented over 90% of the fauna, but only 8 species were present and more than 90% of the worms were immature. From an inspection of the species list, it is clear that all but the three *Limnodrilus* species and *Tubifex tubifex* should have been recognizable from immature specimens. The size spectra proved useful in identifying several environmental factors, such as high chloride and iron content in deep water or Na-rich silicates in sediment, that were possibly structuring the assemblage. The stress mechanism on the benthic fauna was not discussed, and requires further experimental study. The exclusion of the small, asexually reproducing naidids from the data set had no effect on the patterns observed. The authors considered the motility of the naidids to be of significance as the species were found in stressed habitats along with other motile forms (*e.g. Procladius, Chaoborus*).

The use of the size spectrum of the oligochaete fauna rather than species identification could produce misleading results, since species of similar size can be found in oligotrophic (*e.g. Embolocephalus* or *Spirosperma* spp.) or eutrophic (mostly *Limnodrilus* and *Potamothrix* spp.) conditions. Ecological analysis based on taxon identity seems more reliable and the only way to compare results with other sites or with other time periods.

3.5.2 Mesocosms

Mesocosms provide an experimental field approach where communities can be manipulated *in situ* with some degree of control. This provides the advantages of the control of the laboratory approach in field studies, although mesocosms are generally costly in terms of time and resources. We are aware of one study focussing specifically on oligochaetes. A series of experimental ditches were used by Verdonschot and Ter Braak (1994) to study toxic stress by the insectide chlorpyrifos, and eutrophication. Colonization pattern was investigated over a 20-week period, sufficient for the species present. No significant chlorpyrifos effects were revealed by the oligochaete populations.

In a field study on Zn availability by Liber et al. (1996), Zn-spiked sediment was placed in the bottom of a pond and the effects on colonization were studied over a 1-year period. Only naidid oligochaetes were significantly reduced in abundance at high Zn concentrations (10.9 μmol Zn g^{-1} dw sediment) in July, and were absent in October, although the absence could be due to species life-cycle.

A few other studies have included oligochaetes among the macroinvertebrates assessed in the exposure to different chemicals. Dieter et al. (1996) studied the tolerance of different macroinvertebrate taxa to the organophosphorous insecticide phorate (1.2–4.8 kg ha^{-1}) in a wetland mesocosm system in South Dakota (USA). A single application of the pesticide was done in each of the treatment mesocosm units, survival was monitored during the first 3 days, and afterwards recovery was determined by inspection of the number of macroinvertebrates at the end of 28 days. Oligochaete worms and hirudineans proved tolerant to the pesticide, while arthropod taxa such as amphipods and most insect orders (Odonata, Hemiptera, Diptera) were sensitive. The statistical tests used in the data analysis were not reported and the criteria used to classify the taxa sensitivity were related to abundance after treatment. In the case of aquatic oligochaetes, the recovery after treatment showed increments in abundance relative to untreated mesocosms, probably due to reduction of competitors among arthropodan taxa.

3.5.3 Field Bioaccumulation

While measurement of tissue concentration is not a biological effect, it is a line of evidence that indicates that organisms have been exposed to the contaminant. As such these data provide useful supporting evidence in environmental risk

Table 3.9 Tubificids body burden of Cu and Pb in the 4 year seasons, at two different zones of the Jarama River (Spain) (From Hernandez et al. 1988)

Tubificid body burden	Copper ($\mu g\ g^{-1}$ dw) Mean ± sd		Lead ($\mu g\ g^{-1}$ dw) Mean ± sd	
	Zone 1	*Zone 2*	*Zone 1*	*Zone 2*
Autumn	23.1 ± 18.4	33.0 ± 17.1	14.8 ± 8.0	16.0 ± 7.4
Winter	31.6 ± 18.4	53.7 ± 29.5	17.9 ± 9.9	21.3 ± 11.2
Spring	24.2 ± 17.1	22.8 ± 15.3	4.6 ± 4.8	8.7 ± 5.8
Summer	7.8 ± 4.9	10.5 ± 9.2	10.0 ± 8.0	8.4 ± 3.6

Zone 1 moderate to high contamination, *Zone 2* very heavy contamination

assessment studies. In Chap. 5, the reader can find more information about field bioaccumulation, mostly the comparison of field and laboratory test bioaccumulation levels and the relationships between bioaccumulation and toxicity.

One of the major advantages of measuring the body burden of chemicals in field organisms is the fact that the tissue concentration indicates the bioavailability of the substance and the capacity of the organisms to regulate uptake, or be able to eliminate either by excretion or transformation by metabolic processes. Ankley (1996) reviewed the metal bioaccumulation literature in freshwater and marine sediments with both spiked and naturally contaminated sediments, and examined metal bioaccumulation in relation to the metal/AVS (acid volatile sulphides) ratios to test the hypothesis that AVS controls metal availability.

The differences in bioaccumulation patterns of essential and non-essential metals by aquatic organisms have been stressed by Chapman et al. (1996). Organisms can actively maintain internal concentration of essential metals (*e.g.* Cu, Zn) required for metabolic purposes and, thus, no relationship between the bioaccumulation factors and pore water or sediment concentrations of the metal is expected to occur. Some studies have concluded that tubificids are unreliable biomonitors of field heavy metal pollution, since similar levels of metals exist in worms from clean and polluted sites (Elbe river, Germany) (Kaiser et al. 1989). However, these results contradict other studies (see Sect. 5.3) and may reflect the effects of local availability and regulatory capacity by organisms on the bioaccumulation process of particular metals.

In field work, daily and seasonal variation in temperature, as well as in sediment organic content, pH, grain size particle distribution and other factors can cause alterations in the bioaccumulation of metals and organic chemicals. Both Chapman et al. (1980) and Kaiser et al. (1989) reported no seasonal variation in tissue metal levels. However, in the River Jarama (Spain) Hernandez et al. (1988) have found that tubificids showed maximum body burdens of Cu and Pb in winter and minimum burdens in summer. In this study, the significance of differences between the different seasonal values was not established, and no correlation between tubificids metal body burden and sediment concentration was found. Seasonal mean levels of Cu and Pb in tubificids are shown in Table 3.9. In none of these cases did the authors discuss the relationships in the context of the proportion of adults or juveniles in the population (*i.e.* differences in the metabolism), or describe any abiotic reasons that would account for seasonal differences in the availability of metals (*e.g.* organic content of sediment, particle size distribution, and so on).

3.5.4 Chaetal Alterations

Changes in morphological characters of organisms exposed to pollutants have been demonstrated to be a useful tool for pollution assessment in some aquatic taxa. Alterations in the symmetry of mouth parts or in development of the antennae of chironomid larvae as a response to pollution is well established (Warwick 1985). In aquatic oligochaetes, there are a few examples showing cause-effect relationships in the alteration of chaetae related to the exposure to metals. Lucan-Bouché et al. (1999) found no morphological alterations in chaetae in short-term (96 h) laboratory experiments with *Tubifex tubifex* exposed to Pb and Cu. However, it is possible that chaetae are shed quite rapidly (Chapman and Brinkhurst 1987) and the development of new chaetae takes more time, therefore, alterations are undetected in short-term experiments. In the field, under long-term exposure together with time enough for development of new set of chaetae, alterations should be easier to find. Worms in Hg-laden sediments in Scandinavia showed enlarged chaetae (Milbrink 1983b). In 34-d experiments, Chapman and Brinkhurst (1987) demonstrated that *Ilyodrilus franzi* exposed to 2.5 and 1.0 mg Hg l^{-1} (in the presence of sediment) lost hair-chaetae. In the case of *T. tubifex,* abnormalities (higher degree of pectination) in dorsal crochets were reported under the same exposure conditions, after 60 days. Similar effects in *T. tubifex* were measured under extreme environmental conditions (salinity 1–5 ppt), in a 60-d exposure.

Some teratological records can be found in the literature, usually related to the displacement in the position of genitalia, the number of gonads or the absence of spermatheca. However, we do not know any of these reports to be associated with contamination. Although field surveys on pollution assessments do not usually examine the morphology of specimens, future studies may identify alterations which could be used as an indicator of pollution.

3.6 Bioturbation

The presence of tubificid communities in the sediment has been reported to affect the physical, chemical and microbiological processes within the uppermost layers of the sediment, and accelerate the diffusion of dissolved substances (*e.g.* Kikuchi and Kurihara 1982). The activity of aquatic oligochaetes through burrowing, feeding, deposition of faecal pellets on the sediment surface, and their respiratory movement, all contribute to the bioturbation effect.

Poddubnaya and Sorokin (1961), Sorokin (1966), Davis (1974) and Wood (1975) have measured vertical transport of particles by oligochaete burrowing and feeding activities. Bioturbation by worms may well cause release of sedimentary contaminants into the water column. For example, Boddington et al. (1979) (cited by Chapman et al. 1980) stated that bioturbation by worms, through defecation, can release both inorganic and methyl Hg into the water column through the resuspension of fine particles thus deposited at the mud – water interface.

Phosphorus release rates in the presence of *Stylodrilus heringianus*, and what was presumed to be *Limnodrilus* spp., were established for Lake Michigan sediments by Gardner et al. (1981). Two temperature regimes were used, 5°C and 20°C, with fed and unfed animals. Release rates were determined for 30-min and 24-h periods. The data obtained were compared with those obtained for P release rates by planktonic organisms, where retention or release of P was related to its availability for phytoplankton production by recycling. In these experiments, there were high release rates of P in faeces in 30-min exposures in comparison with the 24-h period. Worms that had their gut content present at the start did not produce any more P than those that were purged. The authors also examined the rate of P release from intact cores and release rates were determined by extracting P from sand in which worms were placed. They concluded that benthic invertebrate excretion could account for most of the P released from aerobic Lake Michigan sediments. Gardner et al. (1983) also studied the rates of N release by oligochaetes and estimated NH_4: PO_4 ratios in excreted materials to be 35:1 for tubificids. The effect of *Tubifex* on the release of particulate, dissolved organic and orthophosphate phosphorous was determined by Guérin and Labroue (1991). Worms produced an increase in particulate P in the water column when oxygenated, but an equivalent reduction in orthophosphates so that total P was only slightly higher than in controls. At lower oxygen levels, the difference was more marked because of an increase in the particulate fraction due to sediment irrigation but lower defecation rates.

Karickhoff and Morris (1985) examined the effect of bioturbation by *L. hoffmeisteri* and *T. tubifex,* in a mixed assemblage, on the distribution of hexachlorobenzene, pentachlorobenzene, and trifluralin in the sediment. They found that most of the contaminant within the feeding zone of the worms was transported as faeces to the surface in a 30–50 day period. Chemicals moved at a rate of 5 mm d^{-1}. The trifluralin concentration in faecal pellets decreased exponentially, presumably by microbially mediated chemical reduction. While adsorption of chemicals onto the faecal pellets limits pollutant release from them, we would assume that water circulation could redistribute faecal pellets to uncontaminated areas. Once pellets are broken up, release rates become the same as for fine sediments. Transport of sediments upwards by the worms was estimated to reach 1 mm per day, depending on worm population and animal size, whereas diffusion rates would only account for 1 mm per month for the same type of chemicals. Release of chemicals into the water, with continuous stripping, increased from 4 to 6 times in the presence of worms. As a consequence, in shallow waters sediments may be purged of contaminants instead of retaining material and that may be harmful to sensitive species. However, this is achieved at the expense of export and transport of the contaminants to other locations, making detection of source and mitigation difficult. In deeper waters where sediment deposition is occurring, the burial of contaminants is delayed as older material is continually being deposited on the surface by worm activity. The actual burial rate depends on the ratio between sediment deposition and oligochaete re-working.

Sediment re-working by worms has been studied on many occasions in laboratory tests (see McCall and Fischer 1980 for earlier citations, and the papers cited under sediment chronic toxicity in Sect. 4.3.3, both for metals and organic chemicals), but little has been done in the field. However, bioturbation can be of significance since the release of chemicals to the water-column and the transport of chemicals from deeper to upper sediment layers are important factors in environmental risk assessment studies where high worm densities are associated with contaminated sediments.

3.7 Conclusions and Remarks

- There is a long history of using oligochaete community structure in assessment studies. As with more general invertebrate work, the approaches have evolved from simple descriptive studies through the use of various indices and metrics to the use of multivariate approaches.
- There is considerable range in species response both to nutrient contamination and metals and organic chemicals, and some species assemblages have been correlated to stressors other than nutrient enrichment, but the best understood response is to trophic status. Species assemblage composition can change regionally.
- The traditional view of oligochaetes as a group associated to polysaprobic conditions needs reconsideration as most oligochaete species are associated with oligo- and β-mesosaprobic conditions and the species are distributed across the entire range of ecological classes of water quality.
- The use of tolerance values to weight relative abundance data fails to distinguish between stress factors, and the values correspond primarily to tolerance to organic enrichment or other related environmental factors (*e.g.* dissolved oxygen). In many instances, tolerance values do not conform well to existing field survey data as too many species are seen as highly tolerant. Toxicity data clearly show that worm species do not respond in the same way to all types of contaminant which is implied by the general use of a single tolerance value approach.
- The older field data demonstrated that distribution patterns exist associated with pollution. Recent multivariate statistics use those patterns to improve the cost-benefit ratio associated with benthic biology. Pollution biologists need to pay more attention to developments in ecology and less to the search for an indicator species or species-group panacea. Tests of the validity of indices need to be applied to larger numbers of data sets and these should be obtained from wider geographical areas.

- Rapid assessment methods are supposed to be limited to the detection of severe problems with the least expenditure of time and effort, but they are not appropriate to the detailed analysis of complex field situations. To identify specific causative agents and causality requires tailored site specific studies and will often require an experimental approach.
- The measurement of community dominance or diversity addresses only one aspect of a benthic assemblage. As these indices loose information on the identity of the taxa concerned, one could theoretically have two totally different communities with identical indices although they do not have a single taxon in common. Factors unrelated to pollution, such as substratum type, have a profound effect on these indices. It is now possible to contain taxonomic information within the statistics by using cluster analysis, ordination and tests of significance within and between clusters. The use of non-parametric statistics recognizes the limitations in biological data without the need for transformation.
- We do not see focussing on oligochaetes alone in field community-based bioassessment approaches to be a useful strategy, as information on the entire community is just as easily obtained and provides richer information. However, neither do we recommend ignoring oligochaetes or lumping all the species as a single group because of the great response range exhibited by the group, and the fact that in many habitats they comprise the major part of the community. It is necessary to improve the training of benthic zoologists to include knowledge of the most common oligochaete species. At present, this training has been largely neglected.
- Oligochaetes can also be of considerable importance in the contaminant flux between sediment and the water column. A combination of field and laboratory experimental approaches with these organisms may provide valuable insights on the mobility and transport of sediment associated contaminants in the food chain.

References

Adreani L, Bonacina C, Bonomi G (1981) Production and population dynamics in profundal lacustrine Oligochaeta. Verh Internat Verein Limnol 21:967–974

Adreani L, Bonacina C, Bonomi G, Monti C (1984) Cohort cultures of *Psammoryctides barbatus* (Grube) and *Sprirosperma ferox* Eisen: a tool for a better understanding of demographic strategies in Tubificids. Hydrobiologia 115:113–119

Andrews MJ (1984) Thames estuary: pollution and recovery. In: Sheehan RJ, Miller DR, Buffer GC, Bourdeau Ph (eds) Effects of pollutants at the ecosystem level. Wiley, Chichester, pp 195–227

Ankley GT (1996) Evaluation of metal/acid-volatile sulfide relationships in the prediction of metal bioaccumulation by benthic macroinvertebrates. Environ Toxicol Chem 15:2138–2146

Aston RJ (1968) The effect of temperature on the life cycle, growth and fecundity of Branchiura sowerbyi (Oligochaeta, Tubificidae). J Zool Lond 154:29–40

Aston RJ (1973) Tubificidae and water quality: a review. Environ Pollut 5:1–10

Bailey RC, Norris RH, Reynoldson TB (2004) Bioassessment of freshwater ecosystems. Using the reference condition approach. Kluwer Academic, Boston, 170 pp

Barbour MT, Gerritsen J, Snyder BD, Stribling JB (1999) Rapid bioassessment protocols for use in streams and wadeable rivers: periphyton, benthic macroinvertebrates and fish, 2nd edn. EPA 841-B-99-002. US Environmental Protection Agency, Office of Water, Washington, DC

Barton DR (1986) Benthic fauna from Great Lakes Institute cruises on Lake Huron and Georgian Bay 1963 and 1964. Canadian technical report fisheries aquatic science, vol 1487, Minister of Supply and Services Canada, Ottawa

Belyaev GM, Vinberg GG,Gaevskaya NS, Zhadin VI, Zenkevich LA, Kashkin NI, Reznichenko OG, Shcherbakov AP (eds) (1972) Aquatic Oligochaeta worms. Taxonomy, ecology and faunistic studies in the USSR. In: Proceedings of the symposium on aquatic Oligochaeta, Tartu, 1967. Nauka, Moscow (English translation in 1980, Amerind Publish Co, Pvt Ltd, New Delhi), 211pp

Birtwell JK, Arthur DR (1980) The ecology of tubificids in the Thames Estuary with particular reference to *Tubifex costatus* (Claparède). In: Brinkhurst RO, Cook DG (eds) Aquatic oligochaete biology. Plenum Press, New York, pp 331–381

Bode RW, Novak MA, Abele LE (1996) Quality assurance work plan for biological stream monitoring in New York state. New York State Department of Environmental Conservation, Albany, 89pp

Bode RW, Novak MA, Abele LE, Heitzman DL, Smith AJ (2002) Quality assurance work plan for biological stream monitoring in New York state. New York State Department of Environmental Conservation Report, Albany, 115pp

Brinkhurst RO (1964a) Observations on the biology of lake-dwelling Tubificidae. Arch Hydrobiol 60:385–418

Brinkhurst RO (1964b) Studies on the North American aquatic Oligochaeta. I. Naididae and Opistocystidae. In: Proceeding of the academy of natural science of Philadelphia, vol 116, pp 195–229

Brinkhurst RO (1965a) Observations on the recovery of a British river from gross organic pollution. Hydrobiologia 25:9–51

Brinkhurst RO (1965b) The biology of the Tubificidae with special reference to pollution. In: Proceeding of 3rd seminar on "Biological problems in water pollution", Cincinnati 1962, pp 57–66

Brinkhurst RO (1965c) Studies on the North American aquatic Oligochaeta. II. Tubificidae. In: Proceeding of the academy of natural science of Philadelphia, vol 117. pp 117–172

Brinkhurst RO (1966a) The Tubificidae (Oligochaeta) of polluted waters. Verh Internat Verein Theor Angew Limnol 16:854–859

Brinkhurst RO (1966b) Detection and assessment of water pollution using oligochaete worms. Part 2. Water Sewage Works 113:438–441

Brinkhurst RO (1967) The distribution of aquatic Oligochaeta in Saginaw Bay, Lake Huron. Limnol Oceanogr 12:137–143

Brinkhurst RO (1969) Changes in the benthos of Lakes Erie and Ontario. In: Proceeding of conference changes in biota of lakes Erie and Ontario. Bulletin of the Buffalo Society of Natural Sciences, vol 25, pp 35–65

Brinkhurst RO (1970) Distribution and abundance of tubificid (Oligochaeta) species in Toronto Harbour, Lake Ontario. J Fish Res Bd Can 27:1961–1969

Brinkhurst RO, Austin MJ (1979) Assimilation by aquatic Oligochaeta. Int Rev Ges Hydrobiol 64:863–868

Brinkhurst RO, Cook DG (1966) Studies on the North American aquatic Oligochaeta. III. Lumbriculidae and additional notes and records of other families. In: Proceeding of the Academy of Natural Science of Philadelphia, vol 118, pp 1–33

Brinkhurst RO, Cook DG (1974) Aquatic earth worms (Annelida: Oligochaeta). In: Hart CW, Fuller SLH (eds) Pollution ecology of freshwater invertebrates. Academic, New York, pp 143–156

Brinkhurst RO, Kennedy CR (1965) Studies on the biology of the Tubificidae (Annelida, Oligochaeta) in a polluted stream. J Anim Ecol 34:429–443

Brinkhurst RO, Simmons ML (1968) The aquatic Oligochaeta of the San Francisco Bay system. Calif Fish Game 54:180–194

Brinkhurst RO, Hamilton AL, Herrington HB (1968) Components of the bottom fauna of the St Lawrence Great Lakes. University of Toronto Great Lakes Institute Report PR 33, 49 pp

Brinkhurst RO, Chua KE, Kaushik NK (1972) Interspecific interactions and selective feeding by tubificid oligochaetes. Limnol Oceanogr 17:122–133

Burd BJ, Nemec A, Brinkhurst RO (1990) The development and application of analytical methods in benthic marine infaunal studies. Adv Mar Biol 26:162–247

Carr JF, Hiltunen JK (1965) Changes in the bottom fauna of western Lake Erie from 1930 to 1961. Limnol Oceanogr 10:551–569

Chapman PM, Brinkhurst RO (1981) Seasonal changes in interstitial salinities and seasonal movements of subtidal nethic invertebrates in the Fraser River Estuary, B.C. Estuar Coast Mar Sci 12:49–66

Chapman PM, Brinkhurst RO (1987) Hair today, gone tomorrow: induced chaetal changes in tubificid oligochaetes. Hydrobiologia 155:45–55

Chapman PM, Churchland LM, Thomson PA, Michnowsky E (1980) Heavy metal studies with oligochaetes. In: Brinkhurst RO, Cook DG (eds) Aquatic oligochaete biology. Plenum Press, New York, pp 477–502

Chapman PM, Farrell MA, Brinkhurst RO (1982a) Effects of species interactions on the survival and respiration of *Limnodrilus hofffmeisteri* and *Tubifex tubifex* (Oligochaeta, Tubificidae) exposed to various pollutants and environmental factors. Water Res 16:1405–1408

Chapman PM, Farrell MA, Brinkhurst RO (1982b) Relative tolerances of selected aquatic oligochaetes to individual pollutants and environmental factors. Aquat Toxicol 2:47–67

Chapman PM, Farrell MA, Brinkhurst RO (1982c) Relative tolerances of selected aquatic oligochaetes to combination of pollutants and environmental factors. Aquat Toxicol 2:69–78

Chapman PM, Allen HE, Godfredsen K, Z'graggen MN (1996) Evaluation of bioaccumulation factors in regulating metals. Environ Sci Technol 30:448A–452A

Coates KA (1995) Widespread poplyploid forms of *Lumbricillus lineatus* (Müller) (Enchytraeidae, Oligochaeta): comments on polyploidism in the enchytraeids. Can J Zool 73:1727–1734

Coates KA, Ellis DV (1980) Enchytraeid oligochaetes as marine pollution indicators. Mar Pollut Bull 11:171–175

Cook DG, Johnson MG (1974) Benthic macroinvertebrates of the St. Lawrence Great Lakes. J Fish Res Bd Can 31:763–782

Davis RB (1974) Statigraphic effects of tubificid worms in profundal lake sediments. Limnol Oceanogr 19:466–488

Diaz RJ (1989) Pollution and tidal benthic communities of the James River Estuary, Virginia. Hydrobiologia 180:195–212

Diaz RJ, Erséus C (1994) Habitat preference and species associations of shallow-water marine Tubificidae (Oligochaeta) from the barrier reef ecosystems of Belize-Central America. Hydrobiologia 278:93–105

Diaz RJ, Erséus C, Boesch DF (1987) Distribution and ecology of Middle Atlantic Bight Oligochaeta. Hydrobiologia 155:215–225

Dieter CD, Duffy WG, Flake LD (1996) The effect of phorate on wetland macroinvertebrates. Environ Toxicol Chem 15:308–312

Engle VD, Summers JK, Gaston GR (1994) A benthic index of environmental condition of Gulf of Mexico. Estuaries 17:372–384

Furse MT, Moss D, Wrigth JF, Armitage PD (1984) The influence of seasonal and taxonomic factors on the ordination and classification of running-water sites in Great Britain and on the prediction of their macro-invertebrate communities. Freshwater Biol 14:257–280

Gamenick I, Jahn A, Vopel K, Giere O (1996) Hypoxia and sulphide as structuring factors in a macrozoobenthic community on the Baltic Sea shore: colonisation studies and tolerance experiments. Mar Ecol Prog Ser 144:73–85

Gardner WS, Nalepa TF, Quigley MA, Malczyk JM (1981) Release of phosphorous by certain invertebrates. Can J Fish Aquat Sci 38:978–981

Gardner WS, Nalepa TF, Slavens DR, Lair D (1983) Patterns and rates of nitrogen release by benthic Chironomidae and Oligochaeta. Can J Fish Aquat Sci 40:259–266

Ghetti PF, Bonazzi G (1977) A comparison between various criteria for the interpretation of biological data in the analysis of the quality of running waters. Water Res 11:819–831
Giani N (1983) Le Riou Mort, affluent du Lot, pollué par métaux lourds, III Étude faunistique générale. Annls Limnol 19:29–43
Giani N (1984) Le Riou Mort, affluent du Lot, pollué par métaux lourds. IV. Étude des oligochètes. Annls Limnol 20:167–181
Giere O (1979) The impact of oil pollution on intertidal meiofauna. Field studies after the La Coruña-Spill, May, 1976. Cah Biol Mar 20:231–251
Giere O (1980) Tolerance and preference reactions of marine Oligochaeta in relation to their distribution. In: Brinkhurst RO, Cook DG (eds) Aquatic Oligochaeta biology. Plenum Press, New York, pp 385–405
Giere O (2006) Ecology and biology of marine Oligochaeta – an inventory rather than another review. Hydrobiologia 564:103–116
Giere O, Hauschildt D (1979) Experimental studies on the life-cycle and production of the littoral oligochaete *Lumbricillus lineatus*, and its response to oil pollution. In: Naylor E, Hartnall RG (eds) Cyclic phenomena in marine plants and animals. Pergamon, Oxford, pp 113–122
Giere O, Pfannkuche O (1982) Biology and Ecology of marine Oligochaeta, a review. In: Barnes M (ed.) Oceanography and marine biology: an annual review, vol 20. Aberdeen University Press, Aberdeen, pp 173–308
Goodnight CJ, Whitley LS (1960) Oligochaetes as indicators of pollution. Water Sewage Works 107:311
Goodnight CJ, Whitley LS (1961) Oligochaetes as indicators of pollution. In: Proceeding 15th industrial waste conference, Purdue University Engineering Extension Series, vol 106. pp 139–142
Grapentine L, Anderson J, Boyd D, Burton GA, DeBarros C, Johnson G, Marvin C, Milani D, Painter S, Pascoe T, Reynoldson TB, Richman L, Solomon K, Chapman PM (2002) A decision making framework for sediment assessment developed for the Great Lakes. Hum Ecol Risk Assess 8:1641–1655
Gray JS (1971) The effects of pollution on sand meiofauna communities. Thalassia jugosl 7:79–86
Grigelis AI (1980) Ecological studies of aquatic oligochaetes of the USSR. In: Brinkhurst RO, Cook DG (eds) Aquatic Oligochaeta biology. Plenum Press, New York, pp 225–240
Gross F (1976) Les communautes d'oligochètes d'un ruisseau de plaine. Leur utilisation comme indicateurs de la pollution organique. Annls Limnol 12:75–87
Guérin C, Labroue L (1991) A laboratory study of the effect of *Tubifex tubifex* Müll. (Tubificidae, Oligochaeta) on the release of phosphorus through the sediment. Annls Limnol 27:253–265
Henderson AR (1983) Long term monitoring of the macrobenthos of Upper Clyde Estuary. Water Sci Tech 16:359–373
Hering D, Moog O, Sandin L, Verdonschot PFM (2004) Overview and application of the AQEM assessment system. Hydrobiologia 516:1–21
Hernandez M, Rovira JV, Antonio MT (1988) Relationship between Cu and Pb levels in sediments and dwelling Tubificidae in the Jarama River (Madrid, Spain). In: Astruc M, Lester JN (eds) Heavy metals in the hydrological cycle. Selper Ltd, London, pp 531–538
Herricks EE, Cairns J Jr (1982) Biological monitoring. III. Receiving system methodology based on community structure. Water Res 16:141–153
Hiltunen JK (1969) Distribution of oligochaetes in western Lake Erie, 1961. Limnol Oceanogr 14:260–264
Howmiller RP (1974a) Composition of the oligochaete fauna of central Lake Michigan. In: Proceeding of the 17th conference on Great Lakes Research, pp 589–592
Howmiller RP (1974b) Studies on aquatic Oligochaeta in inland waters of Wisconsin. Trans Wisc Acad Sci Arts Lett 62:337–356
Howmiller RP, Beeton AM (1970) The Oligochaeta fauna of Green Bay, Lake Michigan. In: Proceeding of the 13th conference on Great Lakes Research, pp 15–46

Howmiller RP, Beeton AM (1971) Biological evaluation of environmental quality, Green Bay, Lake Michigan. J Water Pollut Control Fed 43:123–133
Howmiller RP, Scott MA (1977) An environmental index based on relative abundance of oligochaete species. J Water Pollut Control Fed 49:809–815
Hunter J (1981) Survival strategies in Tubificids in the Thames and other estuaries. In: Jones NV, Wolff WJ (eds) Feeding and survival strategies of estuarine organisms. Marine sciences vol 15. Plenum Pub Corp, New York, pp 53–63
Johnson MG, Brinkhurst RO (1971a) Benthic community metabolism in Bay of Quinte and Lake Ontario. J Fish Res Bd Can 28:1699–1714
Johnson MG, Brinkhurst RO (1971b) Benthic community metabolism in Bay of Quinte and Lake Ontario. J Fish Res Bd Can 28:1715–1725
Johnson MG, Matheson DH (1968) Macroinvertebrate communities of the sediments of Hamilton Bay and adjacent Lake Ontario. Limnol Oceanogr 13:99–111
Johnson RK, Wiederholm T (1989) Classification and ordination of profundal macroinvertebrate communities in nutrient poor, oligo-mesohumic lakes in relation to environmental data. Freshwater Biol 21:375–386
Juget J (1958) Recherches sur la faune des fonds du Leman et Lac d'Annecy. Ann Stn Cent Hydrobiol Appl 7:7–95
Kachalova OL, Parele EA (eds) (1987) Aquatic Oligochaeta. In: Proceedings of the 6th all-union symposium, Salaspils, 27–30 April 1987, Riga, 160 pp
Kaiser M, Irmer U, Weiler K (1989) Monitoring of water quality: seasonal variation of heavy metals in sediments, suspended particulate water and tubificids of the Elbe River. Environ Technol Lett 10:845–855
Kangur K, Timm H, Timm H, Timm V (1998) Long-term changes in the macrozobenthos of Lake Vortsjärv. Limnologica 28:75–83
Karickhoff SW, Morris SW (1985) Impact of tubificid oligochaetes on pollutant transport in bottom sediments. Environ Sci Technol 19:51–56
Kathman RD, Brinkhurst RO (1991) A review of interpretation methods for freshwater benthic invertebrate survey data used by selected state and federal agencies. Young-Morgan and Associates, Franklin, 45 pp
Kennedy CR (1966a) The life history of *Limnodrilus udekemianus* Clap. (Oligochaeta, Tubificidae). Oikos 17:10–18
Kennedy CR (1966b) The life history of *Limnodrilus hoffmeisteri* Clap. (Oligochaeta, Tubificidae) and its adaptative significance. Oikos 17:158–168
Kikuchi E, Kurihara Y (1982) The effects of the oligochaete *Branchiura sowerbyi* Beddard (Tubificidae) in the biological and chemical characteristics of overlying water and soil in a submerged ricefield soil system. Hydrobiologia 97:203–208
King DL, Ball RC (1964) A quantitative biological measure of stream pollution. J Water Pollut Control Fed 36:650–653
Kolkwitz R, Marsson M (1908) Oekologie der pflanzlichen Saprobien. Ber Dtsch Bot Ges 26:505–519
Krieger KA (1984) Benthic macroinvertebrates as indicators of environmental degradation in the southern nearshore zone of the central basin of Lake Erie. J Great Lakes Res 10:197–209
Kröncke I (1996) Impact of biodeposition on macrofaunal communities in intertidal sandflats. Mar Ecol 17:159–174
Kurashvili BE (ed.) (1983) Aquatic Oligochaeta. In: Proceedings of the 4th all union symposium Tbilisi 5–7 October 1983, 136 pp (Canadian Translation of Fisheries and Aquatic Sciences 5091, 1984, National Research Council, Ottawa, Ontario,143 pp)
Lafont M (1977) Les Oligochètes d'un cours d'eau montagnard pollué: le Biet Rouge. Annls Limnol 13:157–167
Lafont M (1984) Oligochaete communities as biological descriptors of pollution in the fine sediments of rivers. Hydrobiologia 115:127–129
Lafont M, Juget J (1985) Les oligochètes de quelques lacs français: proposition en vue de leur utilisation pratique pour apprécier l'état biologique des sédiments profonds. Verh Internat Verein Limnol 22:3019–3023

Lafont M, Juget J, Rofes G (1991) Un indice biologique lacustre basé sur l'examen des oligochètes. Rev Sci de l'Eau 4:253–268

Lang C (1978) Factorial correspondence analysis of Oligochaeta communities according to eutrophication level. Hydrobiologia 57:241–247

Lang C (1985) Eutrophication of Lake Geneva indicated by oligochaete communities of the profundal. Hydrobiologia 126:237–243

Lang C (1986) Eutrophisation du Léman indiquée par les communautés d'oligochètes: campagnes 1982–1985. Schweiz J Hydrol 48:230–239

Lang C (1989a) Eutrophication of Lake Neuchâtel indicated by the oligochaete communities. Hydrobiologia 174:57–66

Lang C (1989b) Effects of small-scale sedimentary patchiness on the distribution of tubificid and lumbricid worms in Lake Geneva. Freshwater Biol 21:477–481

Lang C (1990) Quantitative relationships between oligochaete communities and phosphorous concentrations in lakes. Freshwater Biol 24:327–334

Lang C (1991) Decreasing phosphorus concentration and unchanged oligochaete communities in Lake Geneva: how to monitor recovery? Arch Hydrobiol 122:305–312

Lang C (1992) Reversal of eutrophication in lake Geneva - evidence from the oligochaete communities. Freshwater Biol 28(2):145–148

Lang C (1993) Eutrophication of Lake Morat indicted by the oligochaete communities. 1980–1991. Trend Rev Suisse Zool 100:11–18

Lang C (1999) Contrasting responses of oligochaetes (Annelida) and chironomids (Diptera) to the abatement of eutrophication in Lake Neuchâtel. Aquat Sci 61:206–214

Lang C, Lang-Dobler B (1979) The chemical environment of tubificid and lumbricid worms according to the pollution level of the sediment. Hydrobiologia 65:273–282

Lang C, Reymond O (1995a) Recovery from eutrophication delayed in the profundal of Lake Neuchâtel: evidence from oligochaete communities. Rev Suisse Zool 102:907–912

Lang C, Reymond O (1995b) Empirical relationships between oligochaete communities and phosphorous in lakes before and after the abatement of eutrophication. Arch Hydrobiol 134:491–497

Lang C, Reymond O (1995c) Contrasting responses of oligochaete communities to the abatement of eutrophication in lake Geneva. Hydrobiologia 308:77–82

Lauritsen DD, Mozley SC, White DS (1985) Distribution of oligochaetes in Lake Michigan and comments in their use as indices of pollution. J Great Lakes Res 11:67–76

Lenat DR (1993) A biotic Index for the southwestern United States: derivation and list of tolerance values, with criteria for assigning water quality ratings. J N A Benthol Soc 12:279–290

Leppäkoski EJ (1975) Assessment of the degree of pollution on the basis of macrozoobenthos in marine and brackish-water environments. Acta Acad Aboensis B 35:1–90

Leppäkoski EJ, Lindstrom LS (1978) Recovery of benthic macrofauna from chronic pollution in the sea area off a refining plant, southwest Finland. J Fish Res Bd Can 35:766–775

Liber K, Call DJ, Markee TP, Schmude KL, Balcer MD, Whiteman FW, Ankley GT (1996) Effects of acid-volatile sulfide on zinc bioavailability and toxicity to benthic macroinvertebrates: a spiked-sediment field experiment. Environ Toxicol Chem 15:2113–2125

Lucan-Bouché ML, Biagianti-Risbourg S, Arsac F, Vernet G (1999) Autotomy as a mechanism of decontamination used by the oligochaete *Tubifex tubifex*. Bull Soc Zool Fr 124:383–387

Marchese MR (1987) The ecology of some benthic Oligochaeta from the Paraná River, Argentina. Hydrobiologia 155:209–214

Martinez-Ansemil E (1990) Étude biologique et ecologique des Oligochètes aquatiques de la rivière Tambre et de ses milieux associés (Galice, Espagne). Annls Limnol 26:131–151

McCall PL, Fischer JB (1980) Effects of tubificid oligochaetes on physical and chemical properties of lake Erie sediments. In: Brinkhurst RO, Cook DG (eds) Aquatic oligochaete biology. Plenum Press, New York, pp 253–317

McLusky DS, Teare M, Phizacklea P (1980) Effects of domestic and industrial pollution on distribution and abundance of aquatic oligochaetes in the Forth Estuary. Helgolander wiss Meereunters 33:384–392

McLusky DS, Hull SC, Elliot M (1993) Variations in the intertidal and subtidal macrofauna and sediments along a salinity gradient in the upper Forth Estuary. Neth J Aquat Ecol 27:101–109

Milbrink G (1973) On the use of indicator communities of Tubificidae and some Lumbriculidae in the assessment of water pollution in Swedish lakes. Zoon 1:125–139

Milbrink G (1978) Indicator communities of oligochaetes in Scandinavian lakes. Verh Internat Verein Limnol 20:2406–2411

Milbrink G (1983a) An improved environmental index based on the relative abundance of oligochaete species. Hydrobiologia 102:89–97

Milbrink G (1983b) Characteristic deformities in tubificid oligochaetes inhabiting polluted bays of Lake Vänein, southern Sweden. Hydrobiologia 106:169–185

Milbrink G, Timm T (2001) Distribution and dispersal capacity of the Ponto-Caspian tubificid oligochaete *Potamothrix moldaviensis* Vejdovsky et Mrázek, 1903 in the Baltic Sea region. Hydrobiologia 463:93–102

Milbrink G, Timm T (2002) Indicative profundal oligochaete assemblages in selected small Swedish lakes. Hydrobiologia 468:53–61

Montanholi-Martins MC, Takeda AM (2001) Spatial and temporal variations of oligochaetes of the Ivinhema River and Patos Lake on the Upper Paraná River Basin, Brazil. Hydrobiologia 463:197–205

Mozley SC, Howmiller RP (1977) Environmental status of the Lake Michigan region: zoobenthos of Lake Michigan. Argonne National Laboratory, Report N° ANL/ES-40, vol 6, US Energy Research and Development Administration

Nalepa TF, Thomas NA (1976) Distribution of macrobenthic species in Lake Ontario in relation to sources of pollution and sediment parameters. J Great Lakes Res 2:150–163

Nichols SJ (1981) Effect of thermal effluents on oligochaetes in Keowee Reservoir, South Carolina. Hydrobiologia 79:129–136

Nijboer R, Wetzel MJ, Verdonschot PFM (2004) Diversity and distribution of Tubificidae, Naididae, and Lumbriculidae (Annelida, Oligochaeta) in the Netherlands: an evaluation of twenty years of monitoring data. Hydrobiologia 520:127–141

Norris RH, Georges A (1993) Analysis and interpretation of benthic macroinvertebrates surveys. In: Rosenberg DM, Resh VH (eds) Freshwater biomonitoring and benthic invertebrates. Chapman & Hall, New York, pp 234–286

Patrick R (1949) A proposed biological measure of stream condition. Proc Acad Nat Sci Philad 101:277–341

Peterson CH, Kennicutt MC II, Green RH, Mointagna P, Harper DE, Powell EN, Roscigno PF (1996) Ecological consequences of environmental perturbations associated with offshore hydrocarbon production: a perspective on long-term exposures in the Gulf of Mexico. Can J Fish Aquat Sci 53:2637–2654

Poddubnaya TL, Bakanov AJ (1983) Spatial distribution of aquatic oligochaetes. In: Kurashvili BE (ed) Aquatic Oligochaeta. Proceedings of the 4th all-union symposium, Tbilisi 5–7 October 1983 (Canadian translation of Fisheries and Aquatic Sciences 5091, 1984, National Research Council, Ottawa, Ontario), pp 5–11

Poddubnaya TL, Sorokin YI (1961) The depth of the layer of tubificid worm's optimal feeding in connection with their movement in the sediment. Bull Inst Biol Reserv Acad Sci USSR 10:14–17

Probst L (1987) Sublittoral and profundal Oligochaeta fauna of the Lake Constance (Bodensee -Obersee). Hydrobiologia 155:277–282

Qi S (1987) Some ecological aspects of aquatic oligochaetes in the Lower Pearl River (People's Republic of China). Hydrobiologia 155:199–208

Real M, Sabater F, Morguí JA (1992) Significant physiographic disturbances in the Ebro basin reservoirs (NE Spain) reflected by Oligochaeta size spectra. Hydrobiologia 235(236):363–374

Real M, Prat N, Riera JL (1993) Abundance and distribution of profundal zoobenthos in Spanish reservoirs: differences between 1973/75 and 1987/88 surveys. Water Sci Tech 28:45–53

Reynoldson TB (1990) Distribution patterns of oligochaetes in the English Lake District. Arch Hydrobiol 118:303–339
Reynoldson TB, Zarull MA (1993) An approach to the development of biological sediment guidelines. In: Woodley S, Kay J, Francis G (eds) Ecological integrity and management of ecosystems. St. Lucie Press, Del Ray Beach, pp 177–200
Rieradevall M, Real M (1994) On the distribution patterns and population dynamics of sublittoral and profundal Oligochaeta fauna from Lake Banyoles (Catalonia, NE Spain). Hydrobiologia 278:139–149
Rodriguez P (1984) Los Oligoquetos acuáticos del río Nervión: Resultados faunísticos generales. Limnetica 1:169–178
Roff JC, Kwiatkowski RE (1977) Zooplankton and zoo-benthos communities of selected northern Ontario lakes of different acidities. Can J Zool 55:899–911
Rosso A, Lafont M, Exinger A (1994) Impact of heavy metals on benthic oligochaete communities in the River Ill and its tributaries. Water Sci Tech 29:241–248
Sanders HL (1960) Benthic studies in Buzzards Bay. III. The structure of the soft-bottom community. Limnol Oceanogr 5:138–153
Sardá R, Valiela I, Foreman K (1996) Decadal shifts in a salt marsh macroinfaunal community in response to sustained long-term experimental nutrient enrichment. J Exp Mar Biol Ecol 205:63–81
Särkkä J (1987) The occurrence of oligochaetes in lakes chains receiving pulp mill waste and their relation to eutrophication on the trophic scale. Hydrobiologia 155:259–266
Särkkä J (1989) Meiobenthic naidid and aelosomatid oligochaetes from the profundal zone, and relations of species to eutrophication. Hydrobiologia 180:185–190
Särkkä J (1994) Lacustrine profundal meiobenthic oligochaetes as indicators of trophy and organic loading. Hydrobiologia 278:231–241
Särkkä J (1996) Meiofauna of the profundal zone of the northern part of Lake Ladoga as an indicator of pollution. Hydrobiologia 322:29–38
Särkkä J, Aho J (1980) Distribution of aquatic Oligochaeta in the Finnish Lake District. Freshwater Biol 10:197–207
Schwank P (1982) Turbellarien, Oligochaeten und Archianneliden des Breitenbach und anderer oberhessischer Mittelgebirgsbache. III. Die Taxozonosen der Turbellarien und Oligochaeten in Fliessgewassern – eine synokologische Gliederung. Schlitzer produktionsbiologische Studien (43–3). Arch Hydrobiol Suppl 62:191–253
Siegfried CA, Kopache ME, Knight AW (1980) The benthos of a portion of the Sacramento River (San Francisco Bay estuary) during a dry year. Estuaries 3:296–307
Slepukhina T (1984) Comparison of different methods of water quality evaluation by means of oligochaetes. Hydrobiologia 115:183–186
Sorokin YI (1966) Carbon 14 method in the study of nutrition of aquatic animals. Int Rev Ges Hydrobiol 51:209–226
Spencer DR (1980) The aquatic Oligochaeta of St. Lawrence Great Lakes region. In: Brinkhurst RO, Cook DG (eds) Aquatic oligochaete biology. Plenum Press, New York, pp 115–164
Spies RB, Davis PH, Stuermer DH (1980) Ecology of a submarine petroleum seep off the California coast. In: Geyer RA (ed) Marine environmental pollution. Hydrocarbons vol 1. Elsevier, Amsterdam, pp 229–263
Steichen DJ, Holbrook SJ, Osenberg CW (1996) Distribution and abundance of benthic and demersal macrofauna within a natural hydrocarbon seep. Mar Ecol Prog Ser 138:71–82
Stimpson KS, Brice JR, Barbour MT, Howe P (1975) Distribution and abundance of inshore oligochaetes in Lake Michigan. Trans Am Microsc Soc 94:384–394
Takeda AM (1999) Oligochaete community of alluvial Upper Parana River, Brazil: spatial and temporal distribution (1987–1988). Hydrobiologia 412:35–42
Timm T (1970) On the fauna of the Estonian Oligochaeta. Pedobiologia 10:52–78
Timm T (1994) Production of some oligochaete species in large Estonian lakes. Hydrobiologia 278:309–313

Timm T, Kangur K, Timm H, Timm V (1996a) Macrozobenthos of lake Peipsi-Pihkva: taxonomical composition, abundance, biomass and their relations to some ecological parameters. Hydrobiologia 338:139–154

Timm T, Erséus C, Lundberg S (1996b) New and unusual records of freshwater Oligochaeta from Scandinavian Peninsula Nordic. J Freshwater Res 72:15–29

Uzunov JI (1982a) Statistical assessment of the significance of both bottom substrata and saprobity for the distribution of aquatic Oligochaeta in rivers. Limnologica 14:353–361

Uzunov JI (1982b) Dissolved oxygen and saprobity as factors for the distribution of aquatic Oligochaetes in rivers. Hydrobiology 16:19–31

Uzunov JI, Kosel V, Sladecek V (1988) Indicator value of freshwater Oligochaeta. Acta Hydrochim Hydrobiol 16:173–186

Verdonschot PFM (1981) Some notes on the ecology of aquatic oligochaetes in the Delta Region of the Netherlands. Arch Hydrobiol 92:53–70

Verdonschot PFM (1987) Aquatic oligochaetes in ditches. Hydrobiologia 155:283–292

Verdonschot PFM (1989) The role of oligochaetes in the management of waters. Hydrobiologia 180:213–228

Verdonschot PFM (2006) Beyond masses and blooms: the indicative value of oligochaetes. Hydrobiologia 564:127–142

Verdonschot PFM, Ter Braak CJF (1994) An experimental manipulation of oligochaete communities in mesocosms treated with chlorphyrifos or nutrient additions: multivariate analyses with Monte Carlo permutation tests. Hydrobiologia 278:251–266

Warwick WF (1985) Morphological abnormalities in Chironomidae (Diptera) larvae as measures of toxic stress in freshwater ecosystems: indexing antennal deformities in *Chironomus* Meigen. Can J Fish Aquat Sci 42:1881–1914

Watt WD, Harding GH, Caldwell J, McMinn A (1973) Ecology of the Saint John River Basin VI. Oligochaetes as water pollution indicators. Department of Environment, Canada, Report, 29 pp

Weinstein JE, Sanger DM (2003) Comparative tolerance of two estuarine annelids to fluoranthene under normoxic and moderately hypoxic conditions. Mar Environ Res 56:637–648

Wentsel R, McIntosh A, Atcinson G (1977) Sediment contamination and benthic macroinvertebrate distribution in a metal-impacted lake. Environ Pollut 14:187–193

Wiederholm T (1980) Use of benthos in lake monitoring. J Water Pollut Control Fed 52:537–547

Wood LW (1975) Role of oligochaetes in the circulation of water and solutes across the mud-water interface. Verh Internat Verein Limnol 19:1530–1533

Wright S, Tidd WM (1933) Summary of limnological investigations in western Lake Erie in 1929 and 1930. Trans Am Fish Soc 63:271–285

Wright JF, Moss D, Armitage PD, Furse MT (1984) A preliminary classification of running-water sites in Great Britain based on macro-invertebrate species and the prediction of community type using environmental data. Freshwater Biol 14:221–256

[illegible]

[illegible] Oligochaeta [illegible]

[illegible]

[illegible]

Chapter 4
Toxicology and Laboratory Studies

Abstract This chapter provides a chronological synthesis of the available literature and information on laboratory based studies of aquatic oligochaetes and has been organized into water-only vs. sediment toxicity tests, and acute vs. chronic toxicity tests. Toxicological studies have employed relatively few species. Absence of sediment in tests with aquatic oligochaetes is a source of stress and reduces realism in exposure conditions and most current toxicity work is performed in the presence of sediment. Thus, unless there is a requirement for water-only tests, for interspecies comparison of toxicity of chemical compounds, or for routine control of species sensitivity in laboratory cultures, these tests are considered inappropriate for oligochaete worms in ecological risk assessment. Field populations of several species have been demonstrated to vary genetically, and may have different tolerances to both environmental and anthropogenic stress. This suggests that test organisms should be obtained from cultures of a defined genetic strain, or alternatively the sensitivity of the populations should be intercalibrated with reference toxicants. Oligochaete worms are not uniformly more tolerant to contaminants than other test organisms. Comparative data for worms and other benthic invertebrates show that responses are species-specific and also contaminant-specific. Among the sublethal measurements, sexual reproduction seems to be the most easily standardised and informative endpoint, and has high ecological relevance, since reproductive impairment has the same long-term effect as mortality on the population. Sublethal, short-term responses such as behavioural responses (avoidance behaviour, sediment reworking and sediment borrowing) need further standardization. In the future, *in situ* bioassays are promising tools in environmental risk assessment but will require the use of species characteristic of particular habitats or that are suitable for reproduction and toxicity assessment under different environmental conditions.

P. Rodriguez and T.B. Reynoldson, *The Pollution Biology of Aquatic Oligochaetes*,
DOI 10.1007/978-94-007-1718-3_4,

4.1 Introduction

Aquatic toxicology has been defined as the study of the effects of manufactured chemicals, natural materials and other anthropogenic activities on aquatic organisms at various levels of biological organisation, from the sub-cellular through individuals to communities and ecosystems (Rand et al. 1995). While this is a useful definition, there is a clear separation between laboratory based studies that use more traditional toxicological approaches where the test organisms are exposed to a toxic medium in a dose/response type experiment, and ecological studies conducted in the field where either populations or communities are the focus of study (see Chap. 3). While we have elected to examine the literature on aquatic oligochaetes using this dichotomy, this does not mean, however, that we support one approach over the other; in fact, we would strongly encourage those involved in pollution biology to use both approaches.

In this chapter, we have chronologically synthesised the available literature and information on laboratory based studies of aquatic oligochaetes. The chapter has been organized into water-only vs. sediment toxicity tests, and acute vs. chronic toxicity tests, although some publications provide information on several types of toxicity tests. As acute effects we have included information on short-term toxicity tests (usually less than 96-h exposure) mostly measuring mortality as the endpoint, although other sublethal measurements are also described including respiration and behaviour. The sections describing chronic toxicity tests report the effects of longer exposures (usually more than 10 days up to several months) and the endpoints include mortality as well as sublethal effects related to growth and reproduction, and less frequently behavioural response.

Since the seventies, there have been several major ecotoxicological reviews of both general toxicity assessment and specifically for sediment toxicity testing (Table 4.1). At the beginning of that period most of the available methods used fish as the standard test organism, however, in response to the increasing need by regulatory agencies for standardised "cookbook" methods Buikema and Cairns (1980) reviewed the state of knowledge on invertebrate aquatic toxicity testing at that time. Maciorowski and Clarke (1980) found that most (76%) toxicity tests with freshwater organisms were acute (<100 h) and measured lethality (89%). These authors also reported that of 86 taxa suggested for freshwater testing none were annelids, and of 72 taxa used for marine testing only 5 were polychaetes. In fact, at that time few marine oligochaetes had even been identified, at present approximately 600 species are known (Erséus 2005). In the same publication, Benfield and Buikema (1980) synthesised the state of invertebrate toxicity testing and presented the commonly held view of freshwater oligochaetes as "*notorious for being tolerant to polluted conditions*". However, they did acknowledge that not all oligochaete species were pollution tolerant. They also cited some examples of testing with oligochaetes (*Lumbricillus rivalis*) and the aphanoneuran *Aelosoma headleyi*, and suggested that some species could be suitable for bioassay work. Clesceri et al. (1989) provided a description of standard methods for aquatic toxicity tests and suggested a number of

Table 4.1 Oligochaete species reported in selected invertebrate toxicity reviews or international organisations

Author	Oligochaete species identified	Media	Comments
Benfield and Buikema (1977)	*Lumbricillus rivalis* (f)	Sewage beds	For screening potential toxicity of industrial effluents, metals and cutting oils
APHA: Clesceri et al. (1989)	*Limnodrilus hoffmeisteri* (f) *Tubifex tubifex* (f) *Branchiura sowerbyi* (f) *Stylodrilus heringianus* (f) *Lumbriculus variegatus* (f) *Quistadrilus multisetosus* (f) *Spirosperma ferox* (f) *Spirosperma nikolskyi* (f) *Rhyacodrilus montana* (f) *Varichaetadrilus pacificus* (f) *Paranais frici* (f) *Monopylephorus cuticulatus* (m) *Tubificoides fraseri* *(as T. gabriellae)* (m) *Tectidrilus verrucosus* (m) *Paranais litoralis* (m)	Water	No recommendations are provided
Giesy and Hoke (1990)	None	Sediment	Not included in the battery of tests proposed
Burton et al. (1992)	*Limnodrilus hoffmeisteri* (f) *Tubifex tubifex* (f) *Stylodrilus heringianus* (f)	Sediment	–
Persoone and Janssen (1993)	*L. hoffmeisteri* (f) *T. tubifex* (f) *B. sowerbyi* (f) *S. heringianus* (f)	Water	*L. hoffmeisteri*, *T. tubifex* and *B. sowerbyi* are also reported from FAO (Reish and Oshida 1986)
Reynoldson and Day (1993)	*Tubifex tubifex* (f) *Limnodrilus hoffmeisteri* (f) *Stylodrilus heringianus* (f) *Enchytraeus* spp. (f) *Lumbriculus variegatus* (f) *Ilyodrilus frantzi* (f) *Branchiura sowerbyi* (f)	Sediment	Most tests describe acute toxicity. No single invertebrate test can be relied upon for the detection of impact
Burton and MacPherson (1995)	*Lumbriculus variegatus* (f) *Tubifex tubifex* (f)	Sediment	–
Cooney (1995)	None	Water	Water tests only

(continued)

Table 4.1 (continued)

Author	Oligochaete species identified	Media	Comments
Ingersoll (1995)	*Lumbriculus variegatus* (f) *Tubifex tubifex* (f)	Sediment	Their tolerance may be a positive attribute for assessing toxicity or bioaccumulation of severely contaminated sites
Traunspurger and Drews (1996)	*Lumbriculus variegatus* (f) *Tubifex tubifex* (f)	Sediment	Eight species identified but only two widely used
Rodriguez and Reynoldson (1999)	*Stylodrilus heringianus* (f) *Lumbriculus variegatus* (f) *Limnodrilus hoffmeisteri* (f) *Branchiura sowerbyi* (f) *Tubifex tubifex* (f) *Pristina longiseta* (as *P. leidyi*) (f) *Monopylephorus cuticulatus* (m)	Sediment Sediment, water Sediment, water Sediment Sediment, water Sediment, water Water	Benthic species have more ecological relevance in sediment toxicity bioassessment and should not be used in water-only toxicity tests. A battery of tests with other taxa is recommended

f freshwater, *m* marine

freshwater oligochaetes as being suitable for toxicity testing: the tubificids *Limnodrilus hoffmeisteri*, *Tubifex tubifex*, and *Branchiura sowerbyi*; and the lumbriculid *Stylodrilus heringianus*. Among the marine/estuarine annelid species, these authors considered three tubificid oligochaetes (*Monopylephorus cuticulatus*, *Tubificoides fraseri*, and *Tectidrilus verrucosus*[1]). They also indicated that other species had been used in toxicity testing: *Lumbriculus variegatus, Quistadrilus multisetosus, Spirosperma ferox, Spirosperma nikolskyi, Rhyacodrilus montana, Varichaetadrilus pacificus* (reported as *Varichaeta pacifica*), *Ilyodrilus frantzi, Nais communis* and *Paranais frici* for freshwater, and *Paranais litoralis* for estuarine testing. However, this review was based on the work of only two research groups (Chapman et al. 1982a, b, c; Chapman and Mitchell 1986; Bailey and Liu 1980) and their studies did not deal with standardisation of test procedures.

In the late eighties, there was an increasing awareness of the importance of sediments as both a sink and source for chemical contaminants. This awareness arose as a result of concern by the USEPA (1988) and in the North American Great Lakes

[1] North American specimens of *Tubificoides gabriellae* have been recognized to belong to a variety of other species (Brinkhurst and Wetzel 1984). The specimens reported by Chapman and Brinkhurst (1980, 1981, 1984) and Chapman et al. (1982b) are classified as *T. fraseri* (see Brinkhurst 1986). This must not be confounded with *Peloscolex gabriellae* Marcus, a South American species, at present classified as *Tectidrilus* (see Erséus 1982). *Limnodriloides verrucosus* was also transferred to *Tectidrilus* by Erséus (1982).

where disposal of dredged materials containing contaminated sediments became a contentious issue (IJC 1988). As a consequence, several reviews by Giesy and Hoke (1990), Burton (1992), Calow (1993), Ankley et al. (1993), Hoffman et al. (1995), Rand (1995), Traunspurger and Drews (1996) and Mudroch et al. (1999) examined the state of the art in sediment toxicity and ecotoxicological assessment. Giesy and Hoke provided no further information on the utility of oligochaetes than in earlier reviews, however, they did give a useful guide to ranking the attributes of toxicity tests. In their review of sediment toxicity assessment, Burton et al. (1992) provided a valuable history, state of the art and practical guide to sediment toxicity assessment. They summarised the expansion in the literature addressing sediment contamination (Burton 1992) and noted the activity of the American Society for Testing and Materials (ASTM) in providing guides for whole-sediment toxicity testing with species of the midge *Chironomus*, the amphipod *Hyalella* as well as the cladocerans *Daphnia* and *Cerodaphnia and* the ephemeropteran *Hexagenia.* However, their review of the usefulness of oligochaetes still suggested that this group had limited utility. This despite the production of standard protocols for *Lumbriculus variegatus* (Dermott and Munawar 1992) and *Tubifex tubifex* (Reynoldson et al. 1991) and behavioural tests with *Stylodrilus heringianus* and *Limnodrilus hoffmeisteri* (Keilty et al. 1988a). Burton et al. (1992) concluded that while benthic organisms as a group were the best overall indicators of toxic sediments, most common tests available at the time were not ideal toxicity indicators, and problems such as laboratory culturing and recovery of early instars or young animals would be difficult to overcome.

Two reviews on the use of invertebrates (including annelids) in toxicity testing appeared in the *Handbook of Ecotoxicology* (Calow 1993). Examination of the use of invertebrates in freshwater toxicity testing by Persoone and Janssen (1993) showed that since 1979 there had been a large increase (25%) in the use of invertebrates as test organisms, including a doubling (from 6% to 12%) in the use of oligochaetes. However, *Daphnia* tests were, at that time, the only freshwater invertebrate bioassay endorsed by international organizations such as the EEC and OECD. Persoone and Janssen (1993) identified four oligochaete species in use at that time in freshwater toxicity tests (*L. hoffmeisteri*, *T. tubifex*, *B. sowerbyi*, and *S. heringianus*). In the same volume, Reynoldson and Day (1993) reviewed the use of benthic invertebrates in assessing freshwater sediments and saw the utility of laboratory tests in providing the only way of comparing sediments from different sources, but were sceptical of the potential of extrapolating results from the laboratory to the field or for predicting effects in field populations. They also reviewed the oligochaete tests available at the time of writing which included both sub-organism and whole-organism tests with several species.

Adams (1995) provided a useful review of toxicity testing in the USA, including a background on the regulatory framework for toxicity testing; however, there was no specific mention of oligochaetes in a regulatory testing context. In the same volume (Hoffman et al. 1995), Burton and MacPherson (1995) provided an update on sediment toxicity testing, pointing out that few standard methods existed. However, they recommended ten tests for assessment of freshwater sediment toxicity, including two for the oligochaete species *L. variegatus* and *T. tubifex*. In an extensive review of aquatic toxicology published the same year (Rand 1995), Ingersoll (1995) reviewed the available freshwater and marine sediment tests and of eight commonly used freshwater species two were oligochaetes (*T. tubifex* and *L. variegatus*), whereas the only annelids in common use in marine sediment testing were polychaete worms. However, Ingersoll still quoted Dillon and Gibson (1986) in suggesting that difficulties in culturing oligochaetes made them difficult to use in toxicity tests, even if experience at that time (see culturing, in Sect. 6.5) suggested that culturing oligochaetes was straightforward for some species in a number of different substrates, both natural and artificial. Finally, in reviewing saltwater tests, the only annelid Ward (1995) identified as a test species was the polychaete worm *Capitella capitata*, although various papers by Brinkhurst and Chapman had described the use of estuarine/marine oligochaetes (*e.g. Monopylephorus cuticulatus:* Brinkhurst et al. 1983; *M. irroratus* and *Tubificoides fraseri*: Chapman and Brinkhurst 1980; *Tectidrilus verrucosus*: Chapman et al. 1982b, c).

Traunspurger and Drews (1996) conducted an extensive review of freshwater and marine sediment toxicity testing and identified a total of 101 benthic invertebrate species used in bioassays. They correctly pointed out that a simple statement about the sensitivity of oligochaetes to contaminants is unlikely to be valid, for differences between species and between toxicants can be significant. In this review, eight oligochaete species were identified although they only described two species as widely used: *L. variegatus*, using the method developed by Carlson et al. (1991) and Phipps et al. (1993) who used mortality, growth, reproduction, and bioaccumulation as test endpoints, and *T. tubifex* (Reynoldson et al. 1991) where survival and reproduction were used as endpoints. The only marine oligochaete species identified by Traunspurger and Drews for toxicity tests was *M. cuticulatus,* as described by Chapman (1987).

Most recently, Rodriguez and Reynoldson (1999) described laboratory methods for sediment assessment and reviewed the available test species for marine and freshwater sediment testing. Seven species used in both aquatic and solid-phase tests were identified, although they did not make any recommendation as to which species were more appropriate but rather recommended the use of a battery of tests.

In conclusion, from the eighties on there has been a general trend toward the acceptance of oligochaetes in laboratory toxicity tests (Table 4.1), primarily because of the utility of this group which has a widespread distribution and is abundant in fine-grained sediment communities. In reviewing the use of this group in toxicity tests two trends are apparent. First, the initial increase in the number of freshwater species used has now focused on two major test species: *L. variegatus* and *T. tubifex*. Second, the use of the oligochaetes is primarily related to sediment testing rather than to water-only toxicity tests. Some of the classical drawbacks associated with the use of aquatic oligochaetes are patently not true, such as the former resistance to their use based on the assumption of their pollution tolerance or their difficulty to work with. Chapman (2001) identified general lack of experience with the group as the main reason for not using oligochaetes as test organisms in bioassays. Thus, perceived difficulties in species identification can be overcome as specialists are present around the world and also regional taxonomic keys are now available. Furthermore, genetic tools are becoming more available to confirm the identification of species used in toxicity bioassays (see Sect. 2.3). Finally, the difficulties of culturing, and handling for counting, weighing or other measurements have largely been resolved. An indication of this is the number of studies now using aquatic oligochaetes, not only in the field of toxicology but also in other physiological (respiration, uptake, or assimilation rates) and behavioural studies.

4.2 Water-Only Toxicity Tests

In this section, we focus on acute and chronic tests that have used aquatic oligochaetes for testing contaminants in water (both freshwater and salt water). Acute tests are usually designed for the calculation of short-term (commonly an exposure period of 96 h or less) effects on survival or physiological and behavioural alterations of the test organisms. Chronic tests study the effects after a long-term exposure (usually several weeks or months), and focus usually on sublethal endpoints such as growth or reproduction. Although it is now generally considered inappropriate to use oligochaetes in water-only tests, the first studies in this medium were a necessary step for intercomparison purposes of the sensitivity of different species with other aquatic invertebrates and also between oligochaete species. These studies were the basis of the selection of suitable test species and the demonstration of their value as test organisms sensitive to different chemicals. Water-only acute tests can be now integrated as a part of a test quality control procedure to verify the health and sensitivity of oligochaete worms used in chronic sediment bioassays (ASTM 2005; Maestre et al. 2009). The range of experimental 48-h and 96-h LC_{50} values for different species exposed in water-only and sediment bioassays are provided in Appendix 1 (Chap. 7).

4.2.1 Acute Freshwater Tests

It is not surprising that the majority of short-term tests utilised lethality as the endpoint, since it allows the estimation of LC_{50} values for comparison with values for standardised daphnid species. However, a few authors have described effects on behaviour, respiration and neurological function that can be useful tools for a rapid diagnosis of potential hazard of chemicals to living organisms in aquatic systems or as an early warning system. The following is a brief review of the most significant findings broadly separated into studies investigating impacts of metals, organic contaminants and other stressors.

4.2.1.1 Survival

Most water-only toxicity tests have been performed with tubificids or lumbriculids. One of the earliest records is that of Jones (1939) who showed that *Tubifex tubifex* was killed within hours when exposed to combined water concentrations of Cu and Pb equivalent to over 1,000 mg l^{-1}. Learner and Edwards (1963) studied a way of controlling naidines in water treatment facilities, and demonstrated that *Nais* spp. were killed within 6 h in hard or soft water by 1 mg l^{-1} of copper sulphate. Rehwoldt et al. (1973) also used *Nais* to determine median lethal times for six metals, and reported a 96 h-LC_{50} of 1.7 mg l^{-1} for mercury. Other studies have used *Nais communis, Dero digitata, Pristina longiseta* (as *P. leidyi*) and *Stylaria lacustris* and calculated acute LC_{50} values for metals and pesticides (see Appendix 1, Chap. 7). However, in spite of their proved sensitivity, little work has been done with naidines and pristinines, probably due to their small size, and also to their common asexual reproduction by paratomy, which makes the measurement of the number of survivors problematic. Most other studies used the well-known cosmopolitan tubificine species, *Tubifex tubifex* and *Limnodrilus hoffmeisteri.* Whitley (1963, 1968) performed the first tests with tubificines. He used a mixture of these two species, having noted no differences in their reactions when tested separately (but see later work on respiration of mixed or separate populations of these species by Chapman et al. 1982a). The upper and lower pH levels at which 50% of the worms survived in an artificial culture medium (Knop solution), at 20°C, were 9.7 and 5.8. The worms were shown to be tolerant to Pb, Zn and pentachlorophenol, with toxicity varying in relation to pH. At pH 6.5, 10 ppm and 50 ppm Pb were surprisingly more toxic than intermediate concentrations, but in other experiments the mortality of worms was directly related to increasing concentrations of contaminant.

Brkovic-Popovic and Popovic (1977a) determined 24- and 48-h acute toxicity for *T. tubifex* exposed to Cd, Cu, Hg, Zn, Cr and Ni. For metals other than Hg, hardness of the water was a significant variable, the lowest toxicity being obtained in the hardest water. Tests for Cr and Ni revealed greater differences on survival after 48-h exposure compared to 24 h, but there was little difference in other tests using other metals. The high toxicity reported in these experiments run with distilled water as the test solution (LC_{50} values two to three units lower than in drinking water) was most likely associated with the additional osmoregulatory stress. Thus, when distilled water has been used as dilution water toxicity values should be interpreted cautiously since toxicity of the chemical compounds is almost certainly overestimated (see for instance, Lucan-Bouché et al. 1999a; Veltz-Balatre 2000; Veltz et al. 1996, in Appendix 1 (Chap. 7)).

Chapman et al. (1980) reviewed the literature on toxicity and bioaccumulation of heavy metals by oligochaetes. Through the period of the eighties, the largest body of toxicological work on aquatic worms was that of Chapman and Brinkhurst (Chapman et al. 1982a, b, c; Brinkhurst et al. 1983; Chapman and Brinkhurst 1980, 1984; Chapman and Mitchell 1986). Twelve different freshwater and marine tubificids and one lumbriculid were used, including species thought to be associated with eutrophic, mesotrophic and oligotrophic water bodies. Toxicity was determined in both water-only and sediment tests in standard conditions, and with various combinations of pH, temperature and salinity. Both lethal and sublethal tests were performed, with respiration being used as the sublethal measure of stress. This series of studies had two primary objectives: first, to compare the tolerance of worm species

Table 4.2 Relative tolerances to pollutants under different environmental conditions (From Chapman et al. 1982c, Table II, reproduced by permission of Elsevier Ltd, © conveyed by Copyright Clearance Centre, Inc.)

Environmental conditions			Toxicants			
	Temperature	Salinity				
pH	(°C)	(ppt)	NaPCP	Black liquor	Cadmium	Mercury
7	10	0/20[a]	4>[3 A]>2>1>B	A>4>[2 1]>B	A>B>[4 3 2 1]	A>[1 2 4 B]
7	1	0/20[a]	[4 A]>B>2>1	A>4>[1 2]>B	A>B>4>2>1	A>[4 2]>[B 1]
7	20	0/20[a]	4>A>[1 2]	A>[1 2]>4	A>[4 2]>1	[4 1]>A>2
6	10	0/20[a]	4>[A 2]>1>B	A>4>2>1>B	A>B>4>2>1	[4 1]>[2 A]>B
8	10	0/20[a]	[4 A]>2>B>1	A>[4 2 1]>B	A>B>4>2>1	1>[4 2]>[A B]
7	10	5/10[a]	4>[1 2]>[3 A]	A>4>[1 2]	A>[4 2]>[3 1]	A>4>[1 2]

Significant differences are indicated by >; non significant differences are into brackets
Freshwater species: *1* = *L. hoffmeisteri; 2* = *T. tubifex; 3* = *Q. multisetosus; 4* = *S. heringianus.*
Salt-water species: *A* = *M. cuticulatus: B* = *T. verrucosus*
[a]Freshwater species/salt-water species

and to overcome the then widely held but erroneous belief that worms were all tolerant to any and all pollutants; and second, to demonstrate that toxicity values could vary depending on other environmental variables, supporting the contention that toxicity tests are valuable forensic tools for managing effluents and establishing comparative toxicity of contaminants.

In the first set of experiments with acute tests, LC_{50} values were calculated under fixed conditions using Hg, Cd, sodium pentachlorophenol (NaPCP), sewage effluent, and black liquor (from pulp mill effluent) (Appendix 1, Chap. 7), in various conditions of pH, temperature, oxygen depletion and salinity (Chapman et al. 1982b, c). For example, increasing salinity to 5 ppt increased the tolerance of the freshwater species to the Cd, NaPCP and black liquor, suggesting in the case of cadmium that metal complexation was taking place. In contrast, the toxicity of Hg was highly variable and therefore difficult to predict in the natural environment. In other instances, LC_{50} values could be doubled or more, or reduced, simply by shifting the temperature from 10°C to 20°C or down to 1°C. Similar modifications of the sensitivity of the worms were found when shifting the pH up or down by one pH unit from 7, but no response pattern was observed, likely as a result of different chemical speciation under different conditions (Table 4.2). Worm species associated with oligotrophic or "clean" environments were the least tolerant to sewage but were usually the most tolerant to metals and the other chemicals. This probably results from adaptations in species occurring in montane oligotrophic systems that may be more exposed to metal contaminants through normal erosional processes. By contrast, species found in lowland, eutrophic or contaminated sites with heavy inputs of organic matter that bind such naturally occurring metals, may not have developed such adaptations. In summary, this work demonstrated that rank orders of tolerance among tubificids varied depending on the contaminant used and exposure conditions, illustrating species-specific, not group-wide, sensitivity.

More recent studies have generated data on short-term responses of tubificids to various metals. Khangarot (1991) examined the effects of 32 metals in 96-h bioassays

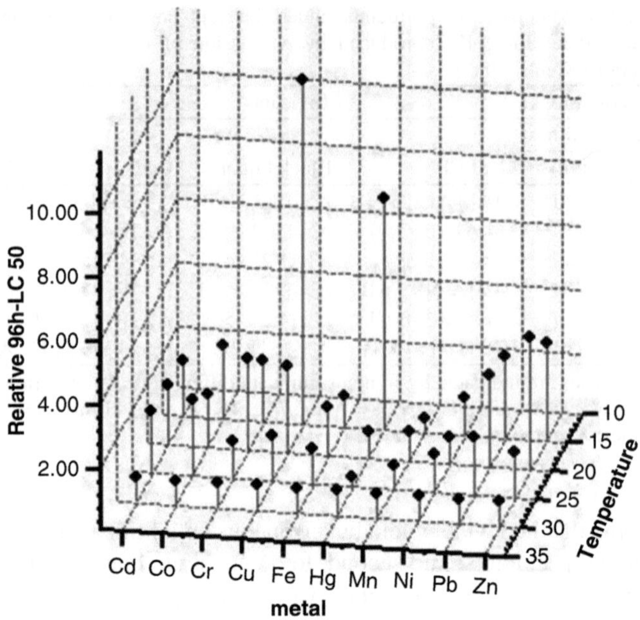

Fig. 4.1 Comparison of 96-h LC_{50} for several metals in *Tubifex tubifex*, exposed at different temperatures (Data from Rathore and Khangarot 2002)

with *T. tubifex,* and among the metals examined, the most toxic were Os, Ag, Pb, Hg, and Pt (see Appendix 1, Chap. 7). Temperature-dependent sensitivity of *T. tubifex* to heavy metals was investigated by Rathore and Khangarot (2002) who described an increase of toxicity in most metals from 15°C to 30°C (Fig. 4.1), concluding that water-quality criteria should consider seasonal (or latitudinal) temperature changes. Further research with dilution water of different hardness has shown, as expected, toxicity to *T. tubifex* worms may increase 6 to more than 60 times from hard to soft water (Rathore and Khangarot 2003), supporting the view (Chapman et al. 1982b) that complexation of metals explains their varying bioavailability. Working with *Branchiura sowerbyi,* Dutta and Kaviraj (1996) showed that acclimation to lime ($Ca(OH)_2$) over a 96-h period would reduce the toxicity of cadmium because of competition of Ca^{2+} ions and Cd^{2+} for binding sites, although the exact location of Ca-Cd interaction in the body wall of the worms was unknown.

The other major category of acute laboratory tests has examined the sensitivity of aquatic oligochaetes to organic compounds. The work described above often combined studies including a suite of metals and organic contaminants. Whitten and Goodnight (1966) used a mixture of *L. hoffmeisteri* and *T. tubifex* species to determine LC_{50} values for six organic insecticides in Knop-solution and demonstrated high tolerance to DDT (96-h $LC_{50} > 100$ mg l^{-1}) and Sevin (96-h $LC_{50} > 100$ mg l^{-1}). Twenty agricultural chemicals were tested by Inoue and Kondo (1962) using *B. sowerbyi, L. hoffmeisteri* (as *L. socialis*) and *L. udekemianus* (as *L. willeyi*) and they found a gradient of response, suggesting NaPCP be used to control worms "damaging" rice terraces. Kabir and Khatoon (1980) used unidentified *Limnodrilus* species (80% *L. hoffmeisteri)* to assay six organophosphorus and chlorinated hydrocarbon insecticides. They tried also to determine the effect of feeding the exposed worms to the guppy *Lebistes reticulatus* (see Sect. 5.5), and they also studied the effect of adding sediment to the worm tests. The LC_{50} values obtained were very similar for 5 of the pesticides (5.5–7.0 ppm for 72-h tests), but Dieldrin produced 100% mortality at the lowest test concentration (1.25 ppm).

Naqvi (1973) used a gilled tubificid species, *Branchiura sowerbyi*, to determine tolerance to 23 insecticides at concentrations with maximum values ranging from 0.5 to 4 ppm, at 21°C, but failed to demonstrate any toxicity after 72-h exposure for 15 of them. Other experiments were run at three different temperatures, toxicity was minimal at 21°C compared with 4.4°C and 32.2°C. Most pesticides produced specific reversible sublethal effects, and the response occurred in relation to insecticide type rather than to concentration. Worms were also used for feeding crayfish, and the author discussed possible resistance mechanisms of the worms to the insecticides, such as through metabolisation or excretion, since worms treated for longer periods were less toxic when fed to crayfish. These results were compared to previous data reported by Inoue and Kondo (1962) working with a population of *B. sowerbyi* from Japan that was as tolerant to DDT, but far more sensitive to endrin, dieldrin and lindane than reported by Naqvi. Dad et al. (1982) used *L. hoffmeisteri* and *T. tubifex* to test two insecticides, Furadan and Malataf. The former was the more toxic of the two, and *L. hoffmeisteri* was more susceptible than *T. tubifex* (see Appendix 1, Chap. 7). These studies show that, in general, there is no predictable response by tubificids to a single class of contaminants.

The tubificids *L. hoffmeisteri* and *B. sowerbyi* were used to test the toxicity of a surfactant (linear alkylbenzensulfonate, LAS) used in household detergents (Casellato and Negrisolo 1989). The worms were more sensitive in water than in sediment, and *L. hoffmeisteri* was more sensitive than *B. sowerbyi.* Verdonschot and Ter Braak (1994) cited Kersting (1990) who reported 48-h LC_{50} values higher than 50 $\mu g\ l^{-1}$ for the pesticide Chlorpyrifos, though other studies did not show any effect on oligochaetes. Most recent studies on the effects of various organic contaminants have been straightforward tests generating LC_{50} values with *T. tubifex*. Fargasová has published several papers (Fargasová and Kizlink 1996; Fargasová 1997, 1998) comparing toxicity of several organic compounds to *T. tubifex* and other aquatic species. Compared with larvae of *Chironomus plumosus*, *T. tubifex* was less sensitive to most organotin compounds tested, but the relative toxicity depended on the species (Fargasová and Kizlink 1996). Fargasová (1997) determined LC_5 and LC_{50} for plant growth regulators using both *Daphnia magna* and *Tubifex tubifex*, resulting in similar sensitivity, and *T. tubifex* was more sensitive to b-indolyl acetic acid and 1-naphtylacetic acid than *D. magna*.

Bailey and Liu (1980) pioneered the use of the lumbriculid oligochaete *Lumbriculus variegatus* as a bioassay organism. Their study involved the determination of 48-h LC_{50} values for trinitrotoluene (TNT) and TNT/RDX, and both 48-h and 96-h LC_{50} values for seven pesticides and six metals. Their work with munitions plant wastewater involved comparative studies with six fish (including rainbow trout) and three arthropodan invertebrates (*Daphnia magna*, *Hyalella azteca* and *Tanytarsus dissimilis*) (see Sect. 4.4). The lumbriculid was more sensitive to metals than the *Tubifex/Limnodrilus* complex used by Whitley (1963, 1968). Sublethal changes were noted, such as the production of a considerable amount of mucus by *L. variegatus* exposed to Pb and Cd, as well as other behavioural changes. In exposures of *L. variegatus* to a series of organophosphate insecticides, the species was found to be relatively insensitive, particularly compared to *Hyalella azteca* and *Chironomus tentans*. Based on a previous study by Brook in 1986, Ankley and Collyard (1995) suggested that this may imply that *L. variegatus* can detoxify the organophosphates via reactions catalyzed by enzymes such as glutathione-S-transferases and / or carboxylesterases.

Ankley's group also examined the toxicity of polycyclic aromatic hydrocarbons (PAHs) to *L. variegatus*. Ankley et al. (1995) registered time-dependent mortality (96 h) that could be accurately predicted through fluoranthene body burden and the light intensity during exposure. Similar experiments on the toxicity of fluoranthene to the naidine *Stylaria lacustris* were conducted by Suedel and Rodgers (1996). In more extensive studies, Ankley et al. (1997) examined the phototoxicity of four PAHs (anthracene, pyrene, fluorene, and fluoranthene), finding anthracene and pyrene to be similarly toxic to *L. variegatus* and four times more potent than fluoranthene, while fluorene showed no phototoxicity. The authors modelled mortality as a function of the product of tissue residue of PAH after 96-h exposure and UV intensity to which organisms were exposed. However, these authors stressed the need to evaluate toxicity to mixtures of PAHs, since this is the most common situation in field exposures. In a similar approach, Erickson et al. (1999) applied an additive model for evaluating the toxicity of binary mixtures of some PAHs suggesting that the additive concentration model adequately described the time-dependent phototoxicity of the mixtures.

Some naidine species are intermediate hosts of sporozoan parasites of farmed fish. This is the case of the worm *Dero digitata* and the catfish *Ictalurus punctatus*. Mischke et al. (2001) identified several toxicants that will eliminate these worms from fish-culture ponds and control proliferative gill disease. Eight chemicals were tested and the LC_{50} values obtained (see Appendix 1, Chap. 7). However, the concentrations required to affect *D. digitata* would also have consequences on fish and become cost-prohibitive.

In an attempt to develop a bioassay test using members of the family Enchytraeidae (mainly of the genus *Enchytraeus*), Römbke and Knacker (1989) found small differences in sensitivity (LC_{50} values) between *Enchytraeus albidus* and *Daphnia magna* when tested with Benomyl, pentachlorophenate (PCP), Parathion, 2,4,5-trichlorophenoxyacetic acid, chloroacetamide, cadmium, and potassium dichromate. The worms were tested in water and in soil. Römbke and Knacker (1989) concluded that enchytraeids of the genus *Enchytraeus* are suitable organisms for toxicity testing of chemicals. Ease of culture and demonstrated sensitivity to different substances make them good candidates as test organisms for soil bioassays. *Enchytraeus albidus* is a medium-sized species often found in compost heaps and in rotting seaweed on beaches, and the *E. buchholzi*-complex is made up of small-size species that

Table 4.3 LC_{50} values (mg l^{-1}) (based on inhibition of movement) for *T. tubifex* populations from Canada and Spain cultures

Reference	48-h LC_{50}	96-h LC_{50}
	Cd^{2+}	
Reynoldson et al. (1996)	6.5 [a]	3.2 [a]
	0.9 [b]	0.4 [b]
	3.6 [c]	1.7 [c]
Maestre et al. (2009) (n=5)	0.5–0.8 [b]	0.2–0.4 [b]
	Cu^{2+}	
Reynoldson et al. (1996)	0.26 [a]	0.09 [a]
	0.18 [c]	0.07 [c]
Maestre et al. (2009) (n=5)	0.07–0.19 [b]	0.03–0.08 [b]
	Cr^{6+}	
Reynoldson et al. (1996)	95.5 [a]	38.1 [a]
	33.2 [b]	9.8 [b]
	54.5 [c]	15.5 [c]
Maestre et al. (2009) (n=5)	11.0–23.8 [b]	4.9–7.2 [b]

[a]Canadian population cultured in sediment from Canadian Great Lakes
[b]Spanish population cultured in sediment from a Spanish mountain river
[c]Spanish population cultured in sediment from the Canadian Great Lakes

inhabit agricultural soils in Europe, and are common in streams. These species could also be used for toxicity testing in sediment toxicity tests. Didden and Römbke (2001) have reviewed the several studies dealing with sensitivity of *Enchytraeus* spp., mostly applied to soil toxicity tests, although some data on exposure in water and agar are also provided (see Appendix 1, Chap. 7).

There are few intercalibration studies of the sensitivity of test species used in acute toxicity tests. Reynoldson et al. (1996) compared tolerances of Canadian and Spanish populations of *T. tubifex* in parallel experiments using three metals (Cu, Cd and Cr) and one organic compound (lindane). Results showed that Spanish populations were somewhat more sensitive than the Canadian, although it was concluded that differences were not sufficient to invalidate the use of either populations in a standard bioassay protocol. Interestingly, this work and recent data (Maestre et al. 2009) have shown that both *T. tubifex* populations (Canadian and Spanish) have apparently more similar LC_{50} values when cultured in the same sediment (Table 4.3). Therefore, acclimation processes likely explain some of the differences in sensitivity in the source populations.

Survival data in laboratory toxicity tests are largely associated with the tolerance of species to a single variable (chemical concentration) and it cannot be used to infer survival in the field as other factors, such as the ability of the species to exploit continuous disturbances through behavioural and reproductive adaptations, as well as the tolerance to abiotic factors play an important role. Nevertheless, these tests are important tools in evaluating the health and sensitivity of the test organisms in culture stocks, in allowing laboratory control charts to be maintained as part of a quality assurance program, in calibrating the sensitivity of different populations of the same species, or in comparing sensitivity with other species. Methods for water-only acute tests are well developed, although decisions on the exposure conditions require standardisation.

4.2.1.2 Respiration

Respiration as an indicator of stress in oligochaetes was first examined by Berg et al. (1962). They looked at the oxygen regulatory ability of four oligochaete species, *Psammoryctides barbatus* (as *Tubifex barbatus*), *Potamothrix hammoniensis* (as *Ilyodrilus hammoniensis*), *Tubifex tubifex*, and *Lumbricillus rivalis,* as well as other benthic invertebrates, as a mechanism for explaining the field distribution patterns of species. Oxygen consumption was measured using the polarometric method of Bartels, in association with variation in temperature, oxygen saturation and starvation (without mud). The authors worked with worms collected from eutrophic Esrom Lake and the polluted river Pøleaa (Denmark), and calculated the critical point at which a marked decrease in oxygen consumption occurred. In a later study, in order to mimic the environmental conditions in the profundal of Lake Esrom, Berg and Jonasson (1964) examined oxygen consumption under conditions of low oxygen content in water (about 0.6 ml l^{-1}) and survival under anoxic conditions.

Fowler and Goodnight (1965) tried to determine the effect of the chemicals (Cu and sodium pentachlorophenate) on oligochaete respiration rates. They placed specimens of *T. tubifex* in the Warburg apparatus, in non-chlorinated water, at 25°C, and claimed that 1 ppm Cu as well as 1 ppm NaPCP caused respiration to increase due to higher worm activity, but that the same concentration of Zn did not. The study also showed no effect of a pH change from 7.5 to 6.1. In contrast, they observed differences in population density to have an effect on respiration, but this was most likely a consequence of either overcrowding at high densities or fragmentation of the worms in the test chambers at low densities. Whitley and Sikora (1970) immersed specimens of a mixed culture of *T. tubifex* and *L. hoffmeisteri* in the test solutions (Pb, Ni and NaPCP) *prior* to placing them in the Warburg respirometer. These authors described a linear drop in respiration rate with increasing Pb concentration from 20 to 60 ppm, but no consistent pattern of change with exposure to NaPCP or Ni. The use of the Warburg respirometer which agitates the worms while in the light, in contrast to their preference for dark, undisturbed conditions, makes these results of limited value. A similar method was used by Brkovic-Popovic and Popovic (1977b) who used *T. tubifex* exposed to six metals. They reported that the direction of change in respiration was not the same for all metals and that concentration affected rates in various ways. Although the error terms of mean values in this study were high, in the acute lethal range of concentrations (48-h and 24-h LC_{50}) Cd, Hg and Cu caused inhibition of oxygen consumption by worms. However, in the 6-h experiments respiratory rates increased at a similar concentration range for both Cd and Hg, while Cu reduced respiration at very low sublethal concentrations. Therefore, respiration rate alone was concluded to be of limited value as a sublethal toxicity measure.

Aquatic organisms are either able to regulate oxygen consumption (oxygen regulators) or not (conformers) in response to reduced ambient oxygen concentration

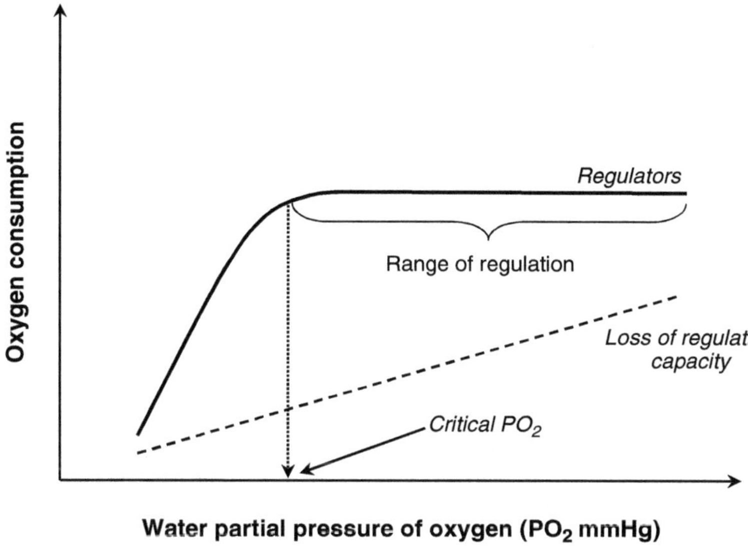

Fig. 4.2 Oxygen consumption plotted against water oxygen concentration (partial pressure of O_2, PO_2 mm Hg). Respiration patterns of oxygen regulators in control conditions and when regulation capacity is lost due to environmental stress

(Fig. 4.2), although the distinction between regulators and conformers is not always clear, particularly when the critical partial pressure of oxygen (PO_2) for a species is high (Willmer et al. 2005). Aquatic oligochaetes have the general reputation of being regulators of oxygen consumption over a wide range of oxygen concentrations and showing the ability to maintain a constant respiratory rate despite falling oxygen levels until a critical point (Critical PO_2) is reached, below which the respiration rate rapidly declines. However, in the absence of sublethal stress, some species have been classified by Chapman and Brinkhurst (1984) as regulators (*e.g. T. tubifex, L. hoffmeisteri* and *Rhyacodrilus montana / Varichaetadrilus pacificus* mixtures), or partial regulators (*e.g. B. sowerbyi, Stylodrilus heringianus, Monopylephorus cuticulatus* and *Spirosperma nikolskyi*).

Brinkhurst et al. (1983) used a Radiometer PHM 73 Blood Gas Analyser to determine respiration rates for *L. hoffmeisteri, S. heringianus* and *M. cuticulatus* exposed to sublethal levels of Cd, Hg, and NaPCP, under various environmental conditions (pH = 6 or 7; salinity = 0, 5, 10, 20 ppt and temperature = 10°C or 20°C). Worms were maintained without sediment in 10-ml syringes filled with water, in the dark, and incubated in water baths. Under the control conditions pertaining in these experiments, *T. tubifex* and *L. hoffmeisteri* were confirmed as respiratory regulators (0.32 and 0.39 μl O_2 mg^{-1} dw h^{-1}, respectively) and *M. cuticulatus* and *S. heringianus* were observed to be partial regulators. All possible changes in the face of exposure to environmental variables and contaminants were observed. Even more complex results were obtained when several stress factors were combined. Strangely, some stress factors actually increased the ability of some species to regulate respiration

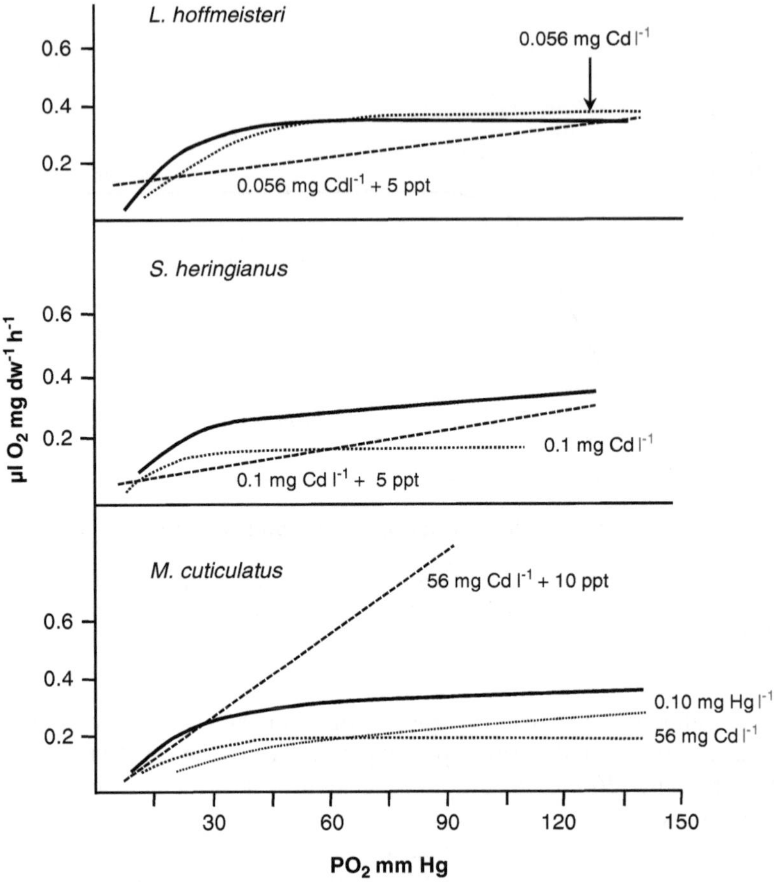

Fig. 4.3 Respiration rates of *L. hoffmeisteri, S. heringianus* and *M. cuticulatus* under standard conditions (*solid line*: 10°C, pH = 7.0 and 0 ppt, except for *M. cuticulatus* 20 ppt salinity), and exposed to Cd and Hg (*dotted lines*). *Dashed line* represents the oxygen consumption when salinity is modified for Cd exposure (Several cases selected from Brinkhurst et al. 1983, redrawn and reproduced with permission of Wiley-VCH Verlag GmbH & Co. KGaA)

rates, that in most instances were elevated due to the presence of organics and decreased under the exposure of metals, or regulation was lost to some degree, or both (Fig. 4.3). In another set of experiments using the same method, Chapman and Brinkhurst (1984) concluded that the ability to regulate respiration is not correlated with tolerance to toxicants.

As part of a complex ecological study that investigated the success of mixed species "flocks" of tubificids, Brinkhurst et al *(*1972*)*, Chua and Brinkhurst (1973) and Brinkhurst and Austin (1979) had shown that *L. hoffmeisteri* and *T. tubifex* grew better and had lower respiration rates in mixed populations than in mono-culture. It seemed likely that if worms were more productive in mixed species flocks than in pure culture, they might also prove to be more resistant to stress under those conditions. Short-term

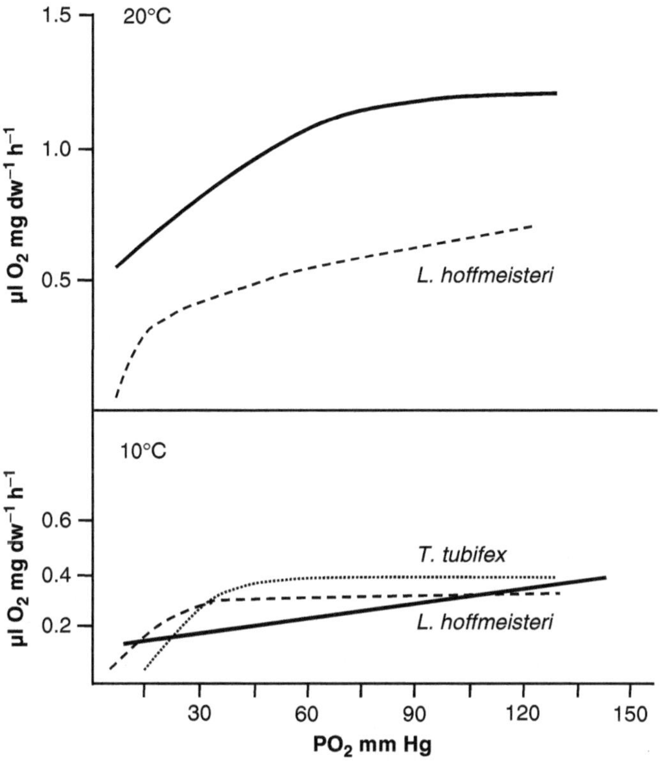

Fig. 4.4 Respiration rates of mixed species (*solid line*) at 10°C and 20°C and for pure cultures of *L. hoffmeisteri* (*dashed line*) and *T. tubifex* (*dotted lines*) (From Chapman et al. 1982a, Fig. 1, redrawn and reproduced by permission of Elsevier Ltd., © conveyed by Copyright Clearance Centre, Inc.)

survival was tested by Chapman et al. (1982a) under the same conditions as was respiration rate, and oxygen consumption was measured in 1-ml water extracted from respiratory vessels, using an IL-pH Blood Gas Analyser. This yielded the unexpected result that the mixture of *T. tubifex* and *L. hoffmeisteri* was less able to regulate respiration (Fig. 4.4) and showed less tolerance to anoxia under the specific conditions of the experiment, than either species alone. Exposure to Hg caused the mixed population to regulate respiration, but not when exposed to Cd or NaPCP. Respiration rates of the mixture were elevated in the face of increased temperature (as expected) and with exposure to phenol, as suggested by Whitley and Sikora (1970).

The conclusion from these studies was that respiration could be used as a sublethal measurement of stress as long as comparisons are made with a standard respiration pattern for the particular species under certain environmental conditions (a control), since toxic effects on respiration rates are species-specific and depend on the nature and concentration of toxicants and the exposure conditions, as in other taxa. While elevation of respiration is predictable for temperature alone, further increases in the presence of a toxicant are less so, and changes in respiration rates in the presence of a toxicant depends on not-always well-understood effects due to the combination of toxicant and environmental variables on the organism.

4.2.1.3 Behaviour, Escape Reflexes and Other Sublethal Responses

Brkovic-Popovic and Popovic (1977a) described the short-term exposure response of *Tubifex tubifex* to heavy metals to be increased mobility and decomposition of the rear portion of the body, with the exception of chromium which provoked weak movement and no decomposition. The degeneration of the posterior region results in a beaded or rosary effect and seems a typical response of worms exposed to heavy metals or pesticides. Naqvi (1973) described several changes in behaviour for *Branchiura sowerbyi* after exposure to insecticides, such as a reversible beaded appearance, and a coiled C-form related to muscular contraction. Naqvi also suggested that changes in morphological appearance depended on substance type rather than concentration. Fragmentation, changes of colour, mucus production, clumping and both local and overall swelling of the body were described by Bailey and Liu (1980) as signs of stress in *Lumbriculus variegatus* when exposed to different chemicals. The worms formed tight, spherical bunches or clumps compared with the controls, which usually were distributed loosely on the bottom of the beaker. Overall swelling was apparent when exposed to 0.056–0.1 mg Cd l^{-1}, the body diameter often doubling in size. The same authors also tested recovery after 5 days in clean water, showing some of the sublethal effects described above to persist in worms previously exposed to Zn, Pb or Cr, whereas worms regained the appearance of the controls after exposure to Cu or Hg. When exposed to Pb (2.5–10.0 mg l^{-1}) and Cd (0.056–0.1 mg l^{-1}), *L. variegatus* also produced a considerable amount of mucus. The production of a mucus/metal complex, which precipitates on the body wall blocking the exchange of oxygen and carbon dioxide, was described by Whitley (1968) as a response of tubificids to Pb and Zn. Bouché et al. (2000) have also reported mucus production together with the loss of the posterior body region of *T. tubifex* exposed to Cd in 96-h water-only toxicity tests. Mucus production was low at 0.01 mg Cd l^{-1} and slightly higher at 0.05 mg Cd l^{-1}. When exposed to cadmium at sublethal levels, worms may also exhibit rupturing of the body wall and degeneration of the digestive epithelium (Thompson et al. 1982; Chapman and Brinkhurst 1984). Worms are able to regenerate their posterior region when it is lost through toxicant exposure, and this supports the suggestion that this process is a decontamination mechanism. It is of interest that under sublethal concentrations of Cu, Pb and Pt, a delay in cephalic regeneration has been described in *L. variegatus*, while mixtures of two metals can cause atypical regeneration processes (Veltz-Balatre 2000).

Rogge and Drewes (1993) examined the neurotoxic and behavioural response of *L. variegatus* to five chemicals (4-aminopyridine, carbofuran, chloroform, diazinon, and cadmium), and median giant nerve fibre (MGF) conduction velocity was considerably reduced after exposure to chloroform and cadmium. Drewes (1997) examined oligochaete escape reflexes in both terrestrial (*Lumbricus terrestris*, *Eisenia fetida*) and aquatic (*L. variegatus*) worms. He found that even short-term cutaneous exposures to a wide variety of toxicants produced sublethal, but reversible, disruptions that could affect the worms rapid escape reflex. More recently, Ding et al. (2001) investigated the effects of the antiparasitic Ivermectin on escape reflexes, swimming, body reversal, and crawling of *L. variegatus*. Ivermectin solutions at 0.3 nM significantly decreased frequency of helical swimming waves and it inhibited swimming, reversal, crawling frequency and crawling speed, with a 3-h median inhibitory concentration (IC_{50}) of 1.1, 16, 91 and 51 nM, respectively. Electrophysiological recordings showed that the drug had no effect on the conduction velocity of the giant fiber system and they concluded that those behaviours controlled by nongiant locomotor pathways are more sensitive to Ivermectin than those controlled by giant interneurons.

The study of behavioural responses is an alternative to measuring lethality in acute toxicity tests and these responses may be ecologically relevant. However, for this to have practical application, the numeric relationship between EC_{50} values measured on lethal and behavioural responses should be established. This task also requires more effort in the standardisation of criteria for measuring specific effects on behaviour.

4.2.2 Acute Estuarine and Marine Tests

Many oligochaete species are known to form large populations in estuaries and some marine coastal habitats and may have potential for future toxicity work. The marine species, especially the truly oceanic species, are harder to identify than their freshwater relatives. Taxonomic keys are required and although a few studies have been done, the selection of suitable species and standardisation procedures for both culturing and testing is an ongoing task. Giere (1980a) and Giere and Pfannkuche (1982) reviewed the temperature and salinity tolerance and optimal values of these variables for certain marine and estuarine oligochaete species in different parts of the world. These authors included notes on acclimation to different environmental conditions, such information is useful for the development of estuarine and marine bioassays.

Data on respiration rate or critical oxygen concentration for marine or estuarine species were reviewed by Giere (2006). The respiration physiology of the tubificine *Tubificoides benedii* has been the subject of study by Bagheri and McLusky (1984) and Giere et al. (1999) and the species is one of the best annelid oxygen regulators (critical oxygen concentration of 2.1 kPa, at 15°C). It also has an unusually low respiration rate and high tolerance to extremely low oxygen levels compared to freshwater species or polychaetes of comparable size. Temperature and salinity tolerance of enchytraeids of the genus *Marionina* have been reviewed by Lasserre (1971). However, the toxicity stress responses of these species are untested.

Median lethal concentration (96-h LC_{50}) to different pollutants with different combinations of pH, temperature and salinity of the North American intertidal species *Monopylephorus cuticulatus* and the subtidal *Tectidrilus verrucosus* (as *Limnodriloides*) were reported by Chapman et al. (1982c). The latter species was not considered for testing as it showed mortality at 20°C and 10 ppt salinity. LC_{50} values changed with environmental factors, *e.g.* tolerances to NaPCP were significantly enhanced at 1°C and 20°C as compared with 10°C, and Cd was more toxic at higher temperatures and lower salinities. The worms were an order of magnitude more tolerant to kraft black liquor than the rainbow trout, to which they were compared. The estuarine species *M. cuticulatus* was included together with freshwater tubificids in a variety of experiments, which measured survival and respiration rates under different environmental stressors (temperature, pH and salinity), both alone and in combination with toxicants (Brinkhurst et al. 1983; Chapman and Brinkhurst 1984) and it was assessed as very tolerant (Brinkhurst et al. 1983). In subsequent work, Chapman (1987) exposed *M. cuticulatus* to elutriates prepared from centrifuged polluted sediment slurries obtained in Puget Sound (Washington, USA), and respiration rates were measured in a Radiometer PHM73 Blood Gas Analyzer, using the same method as the freshwater species. Ninety-seven sediments elutriates were tested (using clean water as the control condition), and 40 produced effects on oligochaete respiration. The respiration rate increased in about half the sites and decreased in the others, and an effect was considered toxic when significant changes in respiration rate occurred in either direction with respect to control. Results showed that there was a high correspondence between toxic responses using respiration tests and tests with other species using reproduction impairment endpoints (oyster larvae, polychaete life-cycle, surf smelt development, and fish cell replication tests). Sediments considered to be potentially contaminated by the various tests were scattered throughout the test area, suggesting very local patchiness. Oligochaete respiration tests using sediment elutriates were therefore considered as a useful tool for assessment of sublethal toxicity of water soluble chemicals from estuarine sediments and, as with other tests, allow sites to be ranked based on toxicity.

The most recent review by Giere (2006) on marine oligochaete biology has reiterated that little work has been done in this field in the last two decades. This is problematic as such data constitutes the fundamental database for the development of new bioassays and the selection of test species based on their ecological relevance, life-cycles, and tolerance to toxicants.

4.2.3 Chronic Freshwater Water-Only Tests

As most oligochaetes are infaunal species, the lack of studies dealing with chronic or subchronic (defined here as >96-hour to <10-day exposure) effects of contaminants in water-only tests is not surprising. Long exposures in aqueous media, with the additional stress imposed on these animals devoid of their natural medium, is likely to confound any result. Nevertheless, a few authors have conducted such studies and we have summarised the results for freshwater and then for marine and estuarine species. One of the few studies on water-only chronic toxicity using a infaunal species was that of Phipps et al. (1995) who compared the toxicity of *L. variegatus* to five metals (Cu, Cd, Ni, Pb, Zn) and to five organics (chlorpyrifos, dieldrin, *p,p'*-DDD, *p,p'*-DDE, *p,p'*-DDT) with *Hyalella azteca* and *Chironomus tentans* in a 10-day flow-through test and found the worm to have generally higher 10-day LC_{50} values.

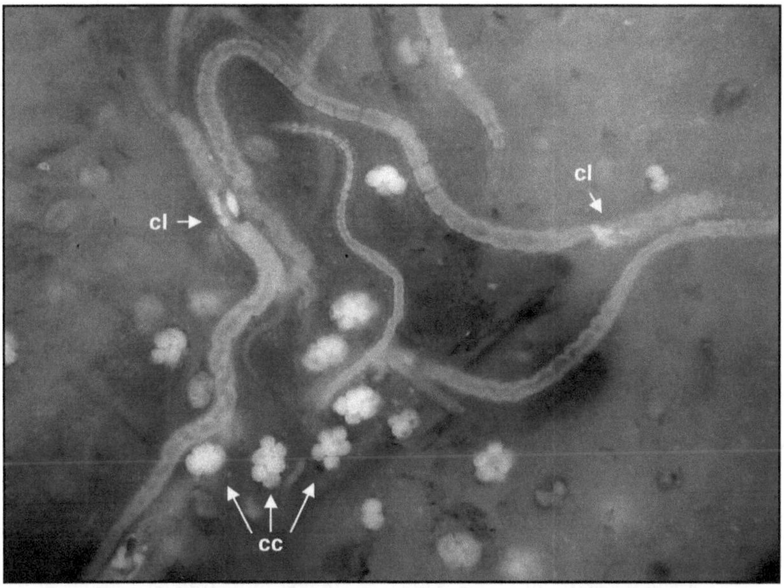

Fig. 4.5 *Enchytraeus coronatus* breeding in an agar culture, showing some mature worms with clitellum (*cl*) and cocoons (*cc*) filled with eggs

Chronic tests in water are appropriate for swimming oligochaete species, such as the many naidines. *Dero digitata* and *Stylaria lacustris* were studied in a semistatic water-only test, and both survival and response to tactile stimuli were measured as endpoints for assessment of toxicity to Derosal (carbendazim 511 g l^{-1}, a benzimidazol fungide) by van Wijngaarden et al. (1998). They used tap water and tests were run with no food, for up to 21 days. *Dero digitata* was only studied in short-term experiments, but effective concentrations for 7-day exposures of *S. lacustris* were calculated (7-day LC_{50} = 232 μg l^{-1}, 7-day EC_{50} = 142 μg l^{-1}, 7-day EC_{10} = 29 μg l^{-1}). After 2-weeks, controls showed effects in more than 10% organisms, probably due to starvation, and the data could not be evaluated, as the authors had intended, in a longer exposure (up to 21 days).

The use of agar is an alternative to water as test medium (Fig. 4.5). Agar is a semi-liquid culture medium that has the advantage of providing food in longer term studies and allows day by day observation of effects on survival, behaviour, and reproduction. Westeheide and Bethke-Beilfuss (1991) described a test guideline that was used successfully with small-sized *Enchytraeus* species. The test duration was 30 days, with inspection every 5 days, and survival, number of cocoons and hatching were measured. Several pesticides were tested including Dimilin 25 WP (diflubenzuron), Benomyl, Cypermethrin, Carbofuran and Parathion. There was no effect in worms exposed to very high concentrations of Dimilin, but effects on cocoon production were observed for the other substances. Ten times the minimum recommended concentration of the insecticide Benomyl reduced cocoon numbers by more than 50% and Cypermethrin was the most toxic substance of the four studied.

Table 4.4 Sensitivity of several *Enchytraeus* species to chemical compounds in laboratory toxicity tests of both short-term and long-term exposure, using agar or water as substrate. *(m) mortality, (r) reproduction*

Species	Substance	Duration (days)	Substrate	LC_{50} (ppm)	NOEC (ppm)	Reference
E. albidus	Chromium	1	Water	20		Didden and Römbke (2001)
E. crypticus	Benomyl	30	Agar	–	11.8 (m)	Didden and Römbke (2001)
	Cadmium	1	Water	0.1	–	
E. doerjesi	Benomyl	30	Agar	–	0.12 (m) 1.22 (r)	Didden and Römbke (2001)
	Carbofuran	30	Agar	–	12,000 (r)	
E. buchholzi	Cadmium	4	Agar	8.4	–	Didden and Römbke (2001)
	Copper	2	Water	0.41	–	
	Zinc	2	Water	57	–	
E. coronatus	Carbendazim	28	Agar	>17.78 (14 days)	0.32 (r)	Arrate et al. (2002)
	4-Nitrophenol	28	Agar	16.79 (14 days)	0.32 (r)	
	Potassium dichromate	28	Agar	17.78 (14 days)	0.32 (r)	

Some authors have used agar tests with *Enchytraeus* species for measuring long-term exposure effects (*e.g.* Westheide et al. 1991; Purschke et al. 1991; Achazi et al. 1995). In the presence of PAHs (fluoranthene and benzo(a)pyrene), Achazi et al. (1995) exposed *Enchytraeus crypticus* for 30 days in agar and showed no effect on cocoon production (up to 500 μmol l^{-1}) and hatching rate was the more sensitive endpoint. However, these same authors showed a significant reduction in the reproduction rate when worms were exposed to 90 μmol Cd l^{-1}. Arrate et al. (2002) exposed *Enchytraeus coronatus* to 4-nitrophenol, Derosal (a.i. carbendazim) and potassium dichromate in a chronic agar test and measured adult survival and the number of cocoons produced per adult after a 2-week exposure, and the number of juveniles produced over 4 weeks. The three substances caused a reduction in the number of juveniles per adult, but they seem to affect different stages of worm reproduction. The reduction in the number of juveniles was explained by a negative effect on cocoon hatching when exposed to carbendazim, a consequence of a decrease in the number of eggs per cocoon when exposed to 4-nitrophenol, and due to a reduction in the cocoon production in worms exposed to potassium dichromate.

In their revision of the sensitivity of enchytraeids to chemicals in laboratory tests, Didden and Rombke (2001) examined *Enchytraeus* species exposed to pesticides (Benomyl and Carbofuran) in agar and water, in both acute and chronic exposure (30 days), although most reported studies were done in soil (Table 4.4). One of the limitations of the agar tests is its lack of ecological relevance (Westeheide and Bethke-Beilfuss 1991); however, their easy and exact recording of several sublethal measures is useful in comparing the relative toxicity of different chemicals.

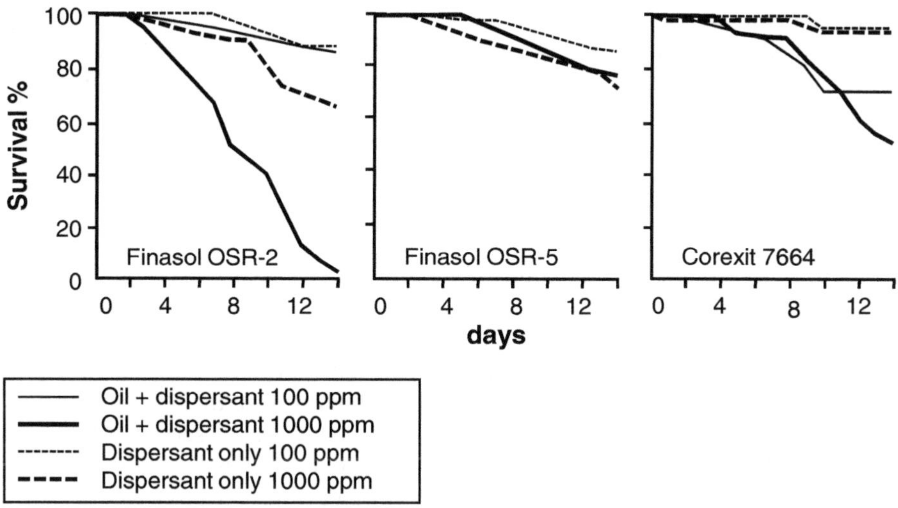

Fig. 4.6 Survival of the interstitial littoral oligochaete *Marionina subterranea* (Knöllner) exposed to different mixtures of oil (crude oil:water = 1:1,000) and three oil dispersants (From Giere 1980b, Fig. 4, redrawn and reproduced by permission of Cahiers de Biologie Marine/ Station Biologique de Roscoff)

4.2.4 Chronic Estuarine and Marine Water-Only Tests

There have been few studies describing estuarine and marine chronic water-only toxicity tests. Giere (1980b) assessed the toxicity of crude oil and oil dispersants, a common and recurrent coastal problem, with the marine enchytraeid oligochaete *Marionina subterranea.* This is a cosmopolitan and abundant meiobenthic species found in the beach zone and adapted to interstitial life. The enchytraeid was exposed to crude oil (both stirred and unstirred with water) and oil dispersants. The studied dispersants in combination with oil (1:10 detergent:oil relation) in 14-day tests caused an additive deleterious effect, compared with the separate components. The use of a combination of two dispersants resulted in synergistic noxious effects (Fig. 4.6). Giere and Pfannkuche (1982) reported results from tests in which the aquatic worm *Lumbricillus lineatus* was exposed for 14 and 60 days to several concentrations of crude oil, either alone or combined with dispersants. The species was highly tolerant to high oil concentrations and worm survival was 100% in all the 14-day tests, at 10–13°C. However, no survivors were observed at 26°C in the 60-day tests, when exposed to 20,000 ppm oil (an oil:water proportion of 1:50), and after only 2 days no survivors were found in the mixture of oil and dispersant. Cocoon production was minimally affected at 10–13°C, but the effect on egg fertility and embryogenesis was severe in mixtures of oil with dispersants, either retarding the development of embryos or preventing hatching. Reduction of egg fertility and hatching success of *L. lineatus* in long-term exposure to high oil concentrations was also reported by Hauschildt-Lillge (1982), although it was compensated by a higher cocoon production. In the presence of oiled *Fucus* algae, *L. lineatus* fed and incorporated large quantities of oil in the gut, chloragocytes became blackened, and apparently transformed the oil through metabolic processes. The tolerance and even the striking ability to feed on oiled substrata should not be surprising since in natural oil seeps in the marine coast some oligochaete species have been reported as abundant taxa, *e.g. Tectidrilus verrucosus* (as *Limnodriloides*) or *Limnodriloides monothecus* (Spies and Davis 1979, Davis and Spies 1980: reported by Giere and Pfannkuche 1982).

Bryant et al. (1985) explored the variability of the toxicity of arsenic to three estuarine invertebrates including the European tubificine *Heterochaeta costata* (as *Tubifex costatus)*. The tubificine survived for more than 384 h at all combinations of temperature and salinity in a concentration of 250 ppm As. Survival time was reduced at higher concentrations (500 and 1,000 ppm As) but the results were very variable though the highest mean survival time (220 h) was observed at 500 ppm As, 5 ppt salinity and 5°C, and the lowest at 1,000 ppm As, 25 ppt salinity and 15°C. Conditions intermediate between these two extremes did not produce consistent changes in survival time, except that survival times at 5°C were always greater than those at 15°C. The oligochaete was the most tolerant of the three species studied. The crustacean *Corophium volutator* had low survival times at 128 ppm As, and the mollusc *Macoma balthica* had shorter survival time periods in many experiments at 500 and 1,000 ppm, and was clearly more sensitive at 250 ppm.

Estuarine and marine oligochaete species have much less information on their general biology (physiology, genetics, life-cycles, or community studies) than freshwater species. Future studies on the sensitivity of some of the common estuarine Holarctic or widely distributed species, such as *Paranais litoralis*, *Heterochaeta costata*, *Monopylephorus rubroniveus*, *Tubificoides benedii* or *T. pseudogaster* to a range of toxicants are required to allow the development of more reliable chronic toxicity tests using oligochaete species for estuarine and coastal toxicity assessment. However, sediment bioassays are more appropriate as water-only chronic tests lack realism and have a very limited probability of success.

In summary, water-only toxicity tests are a necessary step for screening sensitive species to a range of toxicants, or for comparing the sensitivity of oligochaete species to other invertebrates or fishes. Studies from the 1960s to the end of the 1980s focused on estimating median lethal concentration values. Recently, freshwater toxicity tests have mostly chosen three species (*Tubifex tubifex, Limnodrilus hoffmeisteri* and *Lumbriculus variegatus*). These three species have several advantages, including ease to culture and handling and known sensitivity to toxicants. A few tests have also been done with estuarine and marine oligochaete species. The use of water-only toxicological experiments has become less frequent since the late 1990s, but they are helpful in test quality control and assessing the sensitivity of test organisms in the laboratory. The use of water-only chronic tests for the assessment of toxicity is not appropriate for most oligochaete species which are infaunal, but many naidid species inhabit the water column associated with aquatic vegetation (Learner et al. 1978) and more work could be done with species such as *Stylaria lacustris*, *Nais* spp., *Pristina* spp., or *Dero digitata.* They have attracted the attention of a few authors to date and have a high potential for use in both acute and chronic water-only toxicity tests. The more common asexual reproduction of this group limits the use of the reproduction endpoints typical of other chronic tests, however, the population growth rate could be considered as an alternative and relevant sub-lethal chronic measure.

4.3 Sediment Toxicity Tests

In the last two decades, there has been increasing interest by toxicologists in the detection and assessment of toxicity in sediments. The initial approach was to expose standard bioassay animals to sediment elutriates using the toxicity test methods developed for aqueous tests. However, the more common approach now is to test whole sediment. Sediment toxicity testing began in earnest in 1977, with the freshwater benthic invertebrates *Hexagenia limbata* (Ephemeroptera) (*e.g.* Prater and Anderson 1977) and *Chironomus tentans* (Diptera) (*e.g.* Wentsel et al. 1977). Both these studies used acute (96 h) and chronic exposures where the primary focus of the work was testing for the disposal of dredged materials. Organisms initially selected for whole-sediment test tended to be epibenthic (*e.g.* Amphipoda) or benthic but feeding on material deposited in the surface sediment from the water column (*e.g.* Chironomidae, Ephemeroptera). Wiederholm et al. (1987) and Milbrink's (1987) studies did suggest that aquatic oligochaetes would be ideal test animals for sediment toxicity assessment. These worms, particularly the tubificids and some lumbriculids, penetrate deeply into the sediments. They ingest sediment continually to absorb nutrients mainly from the small fraction of organic material and bacteria present on sediment particles, and they complete their entire life-cycle in the sediment. The general hesitation in using worms was a result of the mistaken view that all aquatic oligochaetes are exceedingly tolerant to a whole range of contaminants and stressors such as oxygen shortage, whereas the evidence reviewed here shows that this is not true, and one suspects reflected a general lack of knowledge of the group. In fact, worms are easy to culture (see Sect. 6.4) and are very suitable for long-term exposure in sediment, under laboratory conditions. Sediment bioassay methodologies have been much improved and standardised over the past few years and are included in various international guidance documents (*e.g.* ASTM 2005; OECD 2007). A number of sublethal endpoints have been shown to respond to sediment contaminants, particularly effects on biological activity such as growth and reproduction that are easy to monitor and ecologically relevant.

4.3.1 Acute Freshwater Sediment Tests

While some of the earlier work using oligochaetes as test organisms was conducted with sediment as part of the experimental method, these studies were not designed for sediment testing *per se*. The reduction of toxicity in the presence of sediment

Table 4.5 Acute toxicity of endrin (μg g^{-1} dw sediment) to *Limnodrilus hoffmeisteri* and *Stylodrilus heringianus* in single and mixed species tests, at 10°C (From Keilty et al. 1988a, Table 1, with permission of Springer Publ., © conveyed by Copyright Clearance Centre, Inc.)

Species	96-h LC_{50} μg g^{-1}
Single species test	
Stylodrilus heringianus	2,588
(mean value of four bioassays)	
Single species test	
Limnodrilus hoffmeisteri	2,725
Mean value of 2 bioassays	
Mixed species	
S. heringianus	2,750 (965–7,838)
L. hoffmeisteri	5,600 (2,393–13,104)
Combined response	4,000 (1,493–10,720)

was shown by Naqvi (1973) in short-term (96 h) tests with *Branchiura sowerbyi* exposed to insecticides at different temperatures. The presence of sediment resulted in the absence of mortality and was interpreted as a probable cause of reduction in the (short-term) bioavailability of chlorinated insecticides. The same response was described by Chapman and Brinkhurst (1984) who compared 96-h LC_{50} values for tests run with and without sediment with five different toxicants, and found that sediment modified toxicity, in many cases by an order of magnitude. Nevertheless, oligochaete bioassays in the presence of sediment are generally recommended as the realism of the exposure condition is greater.

An initial series of acute studies on sediment toxicity assessment with aquatic oligochaetes was published in 1988 to assess the toxicity of St. Lawrence Great Lakes sediments with the tubificine species *Limnodrilus hoffmeisteri* and the lumbriculid *Stylodrilus heringianus* (Keilty et al. 1988a, b, c). The pesticide endrin was used in the first study, not because of its importance in lake sediments, but because it represented one of hundreds of highly-sorbed, potentially toxic compounds and was available in a commercially radiolabelled form. Mean 96-h LC_{50} values for the two worm species were similar, and were higher than those determined for fish and crustaceans. There was also evidence of interspecific interactions of the type described above in previous work by Chapman et al. (1982a) in one experiment where LC_{50} values for *L. hoffmeisteri* were significantly higher in the presence of the lumbriculid (Table 4.5).

A more traditional approach to assessing sediment toxicity was reported by Casellato and Negrisolo (1989) who examined the effects of linear alkylbenzensulphonate (LAS), a common surfactant used in household detergent, for two tubificid species, *L. hoffmeisteri* and *B. sowerbyi*. This paper was one of the earliest to use oligochaetes in a straightforward acute sediment test (*B. sowerbyi* 96-h LC_{50} = 10.8 ppm; *L. hoffmeisteri* 96-h LC_{50} = 7.82 ppm). The authors reported that *B. sowerbyi* and *L. hoffmeisteri* exposed to the surfactant LAS in the presence of

Table 4.6 Mean proportion of colonising tubificids in beakers with uncontaminated sediment from beakers with treated sediment, in an avoidance sediment test of 96-h duration, at 14°C (From McMurtry 1984, Table 2, with permission of Elsevier Ltd, © conveyed by Copyright Clearance Centre, Inc.)

	Proportion of worms invading unpolluted sediment	
Treatment (mg l^{-1})	*T. tubifex*	*L. hoffmeisteri*
Control	0.060	0.055
Zn (1.13)	0.152	0.068
Cu (0.57)	0.167	0.130
Zn (1.08)+Cu (0.55)	0.221	0.150

sediment had NOEC values 2.5 and 4–4.5 times higher, respectively, than in water-only toxicity tests.

A series of papers published by scientists at the U.S. Environmental Protection Agency used *Lumbriculus variegatus* as one of a number of test organisms together with a fish, a cladoceran, a chironomid, and an amphipod. These were essentially methodological studies on the evaluation of sediment toxicity (Ankley et al. 1991a; Carlson et al. 1991; Schubauer-Berigan and Ankley 1991) and are described in detail in Sect. 4.4, where the comparative sensitivity with other animal groups is reported. These studies described the bioavailability of metals in association with Acide Volatile Sulphides (AVS), which represents a reactive pool of iron and manganese sulphides that can bind metals reducing their bioavailability. Carlson et al. (1991) determined that metal (Cd) /AVS ratios (molar concentrations)$\leq$1 were related to the absence of toxicity in *L. variegatus*. Results obtained by Ankley et al. (1991a) also supported the importance of AVS normalization in determining metal availability in sediments.

Many of the short-term sediment bioassays were designed to provide rapid test endpoints and therefore considered endpoints other than mortality, and much of that work focused on short-term behavioural responses. Burrowing behaviour was investigated by Fisher and Beeton (1975) using devices to identify both the horizontal and vertical movements of *L. hoffmeisteri* associated with hypoxia. In laboratory experiments, they demonstrated that during hypoxia stress in overlying water, oligochaetes burrowed deeper and when conditions improved they returned to the surface. They speculated on the possibility of using worms as test organisms if this behaviour could be quantified.

McMurtry (1984) replicated some of the Wentsel et al. (1977) assays on sediment avoidance with chironomid larvae but using mixed cultures of predominantly *T. tubifex* and *L. hoffmeisteri* collected from Toronto Harbour, Lake Ontario. He showed that addition of Cu or Zn at sublethal doses increased the migration of tubificids from metal treated cores into the surrounding sediment (Table 4.6). These data suggested that burrowing avoidance could provide a more sensitive bioassay than lethal tests at short-term exposure. The concept was tested using *S. heringianus* exposed to control sediments from open Lake Michigan and one site in the Detroit River, and sediments from three sites in the Detroit River contaminated with a variety of materials (White and Keilty 1988). In clean sediments, the worms burrowed into the sediment in less than one hour and fed normally over a 96-h period. In contrast, after 96 h the contaminated sediments showed no sign of burrows. In the contaminated sediments, the worms returned to the surface a few hours after starting to burrow, the time of return apparently being dependent on the degree of contamination and length of exposure. At the most contaminated site, where the most dramatic results were recorded (94% mortality in 96 h), the experiments were run with a series of sediment dilutions (using sediment from the control site). Burrowing activity was normal in the low concentration mixtures (2% and 10% contaminated mixed with control sediment), and mortality over the 96-h bioassay was 4% and 6%, respectively. Results for the 50:50 mixture were intermediate between these results and those in the 100% contaminated sediments. Air drying and re-wetting the sediments removed the colour and odour of hydrocarbons noticed initially, but the worms still reacted strongly to them. The authors also noted that neither of the tubificids nor the lumbriculid *S. heringianus* adopted what is considered the "normal" posture of worms, with the head in the sediment and the tails extended into the water column to obtain oxygen.

In *T. tubifex* the extent to which the tail is exposed above the sediment has been related to oxygen level in the overlying water (Guérin and Giani 1996). They recorded approximately 1/4 of total body length exposed in the water column under aerated conditions and as much as 3/4 in poor oxygen conditions. This behaviour has been used in toxicity tests as an endpoint. Thus, Meller et al. (1998) described a 72-h test with *T. tubifex* and *L. hoffmeisteri* examining the effects of lindane, hexachlorobenzene and copper through sediment avoidance (defining avoidance when more than an estimated 75% of body length was outside the sediment) together with another three test endpoints: sediment reworking, autotomy, as well as mortality. Autotomy and sediment avoidance EC_{50} values were more than 5 times lower than the LC_{50} values (Table 4.7). However, standardisation of these sublethal endpoints to reduce variability is necessary to reduce operator differences when interpreting the behavioural endpoints. The study is notable also because it used reconstituted water and artificial sediment in the toxicity tests, demonstrating the feasibility of working with aquatic worms under highly standardised laboratory conditions.

Interestingly, Keilty et al (1988a) observed that EC_{50} values for burrowing behaviour in worms exposed to endrin-spiked sediments reached equilibrium in a short-term test. After 16 h, *Stylodrilus heringianus* showed a stable value of $EC_{50} = 15.3$ μg g^{-1}, and *L. hoffmeisteri* after 32 and 48 h showed EC_{50} values of 66 and 59 μg g^{-1}, respectively. In this study, the effective concentrations (EC) for behavioural responses were about 170 times (for the former) and 50 times (for the latter) lower than the lethal concentrations. Coler et al. (1988) described a protocol to assess water-soluble toxicants that integrated respiration and behaviour. Over 4 days, they measured oxygen consumption by tubificid worms (1 g ww) (the name of the species is not given) in a respiratory chamber in the presence of 30-g acid-washed sand, using a flow-through system, at an average rate of 3.46 ml min^{-1} the first day to 0.65 ml min^{-1} after 4 days. Apparently, the tests were not run in the dark and temperature was not controlled. The authors briefly described behavioural changes related to flow regime. Avoidance behaviour, agitation, clumping, beat frequency and degree of extension could be correlated with respiration rate.

Table 4.7 Lethal and sublethal effects of lindane and copper sulphate on tubificids, using artificial sediment as substrate (From Meller et al. 1998, Table 3, with permission of Elsevier Ltd, © conveyed by Copyright Clearance Centre, Inc.)

	Lindane (mg kg^{-1}dw)		Copper sulphate (mg kg^{-1}dw)	
Endpoints	72-h LC_{50}/EC_{50} (95% CI)	LOEC/ NOEC	72-h LC_{50}/EC_{50} (95% CI)	LOEC/ NOEC
T. tubifex				
Mortality	>1,000/–	200/40	>1,000/–	1,000/500
Reworking activity		40/8		125/62.5
Sediment avoidance	–/217 (156–309)	200/40	–/547 (250–1,000)	250/125
Autotomy	–/172 (40–1,000)	200/40	–/601 (500–1,000)	500/250
L. hoffmeisteri				
Mortality	>1,000/–	200/40	516 (458–581)/–	500/250
Reworking activity	–/–	8/1.6	–/–	125/62.5
Sediment avoidance	–/224 (164–314)	200/40	–/392 (250–500)	500/250
Autotomy	–/200	200/40	–/349 (294–403)	125/62.5

However, this protocol is in a very preliminary state for use as a rapid screening tool for contaminated waters.

The oscillation frequency of the worm tail was measured by Guérin and Giani (1996) and the rhythm was slow and regular under well-oxygenated conditions (near oxygen saturation), and some worms did not undulate at all. In comparison, in low-oxygenated conditions (11–38% O_2) the tail (up to 2.2 ± 0.2 cm or 3/4 of total body length) beats regularly at a mean rate of 1.9 oscillations s^{-1}. These data indicate that beat frequency or tail oscillation rhythm in *T. tubifex* may be an additional behavioural variable for assessment of sediment toxicity, providing the level of dissolved oxygen is controlled within a fixed range. In the absence of contaminants and under well-oxygenated conditions, Guérin and Giani also described that after 10 h the proportion of worms in the sediment surface varied between 40% and 50%. The duration of the stay on the surface varied from a few minutes to several hours, thus the variability of this behaviour makes it of limited value in sediment toxicity tests. Regardless, the proportion of worms that did not form burrows was used by Dermott and Munawar (1992) to measure short-term (24 and 48 h) sediment avoidance of *L. variegatus* to contaminated field sediments, showing that batches with marginally the lowest survival rates had the highest number of worms remaining on the surface, whereas in the control sediment all the worms were buried (Fig. 4.7). When the sediment was clean, individual worms usually burrowed quickly (within 24 h) into the sediment. In contaminated sediment many worms remained at the surface of the sediment for several days.

Avoidance response of *L. variegatus* in 12 different sediments was studied by West and Ankley (1998), using a novel water-renewal system developed by Benoit et al. (1993). This approach provided interpretable measures of avoidance compared with a somewhat subjective assessment of 75% of the body being outside the sediment. The assay consists of test chambers where animals can colonise either test sediment, sand or glass beads. Avoidance of sediments was measured by comparing

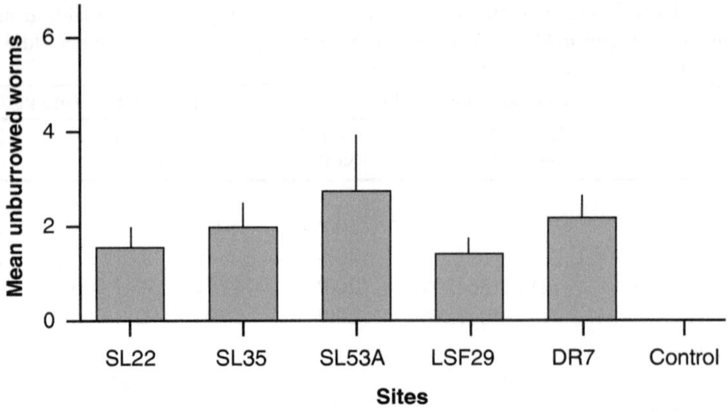

Fig. 4.7 Burrowing avoidance behaviour over 48 h by *Lumbriculus variegatus* in sediments from different sites (*SL* and *LSF*: St. Lawrence, *DR*: Detroit River, *Control*: Lake Ontario). Standard error as a *vertical line* (From Dermott and Munawar 1992, Fig. 5, redrawn and reproduced with permission of Springer Publ., © conveyed by Copyright Clearance Centre, Inc.)

Table 4.8 Sediment avoidance by *Lumbriculus variegatus* in short-term toxicity tests (From West and Ankley 1998, Table 1, with permission of Springer Publ., © conveyed by Copyright Clearance Centre, Inc.)

Test sediment	Test duration (h)	Worms recovered from reference sediment (%)	Worms recovered from test sediment (%)
1	120	71.8±21.5	0*
2	96	97.8±1.8	92.1±10.7
3	72	98.3±2.4	61.9±12.7
4	72	98.3±2.4	80.0±18.9
5	72	98.3±2.4	80.2±14.4
6	72	98.3±2.4	79.0±2.3*
7	72	93.33	83.3±23.6
8	72	93.33	0*
9	72	93.33	28.3±11.8*
10	72	93.33	0*
11	96	79.6 ±11.1	2.0±3.5*
12	96	92.2±1.9	75.5±10.2

*Indicates significant differences of test sediment with reference ($p<0.05$)

worm colonization of test sediments vs. reference sediment. The percent of worms in the test sediments with respect to the total number recovered from all substrates was reported at the end of 72, 96, 120 and 144 h. This showed a significant avoidance behaviour in some of the test sediments (0–79% worms recovery), while reference sediment showed a mean recovery of 71.8–95.8% worms (Table 4.8). West and Ankley (1998) observed the most marked avoidance behaviour in oil/PAH- and DDT-contaminated sediments.

When measuring sediment avoidance in toxicity tests, sediment particle size distribution may constitute a confounding factor. If the sediment grain composition is relatively coarse *T. tubifex* worms may remain at the surface because of the difficulty they may have in burrowing into sediment (P. Rodriguez, pers observ.). Nevertheless, providing that the sediment particle size is not so coarse as to impede normal burrowing activity the approach may have value. Working with *L. hoffmeisteri*, Rodriguez et al (2006) observed the influence of particle size distribution on burrowing behaviour, measured as the length of the galleries in a fixed area of the beaker surface, with more galleries observed in the coarser sediment after 1 and 2 weeks. This behaviour could be attributed to a greater search activity for food in sandy compared with silt-clay sediment, even when the sediments had received a Tetramin® supplement.

In summary, there has been little work in sediment toxicity tests on short-term lethality using oligochaetes. Most studies examined sediment avoidance or other sublethal endpoints. This is mainly because of the higher tolerance of oligochaetes to some contaminants in the presence of sediments, but also because of the increasing emphasis on longer-term testing of effects from sediment associated contaminants. The use of behavioural responses show potential as endpoints in short-term toxicity tests, but their description and measurement require improvement through the production of inter- and intra-laboratory comparable data.

4.3.2 *Acute Estuarine and Marine Sediment Tests*

The oligochaete community can be dominant in many estuarine areas and also in some marine sites, mainly associated with coarser grain sand areas (Diaz et al. 1987). However, their tolerance to toxicants is mostly unknown and only a few studies have measured acute toxicity in oligochaetes exposed to estuarine and marine sediments. These studies were designed for comparing the toxicity of substances under different environmental conditions, such as pH, salinity and temperature. Chapman et al. (1982b, c) reported 96-h LC_{50} values for *Tectidrilus verrucosus* (as *Limnodriloides*), *Tubificoides fraseri* (as *T. gabriellae*) and *Monopylephorus cuticulatus,* and showed that similar to freshwater species, the presence of sediments increased tolerance to both chemicals and environmental factors. Bryant et al. (1985) calculated the Median Lethal Tolerance of *Heterochaeta costata* (as *Tubifex costatus*) to pentavalent As (as sodium arsenate) using sterile sand as a substrate, in a combination of conditions of temperature and salinity. Worms did not survive more than 24 h in salinities above 25‰. The higher the temperature and the salinity, the lower the median survival time in worms exposed to 500 and 1,000 ppm As.

As described for freshwater species, the presence of sediment increases realism but also the tolerance of oligochaetes. In short-term exposures, survival is not a practical endpoint. However, behavioural measurements described above for freshwater toxicity tests, such as burrowing behaviour or sediment avoidance, could likely be used in short-tem toxicity tests with estuarine and marine species.

4.3.3 Chronic Freshwater Sediment Tests

Most toxicological studies with oligochaetes as test organisms have focused on their use in assessing long-term effects associated with sediment contaminants. As such, we would argue that oligochaetes have many useful attributes since they are true sediment dwellers and they are, therefore, exposed to contaminants both through the integument and across the gut wall from ingested sediment. These organisms build galleries in the sediment, can move down through several centimetres of sediment and integrate effects associated with historic deposits. Chronic sediment toxicity tests, in contrast to acute tests, provide information on ecologically important sublethal variables, such as growth or reproduction that may better predict effects in the field at a population level.

Up to 1991, there were few whole-sediment bioassays and most testing was being done with adaptations of water-column assays using sediment elutriates. At the time, the only toxicity tests available using true sediment-dwelling species were based on *Chironomus tentans, Hexagenia limbata* and *Hyalella azteca.* Because oligochaetes are such a widespread group and often the most abundant taxa in fine-grained sediment, their utility as a test organism was investigated by Wiederholm et al. (1987), Milbrink (1987) and Reynoldson et al. (1991). The former developed a method that looked at long-term effects on five tubificine species (*Tubifex tubifex, Limnodrilus hoffmeisteri, L. udekemianus, L. claparedianus, and Potamothrix hammoniensis*). *T. tubifex* and *P. hammoniensis* were concluded to be the most suitable for sediment bioassays because of the ease of handling, and *T. tubifex* in particular as data on growth and reproduction were readily obtained. Their method homogenised the sediment and then sieved it through a 300-μm mesh to remove the macrobenthos. Sediment was renewed every 2 weeks in the 6–18 months of exposure. Experiments were run in series of two or three replicates, in the dark, at 21 ± 1°C, with a sediment depth of 1.5 cm and aerated water (50% tap and 50% distilled water), for 90 days. The influence of food ration was studied, and worms were both unfed and fed weekly with spinach (conditioned in a few drops of water overnight), at two different rates (25 and 75 mg). Results showed that feeding obscured the reactions to contaminants, which were otherwise species-specific. However, the duration of the exposure made the test impractical as a routine method. In addition, concern was raised over change in toxicity of sediment stored for such a long period. Data from a parallel *Daphnia* test (Wiederholm and Dave 1989) suggested a decrease in toxicity over the study period.

Reynoldson et al. (1991) examined three potential test species, *T. tubifex*, *L. hoffmeisteri* and *Quistadrilus multisetosus* for testing Great Lakes sediment, finding *T. tubifex* the most practical and feasible, both because of its ease of culturing, more rapid generation time and it being easier to sort and process. The authors also provided recommendations on culturing, reviewed the methodology including appropriate mesh sizes to be used, test duration, effects of initial density of organisms, effects of animal size, test temperature and the effects of sediment quality. They also determined the numbers of laboratory replicates required to achieve statistical power for different test endpoints including, mortality, hatch rate, young per individual and cocoons per individual. In subsequent papers, Reynoldson (1994) provided information on the field

Table 4.9 Reproduction endpoints, mean (SD), for the chronic sediment bioassay with *T. tubifex* (Reynoldson et al. 1991), a comparison of data for the same bioassay from various authors

	Great Lakes reference sites n=170	Great Lakes control sediments n=46	Great Lakes control sediments n=5	Spain control sediments n=6	Spanish reference sites n=47
Reproduction endpoints	(Reynoldson and Day 1998)	(Reynoldson and Day 1998)	(Marchese and Brinkhurst 1996)	(Martinez-Madrid et al. 1999)	(Rodriguez et al. 2011)
Hatch (%)	58.1 (10)	56.7 (5.8)	39.8 (2.6)	40.2 (16.9)	33.5 (14.2)
Cocoon per adult	9.8 (1.3)	11.1 (0.8)	9.1 (0.6)	7.45 (1.36)	9.0 (1.4)
Young per adult	28.1 (9.1)	36.0 (8.7)	23.3 (3.9)	16.34 (10.3)	24.9 (17.0)

application of the tests with 72 sediment samples taken from 11 contaminated areas in the North American Great Lakes and demonstrated the test could identify six different response groups, including four different groups of toxic sites, based on cluster analysis of five test endpoints measured in the test. Further, Reynoldson and Day (1998) reported on the results of the tests from 170 clean reference sites in the North American Great Lakes and provided numeric targets that define a non-toxic, potentially toxic and toxic response for four test endpoints (survival, number of cocoons per worm, hatch rate of cocoons and number of young per worm). They also demonstrated little sensitivity of the test endpoints to sediment characteristics in clean sediment, finding a similar range in response values in sediments from 170 different reference sites across the Great Lakes, as was observed in their negative control (culture sediment from a single site) tested on 46 separate occasions over a period of several years.

Marchese and Brinkhurst (1996) attempted to reproduce the results of Reynoldson et al. (1991) with *T. tubifex* and also to develop the same tests with an alternative species, *Branchiura sowerbyi,* more appropriate to sub-tropical and tropical regions. They were able to reproduce the results of Reynoldson et al. (1991) with data well within the range reported by Reynoldson and Day (1998) for both natural sediment and their control sediment (Table 4.9).

In addition to the reproduction endpoints of the 28-day *T. tubifex* chronic bioassay (Reynoldson et al. 1991), data on worm production (adult growth and cocoon biomass) and egestion rate were examined by Martinez-Madrid et al. (1999). They showed that a group of very polluted sites were discriminated as well by the number of cocoons per adult as the other measurements, but production data provided a better understanding of some of the physiological processes which affect the reproductive endpoints used in the bioassay (*e.g.* negative growth rate, lower reproductive effort, or low cocoon biomass). In this study, the number of young per adult and hatch percentage were not always a useful measurement, since the control variability was high. The authors attempted to discriminate between somatic and reproductive biomass, and they concluded that measurement of somatic growth in adult tubificids was impractical as it was not possible to separate biomass of the sperm and egg sacs from the somatic biomass. Thus, total growth including both somatic and reproductive

Table 4.10 Instantaneous egestion rates (corrected for 1.5 mg dw) and total production (somatic plus cocoon biomass) in *Tubifex tubifex* exposed to different sediments in the 28-day chronic bioassay (From Martinez-Madrid et al. 1999, data from Tables 4 and 5, with permission of Springer Publ., © conveyed by Copyright Clearance Centre, Inc.)

	Egestion rates ($mm^3\ h^{-1}$)	Total production (mg dw)
Site	Mean (SD)	Mean (SD)
Control Mean (n = 7)	0.120 (0.057)	1.092 (0.351)
P1 (cc)	0.022 (0.004)[a]	0.347 (0.091)[a]
P2 (cc, yg, htc)	0.030 (0.008)[a]	0.311 (0.200)[a]
P3 (htc)	0.104 (0.067)	1.139 (0.202)[a]
P4 (cc, yg)	0.081 (0.051)[a]	–0.272 (0.208)[a]
P5 (htc)	0.056 (0.023)[a]	1.834 (0.253)
P6 (yg, htc)	0.138 (0.042)	1.534 (0.226)
P7 (cc, htc)	0.022 (0.009)[a]	–0.070 (0.192)[a]
P8 (yg, htc)	0.054 (0.009)	0.501 (0.098)
P9 (yg, htc)	0.074 (0.010)	1.049 (0.085)
P10 (htc)	0.036 (0.022)	1.863 (0.825)
R1	0.062 (0.040)	1.786 (0.318)
R2 (yg, htc)	0.058 (0.026)[a]	3.096 (0.351)
R3 (htc)	0.046 (0.037)[a]	2.272 (0.403)

P test sediments, *R* reference sediments. In parentheses, the significantly reduced reproductive endpoints (*cc* number of cocoon per adult, *yg* number of young per adult, *htc* hatch)
[a]Significantly lower than the control ($p < 0.05$)

biomass in the adult and the biomass of cocoons produced during the bioassay was preferred as a measure of production. Martinez-Madrid et al. (1999) also examined the relationships between total growth and egestion rates (Table 4.10). At the end of the bioassay, faecal pellets from each 4-worm replicate were recovered at hourly intervals, disaggregated and the packed volume calculated using a Coulter Multisizer®. The instantaneous egestion rate was calculated as the volume of faeces per hour, normalised for a mean value of worm weight of 1.5 mg dw. These authors found that in half of the contaminated sediments studied the worms had reduced egestion rates (0.022–0.081 $mm^3\ h^{-1}$) with respect to the control (0.095–0.149 $mm^3\ h^{-1}$), and they interpreted the lower production at these sites to be a result of sediment avoidance behaviour that reduced sediment ingestion. At two reference sites significantly ($p < 0.05$) reduced egestion rates were observed, suggesting that other factors (*e.g.* particle size distribution and organic content) may also influence egestion rates (Martinez-Madrid, pers. commun.). Contrary to expected, there was a poor relationship between egestion rates and total growth, suggesting that other energy expenditures such as detoxification processes or mucus production could be important factors reducing worm production (Martinez-Madrid et al. 1999).

In another experiment, Martinez-Madrid et al. (1999) attempted to measure reproduction and total growth in worms exposed to extreme conditions of sediment avoidance that could result in starvation. Thus, worms were starved for 28 days by being maintained in pre-ashed sediment (400°C, for 6 h), thus removing organic material. Results showed that after a 28-day bioassay total growth and reproduction in ashed sediment (–0.349±0.023 mg dw, no cocoons and no young) were similar to the values obtained in some of the most toxic sediments tested (down to –0.272 mg dw, 0.98 cocoons per adult and no young), even if these sites had a normal or high organic content (>4% LOI) and extra food had been added at the beginning of the bioassay. In the same experiment, total growth and reproduction in the control were 0.364±0.309 mg dw, 8.58±1.04 cocoons per adult and 20.17±6.86 young per adult. However, reproduction was almost completely inhibited in some heavily polluted sites but was less affected in other; therefore, sediment avoidance behaviour that prevents animals from feeding on polluted sediments seems to have occurred in some sites but not in others, or to varying degrees.

A new early-life-stage survival test (Vecchi et al. 1999) of 28-day duration using *T. tubifex* was compared with the adult reproduction sediment test (Reynoldson et al. 1991). Results showed the early-life-stage survival test to be less sensitive to copper than the adult reproduction test. Regardless, Vecchi et al (1999) suggested the early-life-stage test as a good alternative as it requires less sediment volume, although it seems to have less discriminatory power. In a later study, Pasteris et al. (2003) compared the standard 28-day *T. tubifex* sediment test in a 6-month experiment on cohorts, again using Cu-spiked sediment (35, 50, 70, 100 and 140 mg kg^{-1} dw, nominal concentrations). The results of both bioassays were very similar regarding the values of LOEC and IC_{50}. It was concluded that the 28-day bioassay provides information relevant to long-term demographic effects.

Dermott and Munawar (1992) adapted the aqueous toxicity test with the lumbriculid oligochaete *Lumbriculus variegatus* to a whole-sediment chronic exposure in a 2-week bioassay, using sediments from the St. Lawrence River and North American Great Lakes. The weight of all prostomium-bearing fragments recovered from the sediment after sieving was recorded for growth/reproduction measurements. A similar standardised method of chronic (4–28 days) test procedures for assessment of both toxicity and bioaccumulation of sediment-associated contaminants using *L. variegatus* has been described by Phipps et al. (1993). Asexual reproduction was shown to be a sensitive endpoint in *L. variegatus* by Hickey and Martin (1995) who studied 10-day survival, reproduction and behavioural endpoints in five different species, including the oligochaete *L. variegatus,* exposed to phenol, pentachlorophenol and cadmium. However, reduced reproduction was also observed in a sandy reference sediment, highlighting the need for using appropriate reference sediments. Many authors have used this test protocol, or later modifications, mostly for bulk-sediment bioassays with *L. variegatus*, mainly by adding chemicals to natural or formulated sediments. However, there are two important problems associated with this bioassay; first, that some expertise and training is needed to both differentiate anterior from posterior parts, and also to distinguish test species of *Lumbriculus* from tubificid species (*e.g. Limnodrilus*, which includes several species with no hair-chaetae) that are commonly present in field collected sediment; second, that asexual reproduction makes it difficult to separate lethal and sublethal effects, and some effects on survival may be masked due to fragmentation of survivors. More recently, synchronization of cultures of *L. variegatus* has been proposed to minimise uncontrolled autotomy and regeneration during the test (OECD 2007).

Table 4.11 Differences in selected physical and chemical attributes of sediment reworked by tubificids in microcosm experiments (population 4×10^4 ind. m^{-2}) (From Karickhoff and Morris 1985, Table I, reproduced with permission of ACS Publ. © conveyed by Copyright Clearance Centre, Inc.)

	Whole sediment	Faecal layer
Volumetric water content (θ)	0.48	0.80[a]
Solids concentration (ρ)	1.25	0.52[a]
Organic carbon fraction	0.008	0.03[a]
Hexachlorobenzene (ppm)	0.96	3.6[b]
Pentachlorobenzene (ppm)	1.0	3.3[b]
Trifluralin (ppm)	1.2	0.5[b]

[a]Determined at day 90
[b]Determined on fresh faecal pellets, at day 29

Bioturbation from burrowing activity of aquatic oligochaetes has also been used as a test variable in chronic toxicity tests (*e.g.* Matisoff et al. 1999; Reible et al. 1996; Wang and Matisoff 1997). It is an important, ecologically relevant effect where oligochaete worms interact with sediments through pumping pore water and the active transport of particulate matter to surficial layers. The reworking activity by worms has been studied in relation to toxicity of chemicals sorbed to the sediment particles and was considered as the most sensitive endpoint because it is an integrated physiological response and incorporates toxicant bioavailability, organism compensatory mechanisms and organism function (Keilty et al. 1988c). Oligochaete species ingest sediment particles from deep layers of sediment and deposit faeces on the surface, and rates at which sediments are transported to the surface have been determined under laboratory conditions. The fact that feeding activity of worms in sediment enhances the solute diffusion in sediments has been demonstrated by Krezoski et al. (1984) and Karickhoff and Morris (1985), among others. In a microcosm experiment (10^4–10^5 ind. m^{-2}), Karickhoff and Morris (1985) measured pollutant (hexachlorobenzene, pentachlorobenzene and trifuralin) transport from deeper sediment layers (typically 2–10 cm depth) to the surface, via ingestion by tubificid worms (90% *L. hoffmeisteri*, 10% *T. tubifex*), over a 30–50 day period, at rates of 0.3–1.0 mm day^{-1}. The rate and depth dependence of worm feeding were followed by periodically introducing fluorescent glass beads onto the sediment surface and observing their subsequent burial. Results suggested that there was a discrete "turnover" of the faecal layer. The transfer of pollutants from the sediment to the water column was enhanced by the presence of the worms 4–6 fold, over a 90-day period. In this study, they also described sediment "pelletization" which "entraps" the sorbed chemical and significantly retards chemical release to the aqueous phase (for hydrophobic chemicals, less than 20% of the chemical concentration contained in an intact pellet was released during the characteristic pellet lifetime of 1–3 days). During the period of faecal layer creation, the effect of particle sorting on both organic content and pollutant concentration was clearly shown (Table 4.11), confirming the intimate link between oligochaete activity and pollutant movement and redistribution in bottom sediment.

The enhancement of a physiological process by exposure to low levels of contaminants is known as hormesis (Laughlin et al. 1981). Several of the studies in this section described higher levels than in the control of the measured endpoints (reworking rates, growth) in worms exposed to low levels of chemicals. However, the concept of hormesis requires better definition, since it is arguable whether it can be applied to all increments of physiological endpoints observed in test organisms when exposed to low doses of toxicants. There has been some debate about this phenomenon and readers are addressed to key papers on the concept of hormesis (Calabrese and Baldwin 1998; Forbes 2000; Stebbing 2000) and its role in ecological risk assessment (Chapman 2002).

The remainder of this chapter examines the response of oligochaetes to different categories of contaminants.

4.3.3.1 Metals

Milbrink (1987) experimented with sediments contaminated with a mixture of heavy metals and sediments from eutrophic and oligotrophic lakes, and measured *Tubifex tubifex* reproduction and growth. These variables were measured using uncontaminated sediment and various mixtures of sediments from a clean and a contaminated lake and showed that growth rate and cocoon and young production were sensitive to metal toxicity in sediments. The author concluded that laboratory test responses could be used to demonstrate sediment toxicity and together with benthic community data would provide valuable information for environmental managers.

This work was also reported by Wiederholm et al. (1987), who used *Limnodrilus hoffmeisteri, L. claparedianus, L. udekemianus* and *P. hammoniensis* as well as *T. tubifex*. The worms were grown in oligotrophic, mesotrophic and eutrophic sediments, and in metal contaminated sediments. Growth and reproduction (cumulative number of young worms) were monitored over long time periods (200–500 days). *L. hoffmeisteri* was the most tolerant of the species in terms of survivorship, but was considered to be the most sensitive species in general. It grew less and failed to reproduce at all in contaminated sediments. *T. tubifex* seemed to produce the most rapid and greatest response in both growth and reproduction, and *L. claparedianus* reproduced better than either of the above species in the contaminated sediments, and was most tolerant. This work confirmed that the absence of healthy worm populations in the contaminated lake was due to metals present in the sediment. In another experiment, sediment was spiked by adding 100 ml of a solution of different quantities of $CuSO_4.5H_2O$ in distilled water to 1 l of sediment, and significant effects in survival and growth in both *T. tubifex* and *L. hoffmeisteri* were reported for high and low doses. The same method was used in a comparison of toxicity of four contaminated lake sediments (with high levels of Pb, Cd, Hg, Zn, Ni, Cr and oil) with a 48-h *Daphnia magna* test and 270-day *T. tubifex* test (Wiederholm and Dave 1989). The two tests gave the same rank order of toxicity of the sediments. The *Daphnia* test was much quicker but the *Tubifex* test was more realistic with regard to field conditions and the effects of pollutants on the benthic community.

Chapman et al. (1999) compared the performance of *L. variegatus* and *T. tubifex* in artificial sediment spiked with copper and cadmium. These authors compared 14-day and 28-day EC_{50} estimated for asexual reproduction in *L. variegatus* toxicity tests and for several reproductive variables (eggs, cocoons and young per adult) in *T. tubifex* tests, respectively, and suggested that the former was more sensitive than the later, although values were similar for Cd and differed by approximately two times for Cu. However, the discussion did not consider the fact that the *T. tubifex* test provided more information (3 reproductive endpoints in this study) than the *L. variegatus* test (1 sublethal endpoint).

A concern regarding metal toxicity evaluation relates to the differential bioavailability of metals dependent on the sediment matrix. The acid-volatile sulphide (AVS) concentration of sediment for normalising sediment bioavailability of metals has been suggested as equivalent to the use of organic content of sediment for normalising the bioavailability of non-polar organic compounds. It has been postulated that AVS is a reactive pool of solid phase sulphide (iron and manganese) that is available to bind metals and render the bound portion unavailable and non-toxic. One difficulty with this hypothesis is that AVS is operationally defined and not a true sediment attribute; in addition, it ignores all the other factors that can affect metal availability. Carlson et al. (1991) investigated the role of AVS in determining Cd availability in freshwater sediments with *L. variegatus*. They spiked sediment with five different concentrations of Cd at three different AVS concentrations in 10-day exposures. When Cd to AVS ratios were ≤ 1.0 no toxicity was observed, which suggests that AVS can play a role in affecting bioavailability, however, it does not exclude the importance of an array of other factors. Worms reproduced, presumably asexually, in sediments in which they were able to survive but changes in population or weight of worms were not used as an endpoint. Mortality of worms was observed in 6 of 7 sediments in which overlying water concentrations exceeded 0.01 mg Cd l^{-1}, and mortality was zero at 0.01 mg l^{-1} and below. A similar abrupt transition occurred in samples in which the pore water Cd levels exceeded about 1.5 mg l^{-1}, which suggests that metal concentration has a threshold effect concentration that may be modified by AVS. Peterson et al. (1996) evaluated the effect of bioturbation by *Lumbriculus variegatus* on surficial sediments showing reductions in AVS concentration and increased pore water concentration of Cd, but not Zn, indicating that bioturbation may increase bioavailability of some metals through oxidation of AVS.

In a long-term study, Ankley et al. (1994a) exposed *L. variegatus* to sediments containing elevated concentrations of Cu, Cd, Zn, Pb, Ni, and Cr. Over a 30-day period, the concentration of the metals in the worms were not significantly greater than in animals exposed in a control medium and this was as predicted by measurements of solid phase metal and AVS concentration or pore water concentration. This was taken by the authors to indicate that metal bioavailability models based on sediment AVS concentration or pore water concentrations may be valid for long-term as well as short-term exposures of benthic animals. In a more comprehensive study, Hansen et al. (1996) examined the relationships between sediment toxicity, AVS concentration, total simultaneously extracted metals (SEM) and interstitial water toxic units (IWTU) from both freshwater and marine locations. *Lumbriculus variegatus* was exposed to 55 different sediments from four freshwater locations and there was no relationship between sediment toxicity and total metals, expressed on a dry weight basis. The authors also found that the difference between molar concentration of the sediment extractable metals (SEM) and the AVS concentration to be a better estimate of metal binding than the ratio of SEM/AVS. In summary, they found that [SEM – AVS] can accurately (99.2%) identify the absence of toxicity but can less accurately (79.1%) identify the presence of toxicity when metals are available.

Occasionally, species from other oligochaete taxa have been used for sediment toxicity assessment. Smith et al. (1991) evaluated the utility of the *Pristina longiseta* (reported as *P. leidyi*) in both short-term aqueous (96 h) (see Sect. 4.2.1) and long-term sediment tests (recommended duration of 15–18 days) and population growth rate was suggested as an endpoint for chronic tests. The test appears to have some potential, the animal has a generation time of 3–7 days in culture fed on rabbit chow suspension and, therefore, rapid multigenerational tests on whole populations can be done in order to calculate demographic parameters. In the reported experiments, the short-term toxicity data suggest the species is equally or more sensitive than other species to cadmium and vanadium. This test is probably worth more consideration, although no further work appears to have been done.

Representatives of the family Enchytraeidae were used by Guérin et al. (1994) who examined the usefulness of *Enchytraeus variatus* as a test species. In exposures to Cd, Cu, and Zn in artificial sediment over 72 hours and 39 days, they measured survival, growth and reproduction by fragmentation and reported that Zn was the most toxic of the metals. They also concluded that fragmentation confounded the test results and did not consider this to be a suitable test organism.

4.3.3.2 Organics

Many studies conducted with organic compounds have examined biological effects related to exposure and bioaccumulation. Where possible, we have attempted to discriminate between these in reporting the current state of knowledge and have selected contributions that show the variety of endpoints used in the chronic testing of organic compounds with aquatic oligochaete worms.

The earliest work in examining effects of specific organic substances on oligochaetes was by Fischer et al. (1980a) who examined the response of chloragocytes in *Tubifex tubifex* to 2,4-dinitrophenol, azide, and monoiode acetate. These compounds are known catabolic inhibitors and the authors observed inhibition in chloragocyte volume with these substances, as observed with metals. This suggests that this may be a very general response in oligochaetes and worth investigating as a sub-cellular biomarker (see Sect. 4.6).

Casellato and Negrisolo (1989) maintained *Branchiura sowerbyi* in sediment spiked with a solution of LAS (linear alkylbenzensulfonate). This study was based on very long time exposures (140 days), with water replacement with a new LAS solution and test beakers being examined every 5 days. At the two highest concentrations of the contaminant (5 ppm, 2.5 ppm), cocoon laying began much earlier

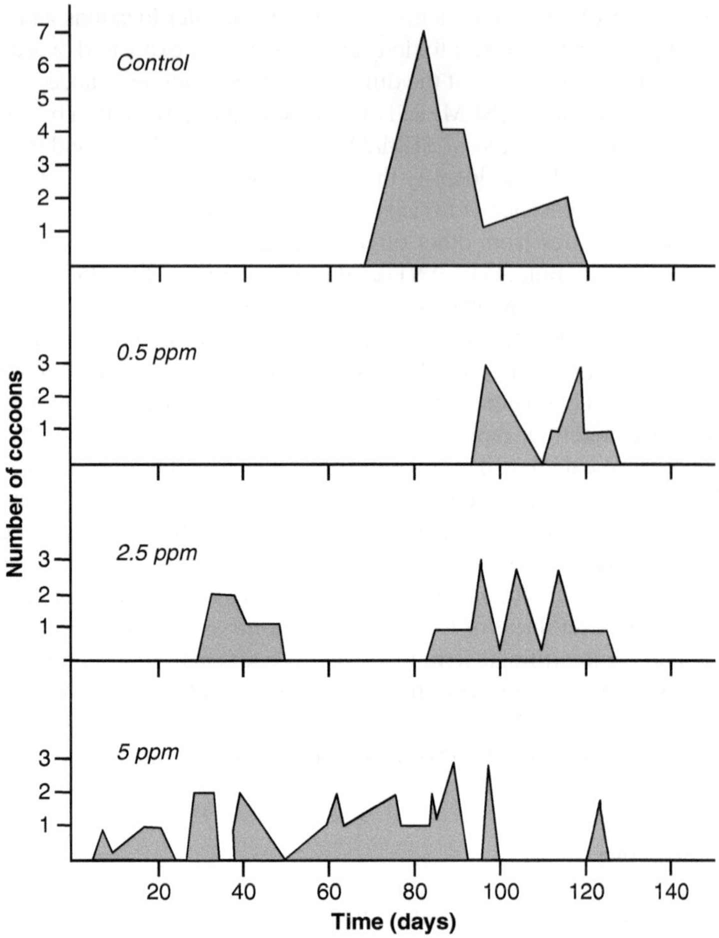

Fig. 4.8 Pattern of cocoon laying activity in *Branchiura sowerbyi* exposed to the anionic surfactant linear alkyl benzensulfonate (LAS) in chronic toxicity tests, at 20°C (From Casellato and Negrisolo 1989, Fig. 4, redrawn and reproduced with permission of Springer Publ., © conveyed by Copyright Clearance Centre, Inc.)

than in controls, but at the lowest concentration cocoons appeared 1 month later. A lower mean number of cocoons and oocytes per cocoon occurred in test sediments compared to controls, but the embryonic period was not affected. Differences in reproduction between controls and the highest concentration were small and, paradoxically, the total number of cocoons increased with increasing concentration of LAS (Fig. 4.8). This fact seems to be related to the precocious laying of cocoons in the most polluted treatments. Evaluation of the number of cocoons per week during the laying period would provide a better estimate of the effects. Subsequently, Casellato et al. (1992) studied the effects of LAS-spiked sediment in *B. sowerbyi* after 45- and 220-day exposures, but observed no effect on reproduction (number of cocoons, number of eggs, % degenerated cocoons, % hatching, mean embryonic

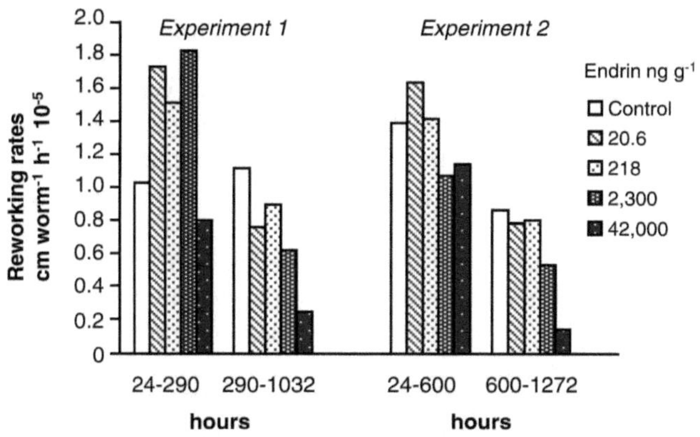

Fig. 4.9 Sediment (mean) reworking rates (cm $worm^{-1}$ h^{-1} 10^{-5}) by *Stylodrilus heringianus* calculated from ^{137} Cs burial rates in two experiments (Data from Table II, in Keilty et al. 1988c)

development time), even when LAS concentration was five times higher than the LC_{50} values calculated in the previous study, where LAS had been added dissolved in water. This study concluded that the bioavailability of LAS sorbed to sediment was very low. The authors did indicate that LAS degrades rapidly (about 20% in 24 h), which explains how LAS concentration in sediment was reduced by 72% at the end of the experiment, as well as the observed lack of effect in this long-term study.

The effect of endrin on sediment reworking rates was determined by following the burial of a submillimeter surface layer of radioactive tracer (^{137}Cs, Keilty et al. 1988b), in microcosms with pure and mixed populations of *Limnodrilus hoffmeisteri* and *Stylodrilus heringianus,* in a long-term (980–1,312 h) exposure. In four experiments, eleven concentrations which ranged in value from 5.5–81,400 ng g^{-1} were investigated. The lower doses did not stimulate sediment reworking in *L .hoffmeisteri* alone, as it had for *S heringianus*. At higher concentrations, reworking rates were initially equal to or higher than control values, but then declined dramatically. Reworking rates for *L. hoffmeisteri* were elevated in the presence of *Stylodrilus*, as were survival and growth. Dry weights of worms at the end of the experiments were inversely proportional to endrin levels in sediment, although worms exposed to lower concentrations tended to weigh more than controls. Reworking rates by *S. heringianus* measured with a gamma-scan system to monitor a marker layer of ^{137}Cs were also reported by Keilty et al. (1988c). The experiments were monitored for 1,300 h and demonstrated that reworking rates were significantly altered by the presence of endrin in concentrations up to 5.5 orders of magnitude lower than the 96-h LC_{50} value. Endrin concentrations ranged from 3.1 to 42,000 ng g^{-1} and although resulting in low mortality (up to 26%) throughout the experiments, this was significantly elevated at the highest endrin concentrations. Moderate levels of endrin contamination stimulated the worms reworking activity so that the contaminant became concentrated in the uppermost sediment layer. In the first 300–600 h of exposure, reworking rates were higher than in controls at lower levels of endrin, whereas rates equalled those observed in control sediments at the higher levels of endrin (Fig. 4.9). However, in the second half of the experiment, reworking rates were slower in all treated sediments. The weights of worms remaining at the end of the experiment were higher at lower endrin levels than in controls, and this corresponded with the faster reworking rates. Weights were very low for high levels of exposure, suggesting that worms decreased feeding activity and/or there was increased stress associated with the presence of the pollutant. Body concentration ranged 34–67 times the sediment concentration, but it was not related to sediment concentration of endrin.

Some evidence of sediment avoidance can be inferred from results by Kukkonen and Landrum (1994) on the exposure of *L. variegatus* to pyrene-spiked sediment. These authors examined the survival and bioaccumulation of worms exposed to pyrene-spiked sediment from Lake Michigan. Pyrene was slightly toxic to the worms and after 7-day exposure mortality was only 3%, 7% and 17%, at 132, 206, and 269 μg g^{-1}, respectively. The low toxicity could be due to sediment avoidance as at high pyrene concentrations the worms did not burrow into the sediment. In fact, the authors reported that 50% and 64% of test (surviving) organisms were on the sediment surface in the two highest treatments.

A study on sublethal effects of 2,4,5-trichlorophenol (TCP) using heat dissipation measured by direct microcalorimetric measurements was conducted by Penttinen et al. (1996). Uncontaminated sediment was spiked with TCP to achieve nominal concentrations of 25, 50, 75 and 100 μg g^{-1} dw sediment. This study was based on the assumption that the sublethal bioenergetic response produced by the additional activity associated with stress from toxicants would be a sensitive measure of that stress. In *L. variegatus,* heat dissipation was of the same order in the control and in those animals with high body residues (>1.5 μmol g^{-1}), despite the fact that at the highest concentration used (83 μg TCP g^{-1} dw) mortality was 20–50%, suggesting that this approach has limited value.

The effect of UV radiation on polycyclic aromatic hydrocarbon (PAHs) chronic toxicity was examined by Ankley et al. (1994b) in field polluted sediments. Solar UV radiation is known to be absorbed by high molecular weight PAHs resulting in excited states that can increase toxicity (Mekenyan et al. 1994). *Lumbriculus variegatus* was exposed to sediments with elevated PAH concentrations in both UV light and fluorescent light. Greater mortality was observed in the animals exposed to PAH-polluted sediments and UV light compared to those exposed only to fluorescent light (12.5–85% higher mortality in UV exposed worms), while mortality was low (less than 10%) and not significantly different in animals irradiated with UV in control and unpolluted sediments. Furthermore, increased mortality was observed in animals exposed in a water-only test to UV light for 2 h, after removal from the sediment, suggesting that bioaccumulated PAHs could be photoactivated. A follow-up field study (Monson et al. 1995) corroborated these results in PAH-contaminated and reference sites, using *in situ* enclosures for comparison of light and dark treatments.

Lotufo and Fleeger (1996) examined the toxicity of pyrene and phenanthrene to *L. hoffmeisteri* in spiked sediment. They used a number of different test endpoints including survivorship, burrowing avoidance, reproduction and daily egestion rates (Fig. 4.10), as measured by faecal pellet production in 5 and 10-day bioassays.

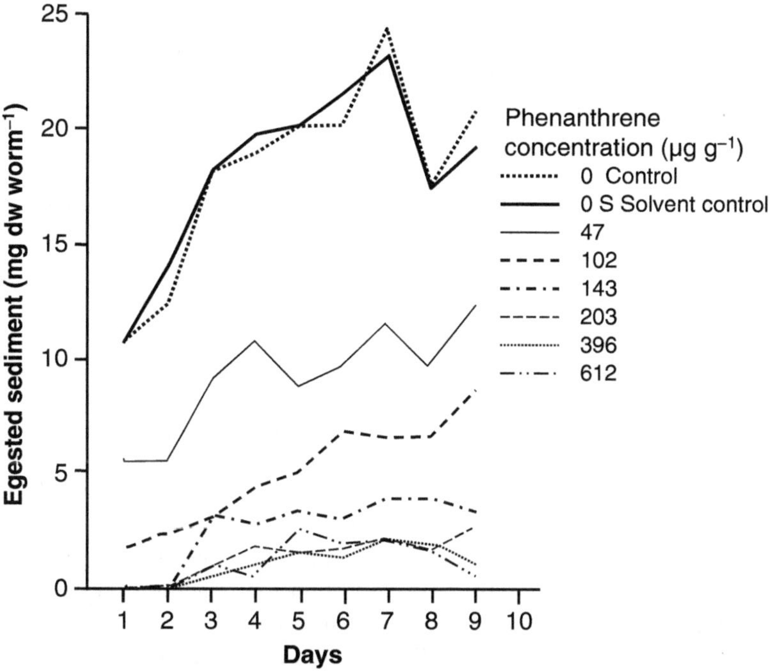

Fig. 4.10 Daily sediment egestion by *Limnodrilus hoffmeisteri* exposed to different sediment concentrations of phenanthrene (From Lotufo and Fleeger 1996, Fig. 2B, redrawn and reproduced with permission of John Wiley & Sons Ltd., © conveyed by Copyright Clearance Centre, Inc.)

Faeces were collected using the method of Kaster et al. (1984) with defecation chambers and expressed on a dry weight basis normalised by averaged post-experimental measurements of body weight (mg dw g^{-1} worm day^{-1}). They observed responses in all the measured endpoints and found the 5-day egestion rates to be particularly sensitive and slightly more sensitive than reproduction.

In summary, a number of oligochaete species have successfully been used for the assessment of chronic toxicity in sediments, and employing a number of sublethal endpoints, such as reproduction (asexual or sexual: number of eggs, cocoons, hatch rate and young), growth (somatic or total), egestion and respiration rates, or behaviour (sediment avoidance and reworking rates). There is clear evidence that oligochaetes constitute good candidates for testing chronic toxicity of sediments polluted either by metals, organics, or both.

4.3.4 Chronic Estuarine and Marine Sediment Tests

Many coastal and estuarine species are tolerant to moderate or wide changes in salinity (*i.e.* euryhaline species), and this is an important consideration when selecting test organisms and bioassay exposure conditions. There are few examples of chronic oligochaete sediment bioassays being used in the ecological risk assessment of estuarine and marine sediments, while other benthic invertebrates such as polychaetes, molluscs and particularly amphipod crustaceans have been widely used. Chapman and Wang (2001) reviewed the use of invertebrates in assessing toxicity of estuarine sediments. For these authors, an important consideration was the fact that most estuarine species do not tolerate all the range of salinities present in estuaries, and none of them reproduce successfully in all estuarine habitats. Therefore, they recommend that a number of species should be selected for different salinity ranges in estuarine assessment. While, at present, no chronic sediment toxicity tests with true estuarine oligochaete species have been recommended by internationally recognized organisations (ASTM, APHA, OECD), this should not be an issue once appropriate test procedures have been standardised for relevant oligochaete species.

Gamenick et al. (1996) performed tolerance experiments with *Heterochaeta costata* (as *Tubifex costatus*) and *Paranais litoralis* on sulphide concentration and hypoxia level, variables that are potential confounding factors in sediment toxicity tests. Results underlined the adaptation of *H. costata* to an endobenthic life under these extreme conditions; whereas the epibenthic and frequently suprabenthic naidine *P. litoralis* showed long-term survival to hypoxia conditions, although its survival time dropped drastically after exposure to sulphide.

Most recently, Rodriguez et al. (2006) have run estuarine sediment bioassays with the euryhaline freshwater species *Limnodrilus hoffmeisteri* as the test organism. Worms were sampled from an estuarine site and acclimated under laboratory conditions to permanent low salinity levels (7–8.5‰), and used for 14-day bioassays to measure survival, growth, reproduction and behaviour when exposed to contaminated sediments from Santander Bay (Northern Spain). The tolerance of *L. hoffmeisteri* up to 15‰ salinity, makes it a useful candidate for the toxicity assessment of oligohaline and β-mesohaline portions of estuaries in Europe and North America. The authors ranked sites according to mortality, number of cocoons and clitellum and gonad resorption. The most toxic sites had up to 65% mortality, and 67% of organisms showed resorption of the clitellum and the reproductive organs together with complete inhibition of reproduction (Table 4.12). Ordination (multidimensional scaling) showed a strong mortality gradient that was associated with Cu and Zn sediment concentration. Burrowing and reproduction did not contribute significantly to the assessment of sites using multivariate analysis, and there was no relationship between survival and clitellum resorption. The bioassay seems to be too short for measuring reproduction as the number of cocoons produced in controls was low (2.4 cocoons per adult), although an effect of ammonia on worm reproduction could not be discounted. Rodriguez et al. (2006) suggested further research with this species and other estuarine species, such as *H. costata* or *Monopylephorus rubroniveus*. Other likely candidate species include *Enchytraeus albidus* which has great osmoregulatory capacity from freshwater to fully marine conditions (after long-term acclimation) (Generlich and Giere 1996), and coastal enchytraeid species (*e.g. Lumbricillus lineatus, L. reynoldsoni*) which inhabit wrack beds on the seashore and are also tolerant to salinity changes.

Table 4.12 Mean values on mortality, clitellum resorption, adult final biomass, number of cocoons and length of galleries measured at the end of 14-day toxicity bioassay with *Limnodrilus hoffmeisteri* exposed to whole-sediment from Santander Bay (Spain) (n=5 for all endpoints, except for length galleries where n=3) (From Rodriguez et al. 2006, Table 4, with permission of Springer Publ., © conveyed by Copyright Clearance Centre, Inc.)

Site	Mortality (%)	Clitellum resorption (% ind)	Adult final biomass (mg dw)	Number of cocoons	Length of galleries (cm)
Control	5.0	10.5	1.271	2.40	7.26
AS3	41.6*	0	1.160	0	8.11
BO2	35.0*	23.1	0.987	0	7.53
BO3	65.0*	57.1	0.681	0	6.85
SO2	60.0*	0	0.993	0	5.30*
SS1	40.0*	25.0	0.900	0.80*	8.65

*Significant differences ($p<0.05$) with respect to control

Table 4.13 Biological characteristics of *Tubificoides benedii* (Udekem)

Issue		Reference
Respiration rates	37 nmol O_2 g^{-1} ww min^{-1}	Bagheri and McLusky (1984)
	63 nmol O_2 g^{-1} ww min^{-1}	Giere et al. (1999)
Habitat preferences	Meso to euhaline sites	Verdonschot (1981); Verdonschot et al. (1982)
Salinity tolerance	2.8–34‰	Birtwell and Arthur (1980)
Temperature tolerance	Up to 28.5°C	Birtwell and Arthur (1980)
Density	Mean density of 26.888 ind m^{-2} in intertidal sands	Zipperle and Reise (2005)
Ecophysiology: effects of sulphide and hypoxia	300 μM sulphide induce anaerobiosis in normoxia, 40 μM sulphide in hypoxia (7% air saturation)	Dubilier et al. (1994, 1995)
Detoxification processes	Precipitation of iron sulphides in body wall	Giere et al. (1988, 1999)

A great deal of field and laboratory experimental work has been performed with *Tubificoides benedii* (Table 4.13) and this species may also be a suitable candidate for sediment chronic bioassays in the risk assessment of marine and estuarine sediments, since it occurs from meso to polyhaline conditions (Giere 2006). Another useful characteristic of *T. benedii* is that it plays an important role in the food chain. Giere (2006) has reported juvenile *Crangon crangon* and gobiid fishes to consume this species in large quantities and in preference to polychaetes of similar size, it also constitutes a significant part of the diet of migratory birds.

Marine whole-sediment toxicity tests are even less developed than estuarine tests and no bioassays have been developed for chronic exposure with marine oligochaetes. Giere (2006) has emphasized the importance of marine oligochaetes as valuable biological models and the need for more research effort on this ecologically relevant group of annelids. Given the high abundance of oligochaetes in some littoral sites (*e.g.* 10^5–10^6 ind. m^2), Giere has suggested the importance of the group in the exchange of pollutants at the water/sediment interface. He has also reviewed the few studies dealing with life-cycles, population dynamics or habitat factors that explain the distribution of the species of marine and estuarine oligochaete species, and concluded that this is also a neglected area of research. This fact handicaps the development of bioassays, since this basic information is required for the culture, test development and interpretation of effects due to chemical exposure.

4.3.5 *In Situ Bioassays*

Laboratory bioassays commonly use sediment that has been processed to eliminate indigenous benthic fauna, and then homogenised to create laboratory replicates. While there are good reasons for these procedures (see Sect. 6.7), they result in the oxidation of the sediment and destroy the geochemical gradients that occur in natural sediments (Luoma and Carter 1993). Thus, laboratory bioassays cannot be seen as replicating field conditions. This deficiency can be addressed through the exposure of invertebrates in field experiments, but there is little work in this area with oligochaetes. Sibley et al. (1999) first used an "in situ" chamber for measuring sediment toxicity and bioaccumulation in *Lumbriculus variegatus,* over a 10-day period. Survival of worms was 40% and 76% in two contaminated sediments, compared with 85% recorded in reference sediment. The authors recommended the new test system, but the methods were not completely optimised and some confounding factors were identified, such as reduction of dissolved oxygen in the chambers to approximately 30% of the outside levels, which can generate problems in environments with low oxygen levels. Predation on oligochaete worms, may also limit the use of *in situ* chambers at some sites.

A further study by Greenberg et al. (2002) conducted *in situ* acute toxicity tests with *Ceriodaphnia dubia* (48 h), *Chironomus tentans* (96 h), *Hyalella azteca* (96 h) and bioaccumulation tests with *L. variegatus* (96 h) in three river sites polluted with chlorobenzenes, and compared results with those obtained from a reference site. Although the study did not address toxicity in the oligochaete, the exposure chambers could be used for this purpose, in a longer exposure period. Transparent core tubes (*ca.* 7-cm outside diameter and *ca.* 13-cm length) closed with polyethylene caps at each end of the tubes, and two windows (4 × 8 cm) covered with polypropylene mesh (74 μm) were buried to approximately half their length into the streambed, with one window in the sediment and the other exposed to the water column. Tubes were left to equilibrate for 24–36 h prior to the addition of test organisms. The use of minipiezometers placed within centimetres of the exposure chambers provided information on the hydraulic characteristics and chemistry of the pore water.

The use of exposure chambers for *in-situ* sediment toxicity has been developed for *Chironomus* spp. (*e.g.* Bervoets et al. 2004) and a similar procedure could be tailored for chronic toxicity tests with *T. tubifex* or *L. variegatus*, or any other oligochaete species appropriate for the *in-situ* conditions.

4.4 Comparative Sensitivity with Other Invertebrates

Several authors have examined the relative sensitivities of oligochaetes to other invertebrates and vertebrate species. In most cases, these cannot be considered as absolute comparisons since the test conditions are frequently inappropriate for one or more species, or the endpoints may not be the most sensitive for the particular species. However, these data do provide a qualitative assessment of the sensitivity of the different species.

A mixture of tubificines (*Tubifex tubifex* and *Limnodrilus hoffmeisteri*) were used by Slooff (1983) among a variety of benthic invertebrates and exposed to 15 chemicals (Hg, Cd and organic compounds). The differences among taxa were not great, but the worms were the fourth most tolerant behind *Erpobdella* (Hirudinea), *Corixa* (Hemiptera) and *Cloeon* (Ephemeroptera). However, on a scale of tolerance from 1 to 12, tubificines ranked anywhere from most to least tolerant depending on the chemical used. In a second comparative study, Green et al. (1985) tested seven freshwater invertebrate species for toxicity with phenol, and in this instance *L. hoffmeisteri* was the most tolerant species with 96-h LC_{50} of 780 mg l^{-1}, a 50-fold increase over that for *Baetis rhodani* (Ephemeroptera). However, the authors stated that, when compared with literature values for taxa related to those that they used, it becomes apparent that there is no standard response within a taxonomic group. In 1994, Fargasová published the results of simple 96-h static toxicity tests with *T. tubifex* exposed to various metals (As, Pb, Cr, Hg, Cd) (see Appendix 1, Chap. 7), and the oligochaete worms were more tolerant than *Daphnia magna*.

Bailey and Liu (1980) conducted comparative studies of *Lumbriculus variegatus* with six fish species (including rainbow trout) and three arthropodan invertebrates (*Daphnia magna*, *Hyalella azteca* and *Tanytarsus dissimilis*) with munitions plant wastewater (for TNT). The lumbriculid was found to be less sensitive to pesticides than most arthropods and fish. It was also less sensitive to munitions waste than the fish at 96-h exposure, but its sensitivity was similar to or greater than that of the arthropods used. The oligochaetes were most sensitive to Cd and had a similar range of sensitivity as the other invertebrate species for Cu, Zn, Pb, Cr, and Hg. A comparison of 96-h LC_{50} values for seven aquatic species was obtained simultaneously by Ewell et al. (1986), who also included *L. variegatus* as a test species. While *Daphnia* was at least as susceptible as the other organisms for all compounds tested, the relative sensitivities of the other species were highly chemical dependent. *Lumbriculus* was as sensitive as *Daphnia* for 19% of the chemicals tested. Most recently, McCrary and Heagler (1997) examined the relative sensitivity of

Table 4.14 Pore water, elutriate and bulk-sediment toxicity test results for fathead minnow, *Ceriodaphnia dubia, Hyalella azteca, Lumbriculus variegatus* and *Chironomus tentans*. Survival in bulk sediment and LC_{50} values (95% confidence interval) in pore water and elutriate tests. Test length: 96 h for all bulk-sediment exposures, 48 h for *C. dubia, H. azteca* and *C. tentans* pore water and elutriate tests, 96 h for fathead minnow and *L. variegatus* pore water and elutriate tests (From Schubauer-Berigan and Ankley 1991, Table 2, with permission of John Wiley & Sons Ltd., © conveyed by Copyright Clearance Centre, Inc.)

	Fathead minnow	*C. dubia*	*H. azteca*	*L. variegatus*	*C. tentans*
Pore water	15 (13–18)	12 (9–15)	10 (9–12)	23 (19–29)	8 (6–11)
Elutriate	57 (47–70)	NT	NT	NT	–
Bulk sediment	0/10[a]	–	0/10	0/10	–

NT non toxic
[a]Test sediment/control sediment, number alive (of 10) at test termination

six species to mercury in simultaneous multiple tests with one oligochaete species, three amphibia and two fishes, showing that *Lumbriculus* was intermediate in sensitivity. However, in a series of papers published by scientists at the U.S. Environmental Protection Agency laboratory at Duluth, *L. variegatus* was the least sensitive organism in tests designed to evaluate the validity of using pore water or elutriates to determine toxicity in place of whole sediments (the other species being fathead minnow, a cladoceran, a chironomid, and the amphipod *Hyalella azteca)* (Table 4.14) (Schubauer-Berigan and Ankley 1991).

In experiments with cadmium, the worm *L. variegatus* and the snail *Helisoma* sp. were exposed to $CdCl_2$ diluted in water, and to a variety of sediments with different AVS spiked with $CdCl_2$, in a flow-through system (Carlson et al. 1991). All worms exposed for 96 h to 0.278 mg Cd l^{-1} died, 40% died at 0.144 mg Cd l^{-1} and all survived at concentrations of 0.074 mg Cd l^{-1}. Compared with the snail, worms were similar or more sensitive to cadmium.

The toxicity of cadmium and nickel to several species was evaluated relative to AVS levels by Ankley et al. (1991b) and showed that *L. variegatus* was less sensitive than *H. azteca* or *Ceriodaphnia dubia*. Water-only toxicity tests gave 10-day LC_{50} values of 158 μg Cd l^{-1} and 12,160 μg Ni l^{-1} for *L. variegatus,* that was approximately 56 and 16 times higher than for *H. azteca*, respectively. In a straightforward test of the sensitivity of three species to Cu-contaminated sediment (West et al. 1993), the sensitivity of the reproductive endpoint in *L. variegatus* was similar to survival in the other two species (*H. azteca* and *Chironomus tentans*). The authors also showed that the relative sensitivity of the three species was not accurately predicted from water-only Cu exposures, which the authors suggested could be related to differences in animal behaviour.

In a comparative study of five benthic invertebrate species using both reference toxicants (phenol and pentachlorophenol), bleached kraft mill effluent (resin-acids) and contaminated sediments, Hickey and Martin (1995) showed that the sensitivity of *L. variegatus* was in the same range as the other species (an amphipod, a tanaid, a cladoceran, an ephemeropteran, and a bivalve). Based on the immobilisation response in acute water-only tests, sensitivity was similar to daphnids for phenol and 2.2-fold lower for Cd (96-h EC_{50} = 35.6 mg Phenol l^{-1}, 0.69 mg PCP l^{-1}, and

0.15 mg Cd l^{-1}). The 10-day survival and reproduction endpoints were also compared for pulp- and paper-mill contaminated sediments, and oligochaete survival was less sensitive than oligochaete reproduction, with amphipod survival and clam reburial as the most sensitive endpoints for assessing effects of this type of contaminant. In a series of papers, the sensitivity of four different invertebrate chronic bioassays to natural or spiked sediments was compared. Tributyltin (TBT) toxicity to four benthic invertebrates was examined in chronic sediment bioassays by Day et al. (1998), using *T. tubifex* as well as *Chironomus riparius*, *Hyalella azteca*, and *Hexagenia limbata.* Their results showed a similar range in sensitivity for most of the endpoints, with reproduction in *T. tubifex* being similarly sensitive to the species considered as more sensitive (*i.e. H. azteca*). Milani et al. (2003) compared sensitivity of the same four taxa, *T. tubifex* was the most tolerant to sediments spiked with Cd, Ni and Cu, and they reported that *T. tubifex* reproduction endpoints were less sensitive than growth in the other three species, but the authors pointed out that the only organism used as an adult was the worm, whereas the other three species were tested as young immatures, which is likely to make them more sensitive to toxicants. These observations were contrary to results obtained in field sediment bioassays in six sites from Collingwood Harbour, in Lake Huron, contaminated by metals, mainly Cu, where only *T. tubifex* reproduction endpoints indicated a toxicity problem (Milani et al. 2003). This agrees with a previous study by Reynoldson et al. (1995), where *T. tubifex* reproduction was more sensitive than growth endpoints of the other three species, although causative agents may be different in the two studies. These authors stressed the importance of using more than one species in toxicity tests since the use of only the most sensitive species can lead to erroneous assessment in particular cases.

In summary, although several publications have shown that worms were not the most sensitive species, they cannot be considered as the most tolerant taxa. Absolute comparisons are not adequate as not only the test conditions but also the endpoints may not be the most adequate for the particular species. Adequate exposure conditions for oligochaete worms require both water and sediment compartments, and the most sensitive endpoints seem to be related with reproduction (asexual or sexual). The range of sensitivities of the various benthic species to the different chemicals suggests that the use of a battery of bioassays with different taxa is most appropriate for assessing toxicity of chemical mixtures that occur in natural sediments (Rodriguez and Reynoldson 1999).

4.5 Acclimation and Genetic Adaptation

The ability to accurately predict or detect toxicity in contaminated sediments depends not only on the species but on the population used as a test organism in bioassays. Field organisms exposed to toxicants can acclimate or can be selected by their genetically determined resistance. Acclimation commonly refers to physiological compensatory changes in the organisms exposed to toxicants that increase internal

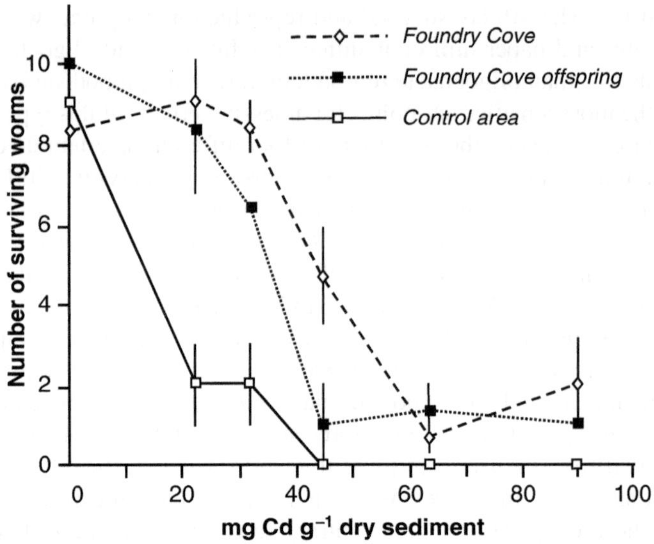

Fig. 4.11 Mean number of survivors (*vertical lines*, SE) of field collected *Limnodrilus hoffmeisteri* from a control area and a polluted site in Foundry Cove, and second generation from Foundry Cove worms exposed to different Cd concentrations (28-day exposure, n=3) (From Klerks and Levinton 1989, Fig. 3, redrawn and reproduced from *The Biological Bulletin* by permission of Marine Biological Laboratory, Woods Hole, MA)

tolerance limits, and responses include behavioural and physiological changes (*e.g.* uptake reduction, increased excretion, or storage in metal granules). On the other hand, genetic adaptation in a population is acquired through natural selection, which results in inheritable changes. Both acclimation and genetic adaptation can result in test organisms becoming more tolerant to toxicants and may explain some of the observed differences among populations used in the toxicological assessment. Few studies have demonstrated inherited tolerance in aquatic oligochaetes, but some clear evidence is described below.

Klerks and Levinton (1989) detected a very rapid evolution of resistance to Cd, Co and Ni levels in *Limnodrilus hoffmeisteri*. Specimens collected from a contaminated site were significantly more resistant to cadmium than those from a reference site, and this persisted after culturing for two generations in clean sediment. The results were based on survival and reproduction data obtained in 28-day bioassays with natural sediment and survival time in metal-spiked water (Fig. 4.11). The authors expected the increased tolerance to be due to a reduction in Cd uptake, but in fact it was associated with the presence of high concentrations a metallothionein-like protein that was genetically determined (Klerks and Bartholomew 1991). Heritability estimates of 0.93 ± 0.12 were obtained by Klerks and Levinton (1993) from the regression of mean offspring survival time on midparent survival time. These authors estimated that the final difference in resistance between the selected lines and the control lines (about 66% of difference in resistance) could have been obtained in four or five generations of natural selection, if the conditions were similar to those used in the laboratory. A more recent contribution by Levinton et al. (2003) has described a rapid loss of resistance to Cd in remediated sediments over 9–18 generations. After a clean up in 1994, the resistance of *L. hoffmeisteri* declined steadily to normal values at a control site by 2002, and Cd body burdens in the remediated site and control were the same. Martinez and Levinton (1996) examined the genetic architecture of metal resistance in *L. hoffmeisteri* and suggested this was controlled by a single gene.

Keilty and Landrum (1990) tested the sensitivity of two populations of *Stylodrilus heringianus*, one from a polluted site offshore of Benton Harbor (Michigan), the other from the less contaminated Grand Haven (Michigan), both in the North American Great Lakes. Benton Harbor sediments were toxic to Grand Haven worms (50.9% mortality), and final worm weights and reworking rates were significantly lower than those obtained in any other experiment in which worm populations from a site were exposed to sediment from the same site. The mortality value had a large error term due to variability, while the sublethal measure of reworking rates was the more sensitive measure. This finding is significant, in that it emphasises the need to use a consistent source of organisms for any type of bioassay.

In genetically different populations of *Tubifex tubifex,* Anlauf (1994) found there may be differences in tolerance to changing temperature, in somatic weight and in reproduction (number of cocoons), all important attributes when selecting a population for conducting a sediment bioassay with this species. In later work, the author measured differences between two genetic lineages of *T. tubifex* through allozyme patterns (Anlauf 1997) and showed that distinct lineages had different habitat preferences in the field (Anlauf and Neumann 1997). Lineage-specific experiments with different European populations of *T. tubifex* by Sturmbauer et al (1999) showed that some lineages were more tolerant to Cd than others. The genetic distances of some of these populations made the authors conclude that several cryptic species constitute the morpho-species *Tubifex tubifex*. However, the possibility that selection pressure has operated on some populations due to the presence of natural and anthropogenic pollutants cannot be discounted as an explanation for the differences in tolerance among populations, and more data from a more rigorous taxonomic study are required before acknowledging the existence of cryptic species (see Sect. 2.3).

Tolerance and acclimation to mercury has been investigated by Vidal and Horne working with *Sparganophilus pearsei* (2003a) and *T. tubifex* (2003b). Differences in tolerance to Hg by *S. pearsei* was interpreted through LC_{50} values in acute water-only toxicity tests and results showed that tolerance depends on the previous exposure history to the metal, *i.e.* whether the source populations were exposed to Hg-contaminated sediment. In a second set of experiments, *T. tubifex* worms raised in Hg-contaminated sediments (0.62 mg kg^{-1} ww) over four generations developed metal resistance, as measured by comparing 96-h LC_{50} in water-only toxicity tests with worms raised in contaminated (1.09–1.57 mg Hg l^{-1}) and uncontaminated sediments (0.17–0.18 mg Hg l^{-1}). Resistance persisted for 3 subsequent generations in offspring moved to clean sediment (96-h LC_{50} values ranged between 1.07–1.48 mg Hg l^{-1} in worms descendants from those exposed to the contaminated sediment, compared to 0.18–0.19 mg Hg l^{-1} in controls). A final experiment was run with worms from the seventh generation, placed in pairs (one worm from each uncontaminated and contaminated treatment), and their offspring then analysed for mercury tolerance. The offspring LC_{50} of 1.39 mg Hg l^{-1} was higher than expected if a mixture of genes from organisms raised in control sediment had occurred (Vidal and Horne 2003b).

In conclusion, differences in resistance to pollutants must be expected when working with the same species, based on their genetic lineage, natural selection, and acclimation processes to the environmental characteristics of the habitat of the source population. These differences can be minimised using organisms from a single source, or by requiring an acclimation period in unpolluted or formulated sediment, together with a regular observation of the sensitivity of the culture population.

4.6 Biomarkers

A biomarker is a biochemical or a cellular response that can be related quantitatively to the extent of exposure to a chemical (or class of chemicals), and can be used as a bioassay of the presence and effects of pollutants (Widdows 1993). The advantage of working at the molecular or subcellular scale of organisation is that the response time is rapid, and effects are often linked to very specific stressors. Responses at molecular or cellular level are also supposed to occur at doses lower than those at the organism scale (Fig. 4.12). The disadvantage, however, is that the ecological significance of the effects is often unclear.

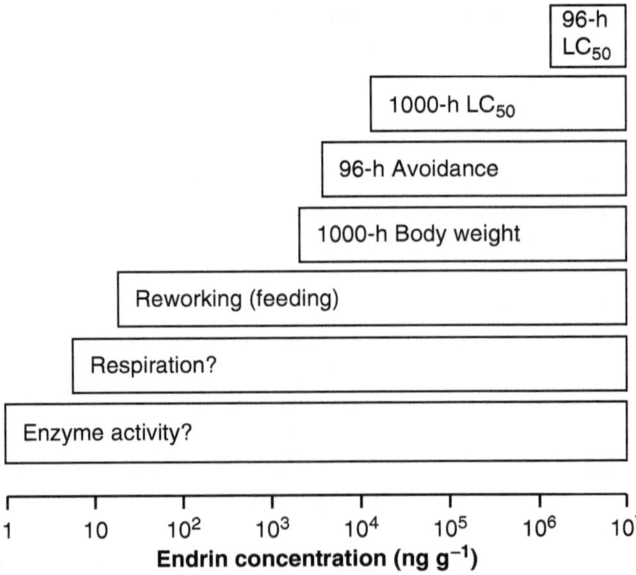

Fig. 4.12 Summary of measured and hypothesized responses by *Stylodrilus heringianus* to endrin-laden sediments (From Keilty et al. 1988c, Fig. 5, redrawn and reproduced by permission of Elsevier Ltd, © conveyed by Copyright Clearance Centre, Inc.). Values are estimations placed along a presumed continuum of response

Table 4.15 Cellular and molecular biomarkers used in aquatic oligochaetes

Biomarker	Species	Chemical	Reference
Chloragocytes: nuclear expansion	*Tubifex tubifex*	Metals, related to hypoxia	Fischer et al. (1980a, b), Fischer and Molnár (1992)
Chloragocytes: Cell volume Nuclear expansion Mitochondrial alterations Lysosomes RER reduction	*Enchytraeus* sp.	Pesticides	Purschke et al. (1991)
Metal-binding proteins	*Monopylephorus cuticulatus*	Cd	Thompson et al. (1982)
MTLP	*Limnodrilus hoffmeisteri* *Stylodrilus heringianus* *Monopylephorus cuticulatus*	Cd	Chapman and Brinkhurst (1984)
MTLP	*Lumbriculus variegatus*	Cd	Bauer-Hilty et al. (1989)
MTLP	*Limnodrilus hoffmeisteri*	Cd	Wallace and Lopez (1996, 1997)
MTLP	*Tubifex tubifex*	Cd	Gillis et al. (2002, 2004a)
MTLP	*Tubifex tubifex*	Cd, Ni, Pb	Gillis et al. (2004b)
MTLP	*Tubifex tubifex*	Isoproturon (pesticide)	Mosleh et al. (2005)
MTLP	*Tubifex tubifex*	Cu	Mosleh et al. (2006)
Depigmentation	*Lumbriculus variegatus*	Cd	Bailey and Liu (1980)
Haemoglobin decrease	*Limnodrilus hoffmeisteri*	Ni	Martínez-Tabché et al. (1999)
Haemoglobin decrease and ACHE activity	*Limnodrilus hoffmeisteri*	Pb	Martínez-Tabché et al. (1999)

MTLP metallothionein-like proteins

There has been relatively little work on biomarkers in aquatic oligochaetes (Table 4.15), but studies with earthworms are likely illustrative of the expected responses and these have been reviewed by Scott-Fordsmand and Weeks (2000). One of the earliest studies on the effects of metals on tubificid oligochaetes was an investigation of the effects of eight metals (Ba, Co, Fe, Mn, Zn, Cu, Cd, and Pb) and hypoxia on the chloragocytes of *Tubifex tubifex* (Fischer et al. 1980b). The chloragocytes are thought to be the main site of haemoprotein synthesis. Therefore, increases in the nuclear volume of chloragocytes under hypoxic conditions, observed by Fischer and Horvath (1979), are probably indicative of an increase in haemoprotein synthesis when trying to regulate oxygen in response to low oxygen tension. However, Fischer et al. (1980b) demonstrated that there was also a metal specific response to nuclear expansion in *T. tubifex*. Under aerated conditions, Cd, Mn, Fe and Pb had a stimulatory effect on nuclear volume, whereas no effect was observed with Ba, Co and Zn. Contrarily, under hypoxic conditions, Ba reduced nuclear expansion, and Co and Cu completely prevented expansion. In another paper, these authors described the stimulatory effect on the nuclear volume of chloragocytes by a 5-day exposure to 10 ppm of 2,4-dinitrophenol, and compared the effects of this toxicant under aerated conditions to those measured under hypoxic conditions, suggesting an uncoupling effect of oxidative phosphorilation (Fischer et al. 1980a). Other catabolic inhibitors (monoiod acetate, azide, fluoride and arsenate) had an indirect effect on retarding phosphorilation, resulting in less or no expansion of the nucleus.

Purschke et al. (1991) described ultrastructural changes in chloragocytes of *Enchytraeus* sp. exposed to the pesticides parathion, cypermethrin and benomyl (benzimidazol) and stated the potential of cellular level characters in detecting early and sublethal effects. The pathological features described in this work included alterations in nuclei (enlargement, reduction of heterochromatin, dispersion of nucleoli), mitochondria (reduction of size or enlargement, shrinkage, swelling and disintegration of cristae), autophagosomes (multiplication and size) containing damaged mitochondria, RER stacks and granules, very large lipid droplets, autolysosomes containing lamellar bodies, increased number of chloragosome-like granules, and RER reduction. The authors reported chloragocyte volume reduction from $175 \pm 48\ \mu m^2$ in the control down to $51 \pm 12\ \mu m^2$ in worms exposed to $0.8\ \mu l\ ml^{-1}$ parathion, and a reduction in the lipid content of chloragocytes from 74.6 ± 6.5 vol % in the control down to 47.2 ± 0.7 vol% in worms exposed to $236\ \mu l\ ml^{-1}$ benomyl. However, there was an unexpected reduction of the histopathologies in survivors after longer periods of exposure with the possible result that damaged cells may renew organelles or completely regenerate cells, although this process probably implies a reduced amount of energy for investment in reproduction and a reduction in life-span.

In a review of metal-containing granules in invertebrates, Brown (1982) discussed the function of Ca-rich granules in detoxification processes in body fluids, whilst storing valuable metabolites such as phosphate for re-use. She pointed out that knowledge on the composition of the granules, particularly the organic components, would provide some understanding of their loading and unloading capabilities with respect to trace metals. In a study published in 1991, Klerks and Bartholomew reported two types of granules present in the digestive duct tissue of *Limnodrilus hoffmeisteri*, most of them were rich in Cd and S (probably CdS), and a lower proportion rich in Ca, Fe and P. Our knowledge on the composition and formation of granules as a response to metal exposure is very limited, and the role of metal-sequestering granules in metal regulation by aquatic oligochaetes needs research.

Metallothionein-like proteins (MTLP) are indicators of metal exposure, and are characterised by their low molecular weight, high cysteine content and an absence of aromatic amino acids (Ireland 1983). They are normally found in traces, but exposure to sublethal levels of heavy metals increases the synthesis of MTLP that are thought to play a role in the metal detoxification processes. Metal-binding proteins in aquatic oligochaetes were first investigated by Thompson et al. (1982) in relation to sublethal exposure of Cd on the estuarine tubificid *Monopylephorus cuticulatus*. Following exposure, there was an increase in Cd and a concomitant decrease of Zn in the higher molecular weight protein pool of a molecular weight

Table 4.16 Cadmium accumulation and subcellular distribution (nmol Cd g^{-1} ww tissue) in *Limnodrilus hoffmeisteri* obtained from a control area and a polluted site (Foundry Cove), and exposed to 8.9 μM Cd in water, for 6 days (From Klerks and Bartholomew 1991, Table 1, with permission of Elsevier Ltd, © conveyed by Copyright Clearance Centre, Inc.)

	Control area	Foundry cove	F_{off}	Scheffé F test
Fraction	Cd concentration (mean ± SE)			(α = 0.01)
Homogenate	1,010 ± 157	2,232 ± 182	2,040 ± 46	C < Foff = F
Particulate	594 ± 98	1,184 ± 142	1,251 ± 83	C < F = Foff
Cytosol	218 ± 10	1,161 ± 238	522 ± 50	C < Foff < F
HMW	26 ± 5	64 ± 7	26 ± 3	C = Foff < F
MT	55 ± 4	415 ± 76	229 ± 13	C < Foff = F
LMW	96 ± 14	388 ± 59	208 ± 26	C < Foff = F

F_{off}: second generation offspring of the Foundry Cove worms, never exposed to elevated metal levels before the experiment

HMW high molecular weight pool, *MT* metallothionein pool, *LMW* low molecular weight pool

similar to metallothioneins, a slight increase in the medium molecular weight pool, and no increase in the low molecular weight protein pool. In Cd-exposed worms, these authors also described the rupture of the body tegument and peritoneum, degeneration of the digestive epithelium and of phagocytic amoebocytes (cytoplasm pale and translucent and nuclei with large clear spaces). Similar results were reported by Chapman and Brinkhurst (1984) in Cd-exposed worms. Metallothionein-like proteins were first characterised by Bauer-Hilty et al. (1989) in *Lumbriculus variegatus* exposed to cadmium chloride (1 mg l^{-1}), for 30 days. Similar MTLP were further described by Klerks and Bartholomew (1991) who showed that *L. hoffmeisteri* from metal-polluted areas were not only more resistant to cadmium by the production of higher levels of metal-binding proteins, but also that this was genetically determined (Table 4.16).

The oligochaete *L. hoffmeisteri* was used as prey in an experiment of cadmium bioavailability to the shrimp *Palaemonetes pugio* by Wallace and Lopez (1996) who demonstrated that ^{109}Cd (0.5–140 μg Cd l^{-1}) bound to MTLP was transferred with high efficiency. In a later set of experiments, Wallace and Lopez (1997) measured Cd concentrations in different subcellular fractions separated by differential centrifugation. The ^{109}Cd sequestered in the cytosolic fraction, which includes the MTLP, constituted the major source of the ^{109}Cd adsorbed by the shrimp (35.8%), while the sources due to intracellular (nuclear, mitochondria and microsomal) fractions or tissue fragments and metal-rich granules represented only 6.7% and 19.0% respectively of the Cd absorbed by shrimp. Most recently, Redeker et al. (2007) have applied a similar procedure to determine Cd and Zn subcellular distribution in *T. tubifex*.

One concern in the studies with biomarkers is the demonstration of a dose-response relationship between exposure level and biomarker response. The relationships between metal doses, MTLP induction and body burden in *T. tubifex* have been investigated by Gillis et al (2002), where *T. tubifex* was exposed to Cd-spiked sediment for up to 4 weeks. A significant correlation between MTLP and Cd in tissue was reported (r = 0.83). Worms from control sediment had a mean MTLP concentration of 2.6 and

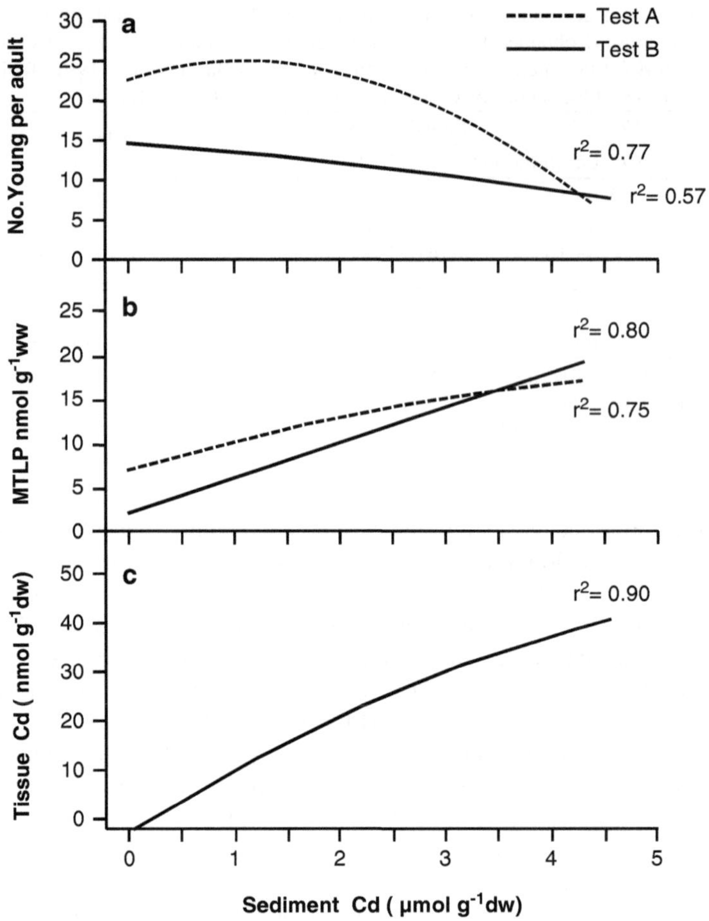

Fig. 4.13 Regression models ($p<0.001$) describing the relationship between the sediment Cd concentration and the mean number of young per adult (**a**), metallothionein-like protein (MTLP) mean concentration (**b**), and Cd mean tissue concentration (**c**) in *Tubifex tubifex* worms exposed to Cd-spiked sediment, for 28 days (From Gillis et al. 2002, Fig. 1, modified and reproduced with permission of John Wiley & Sons Ltd., © conveyed by Copyright Clearance Centre, Inc.)

6.7 nmol g^{-1} ww tissue, measured in two different tests, the concentration increasing significantly when exposed to higher levels of Cd in sediment (Fig. 4.13). The importance of MTLP as an early warning of disorders at high levels of biological organization has also been demonstrated by these authors who measured the relationship between the biomarker response and reproductive output (number of young per adult), in 28-day bioassay. The sediment concentration from which a significant decline in reproduction was observed was 2.78 and 2.68 µmol Cd g^{-1} dw in two experiments, and *T. tubifex* worms with a tissue MTLP concentration above 14 nmol g^{-1} were expected to undergo significant impairment in reproduction.

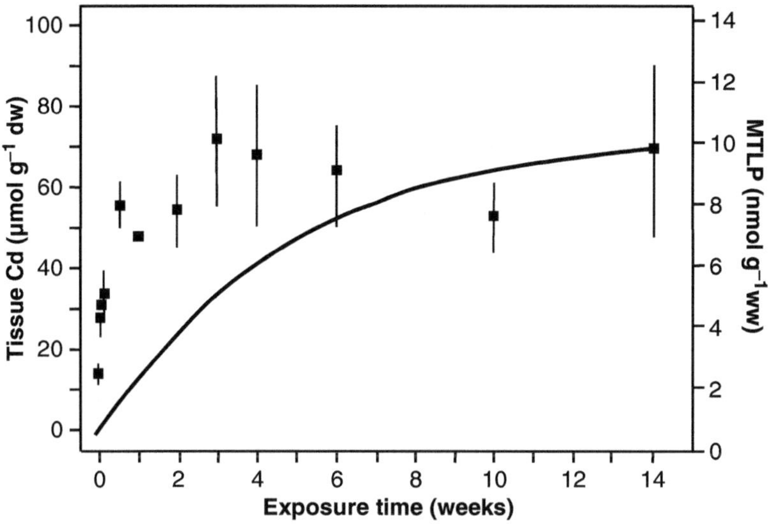

Fig. 4.14 Metallothionein-like protein (*MTLP*) concentration (*black squares*) in gut-cleared *Tubifex tubifex* exposed to sediment spiked with 3.66 μmol Cd g^{-1} dw for up to 14 weeks (n=4). *Solid line* represents the regression model for Cd uptake on exposure time ($r^2=0.87$, $p<0.0001$) (From Gillis et al. 2004b, Fig. 3, redrawn and reproduced with permission of John Wiley & Sons Ltd., © conveyed by Copyright Clearance Centre, Inc.)

If MTLP are used as biomarkers, their natural variation should be known as well as their dependence on worm physiological condition and on environmental variables. Gillis et al. (2004a) examined MTLP concentration in worms with reduced food supply, increased handling, or reduced dissolved oxygen (0–3 mg l^{-1}) and showed no significant differences with control. However, worms exposed to lower temperatures (13°C) showed a significant increase in tissue MTLP concentration, compared to those at 23°C. Gillis et al. (2004a) also examined the response of MTLP throughout the life-cycle of *T. tubifex* suggesting no significant variation associated with the reproductive stage (mean concentration: 2.5 nmol g^{-1} tissue ww), from cocoon to 10-week old worms. This value was slightly higher to background MTLP levels measured by Redeker et al. (2007) of 0.64 nmol g^{-1}. In another series of experiments, Gillis et al. (2004b) investigated the uptake and depuration rates of Cd, Ni and Pb and changes in tissue concentration of MTLP in *T. tubifex*. In a first set of experiments, tissue Ni peaked after only 12 h, whereas Cd and Pb continued to accumulate for the duration of the exposure (two experiments of 6 and 14 weeks, respectively). The resulting MTLP concentrations were highly variable and not correlated with metal tissue concentrations. In a second experiment, worms were exposed to high Cd concentration (3.66 μmol g^{-1} dw) and exhibited rapid Cd uptake in tissue and induced synthesis of MTLP up to a mean concentration of 8.68 μmol g^{-1} ww from day 4 to week 14 (Fig. 4.14). In a third experiment on depuration rates, Gillis et al. (2004b) reported that MTLP declined after a 3-week depuration period in clean sediment, and after 10 weeks the levels were not significantly higher than the pre-exposure levels. This gives us an indication of the period for which MTLP can be used as biomarkers after metal exposure.

Metallothioneins (MTs) and MTLP also participate in cellular functions such as regulation of growth or antioxidative defences, and their synthesis can be induced by exposure to the pesticide isoproturon (Mosleh et al. 2005). In *T. tubifex* worms, 10 mg l^{-1} isoproturon increased the total MTs concentration in the worms, up to a 148%, after 4-day exposure (MT concentration in control worms of 46.9–48.8 μg g^{-1} ww), suggesting that the increase of metallothionein cannot be consider a specific biomarker for metal pollution. In this study, worms exposed to isoproturon suffered clear reduction in growth rates, probably related partly to starvation, since worms were not fed during the exposure period, and partly to investment in detoxication processes. In a later contribution, Mosleh et al. (2006) measured significant increases of MT levels and catalase activities in copper-treated *T. tubifex* worms which the authors interpreted as a development of antioxidative defences.

Pigmentation of worms can also be used as a biomarker. It was first reported by Bailey and Liu (1980) in *L. variegatus* exposed to 3.4–6.2 mg l^{-1} Sevin and 2.5–20.0 mg l^{-1} Zn, turning the animals dark brown and gray, respectively. White and Keilty (1988) reported depigmentation in *Stylodrilus heringianus* after 24-h exposure to sediments heavily polluted by metals, chlorinated hydrocarbons and volatiles, and in 96 h only a mean of 0.6 individuals per replicate (initially 10) were alive. Unburrowed animals were considered healthy if active and bright red. Some of these colour changes can be attributed to changes in the synthesis of blood pigments. Martínez-Tabché et al. (1999) measured haemoglobin (Hb) concentration in *L. hoffmeisteri* exposed to three different sediments spiked with Ni, and reported Hb decrease after treatment (Fig. 4.15). However, these authors detected differences among different sediments for the same concentration of spiked Ni that were attributed to the interaction of Ni with other metals present (Zn) and to sediment characteristics (pH, particle size, and organic matter), concluding that the haemoglobin could be used as an indicator of Ni bioavailability in the sediments but it can be modified by the influence of sediment characteristics. In more recent work, the authors have investigated haemoglobin concentration and acetylcholinesterase (AChE) activity in *L. hoffmeisteri* worms exposed to Pb-spiked sediments, reporting a decrease of 39.9–93.1% Hb after 25-day exposure and usually associated with a decrease in AChE activity (Martínez-Tabché et al. 2001). Thus, haemoglobin levels or pigmentation could also be used as a biomarker but a control level should be established under specific environmental conditions to allow interpretation of the response to xenobiotics. Some environmental stressors, such as cold or salinity, have been demonstrated to provoke pigmentation changes in earthworm chloragocytes (Fischer and Molnár 1992), changing the general colour of the worm.

In the oligochaetes, the haeme molecule is dissolved in the blood plasma and its synthesis is controlled by a complex of enzymes. Reynoldson and Thompson (unpub.) have investigated the utility of porphyrin profiles as a rapid sub-cellular biomarker in *Tubifex tubifex*. The exposure to some toxics (called porphyrinogenic chemicals) may disturb the process and the porphyrins and their precursors may

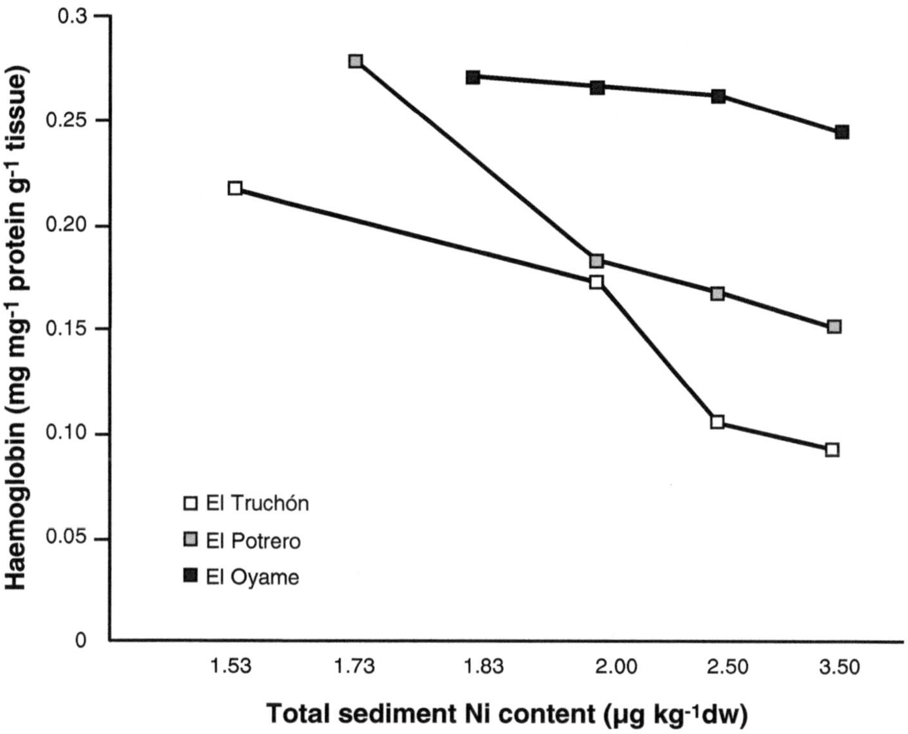

Fig. 4.15 Mean effect (n = 3) on haemoglobin concentration in *Limnodrilus hoffmeisteri* exposed to sediments from three different sites spiked with nickel (From Martínez-Tabché et al. 1999, redrawn and reproduced by permission of Elsevier Ltd, © conveyed by Copyright Clearance Centre, Inc.)

accumulate. Low levels of porphyrins are found under normal circumstances and the coproporphyrinogen to total porphyrin ratios were used as indicator of the toxicity stress in worms exposed to copper. The coproporphyrinogen /total porphyrin ratio appeared to be a good indicator of the stress, and this study showed that a ratio of 1.0 or close to 1.0 was found in control organisms, whereas a ratio >1 indicates that the porphyrin synthesis is being blocked (unpub.).

Chaetal variation has also been studied to determine if pollutants as well as environmental factors could induce changes in morphology. Chapman and Brinkhurst (1987) found that dorsal chaetal bundles had reduced hairs and pectination of the bifid chaetae in *T. tubifex* when exposed to certain pH, salinity and hardness conditions, and the resulting chaetal morphology was that of the *T. tubifex bergi* type (with short, rudimentary hair-chaetae and few pectinates in dorsal bundles) in some but not all survivors. Either high or low pH may cause a total loss of hairs and pectination in *T. tubifex*, resulting in a chaetal morphology characteristic of the *T. tubifex blanchardi* type (only bifid dorsal chaetae present). A similar experiment with the species *Ilyodrilus frantzi* (two forms, with and without hair-chaetae) was performed

Table 4.17 Changes observed in chaetae in *Tubifex tubifex* and *Ilyodrilus frantzi* exposed to different stressors (From Chapman and Brinkhurst 1987, Table 3, with permission of Springer Publ., © conveyed by Copyright Clearance Centre, Inc.)

Species	Treatment	Changes observed	
Tubifex tubifex		**30-day exposure**	**60-day exposure**
	pH = 5	*bergi*-type	–
	pH = 9	*bergi*-type	*bergi*-type
	Salinity 1.0 ppt	–	Abnormalities in dorsal chaetae
	Salinity 2.5 ppt	–	*bergi*-type + abnormalities in dorsal chaetae
	Salinity 5.0 ppt	–	Abnormalities in dorsal chaetae
	Hardness 200 mg l^{-1}	Partial loss of hairs and pectinates	*bergi*-type
	2.5 mg Hg l^{-1}	–	Abnormalities in dorsal chaetae
	1.0 mg Hg l^{-1}	–	Abnormalities in dorsal chaetae
Ilyodrilus frantzi		**34-day exposure**	
	pH = 5	Loss of hairs[a]	–
	pH = 9	Loss of hairs[a]	–
	Salinity 1.0 ppt	Loss of hairs[a]	–
	Salinity 2.5 ppt	Loss of hairs[a]	–
	Salinity 5.0 ppt	With hairs	–
	2.5 mg Hg l^{-1}	Loss of hairs[a]	–
	1.0 mg Hg l^{-1}	Loss of hairs[a]	–
	Hardness 200 mg l^{-1}	With hairs[b] and loss of hairs[a]	–
	Hardness 100 mg l^{-1}	Loss of hairs[a]	–
	Hardness 50 mg l^{-1}	Loss of hairs[a]	–

[a]*Ilyodrilus frantzi (capillatus* form with hair-chaetae)
[b]*Ilyodrilus frantzi* (*typical* form with only bifid chaetae)

with different stressors (pH, metals, hardness, salinity) and resulted in the loss of hair chaetae in the *capillatus* form in soft water (Table 4.17). Other authors have studied the chaetal morphology under exposure to heavy metals but did not find any alteration (Lucan-Bouché et al. 1999b). However, field data on chaetae modifications due to mercury pollution have been reported (Milbrink 1983).

If biomarkers are to be included in risk assessment approaches for monitoring the effects of toxicants, more knowledge is required on their physiological role. Some of the cellular and molecular biomarkers have shown that effects are reversible, and this should be considered when selecting endpoints, since the exclusive use of biomarkers for toxicity assessment of sediments would be inadequate. Better understanding of the relationships of these early-warning alterations with other endpoints that have clear demographic consequences (*e.g.* reproduction) are required to validate their use and adequately interpret their meaning.

4.7 Conclusions

Relatively few oligochaete species have been used in toxicological studies, so that true comparisons among species are hard to make. Only the work of Chapman et al. (1982b, c) enables direct comparison between oligochaete taxa and showed how a small modifications in protocols can produce large changes in LC_{50} values. Presumably this is related to the degree of stress under the particular environmental conditions, even without the additional stress induced by the presence of a contaminant. Absence of sediment in tests with aquatic oligochaetes is a source of additional stress and reduces realism in the exposure conditions, hence most current toxicity work is performed in the presence of sediment. Thus, unless there is a requirement for water-only tests, for interspecies comparison of toxicity of chemical compounds, or for quality control of species sensitivity in laboratory cultures, these tests are considered inappropriate for oligochaete worms in ecological risk assessment.

When selecting the worm population for use in toxicological studies, it is important to know their origin and verify their identity. Taxa purchased from supply houses may not be correctly identified and may contain mixtures of species. Field populations of several species have been demonstrated to vary genetically, and may have different tolerances to both environmental and anthropogenic stress. This suggests that test organisms should be obtained from cultures of a defined genetic strain, or alternatively the sensitivity of the populations should be intercalibrated with reference toxicants. Oligochaete worms are not uniformly more tolerant to contaminants than other test organisms. Comparative data for worms and other benthic invertebrates show that responses are species-specific and also contaminant-specific. Mixed species cultures are often more tolerant than the same species in pure culture. More work on species assemblages is needed as this is what typically occurrs in the field.

Among the sublethal measurements, sexual reproduction (number of cocoons and young per adult) seems to be the most easily standardised and informative endpoint, and has high ecological relevance, since reproductive impairment has the same long-term effect as mortality on the population. Asexual reproduction in *L. variegatus* makes it difficult to resolve effects of fragmentation and mortality that both may occur in exposed populations, although methods have been proposed to synchronize cultures that help clarify interpretation of responses. The use of growth as endpoint for adult *T. tubifex* requires control of the initial fresh biomass of the adult worms and of the cocoons laid in the chronic bioassay. Loss of adult biomass cannot be completely attributed to toxic stress, but includes investment in reproduction. Total growth rate can be measured as the increase in adult biomass plus the biomass of cocoons produced in the bioassay, and used as a total-growth endpoint at the end of the bioassay. This endpoint may be critical for interpretation of reproductive inhibition in cases where a change in energy allocation pattern in the test organisms is occurring in response to environmental stress. Young worms have a higher surface-to-volume ratio and, in consequence, higher sensitivity to pollutants that may increase their uptake rate. Therefore, an adaptive response of the adult to the toxic stress may consist of investing in detoxification processes to assure its own survival, at the expense of the immediate investment in offspring (Forbes 2000).

Sublethal, short-term responses such as behavioural changes (avoidance behaviour, sediment reworking and sediment borrowing) need further standardization. Behavioural responses may produce quicker results in situations requiring rapid, inexpensive tests; however, this should not be achieved at the expense of reliability. Another short-term sublethal endpoint used for oligochaete worms is respiration rate, but this requires experienced technical staff and the results are complex to interpret since they are both contaminant and species-specific. There is probably no advantage to using this approach in contrast to mortality measurements in short-term toxicity tests.

In situ bioassays are promising tools in environmental risk assessment and will require working with species characteristic of certain habitats and that allow reproduction to be estimated under different environmental conditions. Alternatively, acclimation to different field environmental conditions and adequate test vessels should be investigated using tolerant species, such as *T. tubifex* or *L. hoffmeisteri*, which can be used in *in-situ* bioassays in a wide variety of field situations.

References

Achazi RK, Chroszcz G, Duker C, Henneken M, Rothe B, Schaub K, Steudel Y (1995) The effect of fluoranthene (Fla), benzo(a)pyrene (BaP) and Cadmium (Cd) upon survival rate and life cycle parameters of two terrestrial annelids in laboratory test systems. Newsl Enchytraeidae 4:7–14

Adams WJ (1995) Aquatic toxicology testing methods. In: Hoffman DJ, Rattner BA, Burton GA Jr, Cairns J Jr (eds) Handbook of ecotoxicology, Chapter 3. Lewis Publ, Boca Raton, pp 25–46

Ankley GT, Collyard SA (1995) Influence of piperonyl butoxide on the toxicity of organophosphate insecticides to three species of freshwater benthic invertebrates. Comp Biochem Physiol 110:149–155

Ankley GT, Phipps GL, Leonard EN, Benoit DA, Mattson VR (1991a) Acid-volatile sulfide as a factor mediating cadmium and nickel bioavailability in contaminated sediments. Environ Toxicol Chem 10:1299–1307

Ankley GT, Schubauer-Berigan M, Dierkes JR (1991b) Predicting the toxicity of bulk sediments to aquatic organisms with aqueous test fractions: pore water vs. elutriate. Environ Toxicol Chem 10:1359–1366

Ankley GT, Benoit DA, Hoke RA, Leonard EN, West CW, Phipps GL, Mattson VR, Anderson LA (1993) Development and evaluation of test methods for benthic invertebrate and sediments: effects of flow rate and feeding on water quality and exposure conditions. Arch Environ Contam Toxicol 25:12–19

Ankley GT, Leonard EN, Mattson VR (1994a) Prediction of bioaccumulation of metals from contaminated sediments by the oligochaete *Lumbriculus variegatus*. Water Res 28:1071–1076

Ankley GT, Collyard SA, Monson PD, Kosianm PA (1994b) Influence of ultraviolet light on the toxicity of sediments contaminated with polycyclic aromatic hydrocarbons. Environ Toxicol Chem 13:1791–1796

Ankley GT, Erickson RJ, Phipps GL, Mattson VR, Kosian PA, Sheedy BR, Cox JS (1995) Effects of light intensity on the phototoxicity of fluoranthene to a benthic macroinvertebrate. Environ Sci Technol 29:2828–2833

Ankley GT, Erickson RJ, Sheedy BR, Kosian PA, Mattson VR, Cox JS (1997) Evaluation of models for predicting the phototoxic potency of polycyclic aromatic hydrocarbons. Aquat Toxicol 37:37–50

Anlauf A (1994) Some characteristics of the genetic variants of *Tubifex tubifex* (Müller, 1774) (Oligochaeta, Tubificidae) in laboratory cultures. Hydrobiologia 278:1–6

Anlauf A (1997) Enzyme variability of *Tubifex tubifex* (Müller) (Oligochaeta, Tubificidae) and seven other tubificid species. Arch Hydrobiol 139:83–100

Anlauf A, Neumann D (1997) The genetic variability of *Tubifex tubifex* (Müller) in 20 populations and its relation to habitat type. Arch Hydrobiol 139:145–162

Arrate JA, Rodriguez P, Martinez-Madrid M (2002) Effects of three chemicals on the survival and reproduction of the oligochaete worm *Enchytraeus coronatus* in chronic toxicity test. Pedobiologia 46:136–149

ASTM (2005) Standard guide for conducting sediment toxicity tests with freshwater invertebrates. ASTM E1706-05, Philadelphia

Bagheri EA, McLusky DS (1984) The oxygen consumption of *Tubificoides benedeni* (Udekem) in relation to temperature and its application to production biology. J Exp Mar Biol Ecol 78: 187–197

Bailey HC, Liu DHW (1980) *Lumbriculus variegatus*, a benthic oligochaete, as a bioassay organism. In: Eaton JG, Parrish PR, Hendricks AC (eds) Aquatic toxicology. ASTM Special Technical Publication 707, Philadelphia, pp 205–215

Bauer-Hilty A, Dallinger R, Berger B (1989) Isolation and partial characterization of a cadmium-binding protein from *Lumbriculus variegatus* (Oligochaeta, Annelida). Comp Biochem Physiol 94C:373–379

Benfield EF, Buikema AL Jr (1977) Review of macroinvertebrate toxicity test techniques. ASTM Symposium on Invertebrate Bioassays

Benfield EF, Buikema AL Jr (1980) Synthesis of miscellaneous invertebrate toxicity tests. In: Buikema AL Jr, Cairns J Jr (eds) Aquatic invertebrate bioassays. ASTM Special Technical Publication 715, Philadelphia, pp 174–187

Benoit DA, Phipps GL, Ankley GT (1993) A sediment testing intermittent renewal system for the automated renewal of over-lying water in toxicity tests with contaminated sediments. Water Res 27:1403–1412

Berg K, Jonasson P (1964) Oxygen consumption of profundal lake animals at low oxygen content of the water. Hydrobiologia 26:131–143

Berg K, Jonasson P, Ockelmann KW (1962) The respiration of some animals from profundal zone of a lake. Hydrobiologia 19:1–39

Bervoets L, Meregallai G, De Cooman W, Goddeeris B, Blust R (2004) Caged midge larvae (*Chironomus riparius*) for the assessment of metal bioaccumulation from sediments in situ. Environ Toxicol Chem 23:443–454

Bhunia F, Saha NC, Kaviraj A (2003) Effects of aniline- an aromatic amine to some freshwater organisms. Ecotoxicology 12:397–404

Birtwell JK, Arthur DR (1980) The ecology of tubificids in the Thames Estuary with particular reference to *Tubifex costatus* (Claparède). In: Brinkhurst RO, Cook DG (eds) Aquatic oligochaete biology. Plenum Press, New York, pp 331–381

Bouché ML, Habets F, Biagianti-Risbourg S, Vernet G (2000) Toxic effects and bioaccumulation of cadmium in the aquatic oligochaete *Tubifex tubifex*. Ecotoxicol Environ Safe 46:246–251

Brinkhurst RO (1986) Taxonomy of the genus *Tubificoides* Lastockin (Oligochaeta, Tubificidae) species with bifid setae. Can J Zool 64:1270–1279

Brinkhurst RO, Austin MJ (1979) Assimilation by aquatic Oligochaeta. Int Rev Ges Hydrobiol 64:863–868

Brinkhurst RO, Wetzel MJ (1984) Aquatic Oligochaeta of the world: supplement. A catalogue of new freshwater species, descriptions and revisions. Can Tech Rep Hydrogr Ocean Sci 44, 101 pp

Brinkhurst RO, Chua KE, Kaushik NK (1972) Interspecific interactions and selective feeding by tubificid oligochaetes. Limnol Oceanogr 17:122–133

Brinkhurst RO, Chapman PM, Farrell MA (1983) A comparative study of respiration rates of some aquatic oligochaetes in relation to sublethal stress. Int Rev Ges Hydrobiol 68:683–699

Brkovic-Popovic Y, Popovic M (1977a) Effects of heavy metals on survival and respiration rate of tubificid worms: Part I- Effects on survival. Environ Pollut 13:65–72

Brkovic-Popovic Y, Popovic M (1977b) Effects of heavy metals on survival and respiration rate of tubificid worms: Part II- Effects on respiration rate. Environ Pollut 13:65–72

Brown BE (1982) The form and function of metal-containing "granules" in invertebrate tissues. Biol Rev 57:621–667

Brust K, Licht O, Huiltsch V, Jungmann D, Nagel R (2004) Effects of terbutryn on aufwuchs and *Lumbriculus variegatus* in artificial indoor streams. Environ Toxicol Chem 20:2000–2007

Bryant V, Newbery DM, McLusky DS, Campbell R (1985) Effect of temperature and salinity on the toxicity of arsenic to three estuarine invertebrates (*Corophium volulator, Macoma balthica, Tubifex costatus*). Mar Ecol Prog Ser 24:129–137

Buikema AL Jr, Cairns J Jr (1980) Aquatic invertebrate bioassay. ASTM Special Technical Publication 715, Philadelphia, 209 pp

Burton GA Jr (1992) Sediment toxicity assessment. Lewis Publ, Boca Raton, 457 pp

Burton GA Jr, MacPherson C (1995) Sediment toxicity testing issues and methods. In: Hoffman DJ, Rattner BA, Burton GA Jr, Cairns J Jr (eds) Handbook of ecotoxicology. Lewis Publ, Boca Raton, pp 70–103

Burton GA Jr, Nelson MK, Ingersoll CG (1992) Freshwater benthic toxicity tests. In: Burton GA (ed) Sediment toxicity assessment, Chapter 10. Lewis Publ, Boca Raton, pp 213–240

Calabrese EJ, Baldwin LA (1998) Hormesis as a biological hypothesis. Environ Health Persp 106:357–362

Calow P (1993) Handbook of ecotoxicology, vol 1. Blackwell, London, 478 pp

Carlson AR, Phipps GL, Mattson VR, Kosian PA, Cotter AM (1991) The role of acid-volatile sulfide in determining cadmium bioavailability and toxicity in freshwater sediments. Environ Toxicol Chem 10:1309–1320

Casellato S, Negrisolo P (1989) Acute and chronic effects of an anionic surfactant on some freshwater tubificid species. Hydrobiologia 180:243–252

Casellato S, Aiello R, Negrisolo A, Seno M (1992) Long-term experiment on *Branchiura sowerbyi* Beddard (Oligochaeta, Tubificidae) using sediment treated with LAS (Linear Alkylbenzene Sulphonate). Hydrobiologia 232:169–173

Chapman PM (1987) Oligochaete respiration as a measure of sediment toxicity in Puget Sound, Washington. Hydrobiologia 155:249–258

Chapman PM (2001) Utility and relevance of aquatic oligochaetes in ecological risk assessment. Hydrobiologia 463:149–169

Chapman PM (2002) Ecological Risk Assessment (ERA) and hormesis. Sci Total Environ 288:131–140

Chapman PM, Brinkhurst RO (1980) Salinity tolerance in some selected aquatic oligochaetes. Int Rev Ges Hydrobiol 65:499–505

Chapman PM, Brinkhurst RO (1981) Seasonal changes in interstitial salinities and seasonal movements of subtidal nethic invertebrates in the Fraser River Estuary, B.C. Estuar Coast Mar Sci 12:49–66

Chapman PM, Brinkhurst RO (1984) Lethal and sublethal tolerances of aquatic oligochaetes with reference to their use as a biotic index of pollution. Hydrobiologia 115:134–144

Chapman PM, Brinkhurst RO (1987) Hair today, gone tomorrow: induced chaetal changes in tubificid oligochaetes. Hydrobiologia 155:45–55

Chapman PM, Mitchell DG (1986) Acute tolerance tests with the oligochaetes *Nais communis* (Naididae) and *Ilyodrilus frantzi* (Tubificidae). Hydrobiologia 137:61–64

Chapman PM, Wang F (2001) Assessing sediment contamination in estuaries. Environ Toxicol Chem 20:3–22

Chapman PM, Churchland LM, Thomson PA, Michnowsky E (1980) Heavy metal studies with oligochaetes. In: Brinkhurst RO, Cook DG (eds) Aquatic oligochaete biology. Plenum Press, New York, pp 477–502

Chapman PM, Farrell MA, Brinkhurst RO (1982a) Effects of species interactions on the survival and respiration of *Limnodrilus hoffmeisteri* and *Tubifex tubifex* (Oligochaeta, Tubificidae) exposed to various pollutants and environmental factors. Water Res 16:1405–1408

Chapman PM, Farrell MA, Brinkhurst RO (1982b) Relative tolerances of selected aquatic oligochaetes to individual pollutants and environmental factors. Aquat Toxicol 2:47–67

Chapman PM, Farrell MA, Brinkhurst RO (1982c) Relative tolerances of selected aquatic oligochaetes to combination of pollutants and environmental factors. Aquat Toxicol 2:69–78

Chapman KK, Benton MJ, Brinkhurst RO, Scheuerman PR (1999) Use of the aquatic oligochaetes *Lumbriculus variegatus* and *Tubifex tubifex* for assessing the toxicity of copper and cadmium in a spiked-artificial sediment toxicity test. Environ Toxicol 14:271–278

Chua KE, Brinkhurst RO (1973) Evidence of interspecific interactions in the respiration of tubificid oligochaetes. J Fish Res Bd Can 30:617–622

Clesceri LS, Greenberg AE, Trussell RR (eds) (1989) Toxicity test procedures in annelids. In: Standard methods for the examination of water and wastewater, 17th edn. American Public Health Association, APHA, Washington DC: 8/65-8/73

Coler RA, Coler MS, Kostecki PT (1988) Tubificid behavior as stress indicator. *Water Res* 22:263–267

Cooney JD (1995) Freshwater tests. In: Rand GM (ed) Fundamentals of aquatic toxicology: effects, environmental fate and risk assessment, Chapter 2, 2nd edn. Taylor & Francis, Washington, pp 71–102

Dad NK, Qureshi SA, Pandya VK (1982) Acute toxicity of two insecticides to tubificid worms, *Tubifex tubifex* and *Limnodrilus hoffmeisteri*. Environ Int 7:361–363

Day KE, Maguire RJ, Milani D, Batchelor SP (1998) Toxicity of tributyltin to four species of freshwater benthic invertebrates using spiked sediment bioassays. Water Qual Res J Can 33:111–132

Dermott R, Munawar M (1992) A simple and sensitive assay for evaluation of sensitive toxicity using *Lumbriculus variegatus*. Hydrobiologia 235(236):407–414

Diaz RJ, Erséus C, Boesch DF (1987) Distribution and ecology of Middle Atlantic Bight Oligochaeta. Hydrobiologia 155:215–225

Didden W, Rombke J (2001) Enchytraeids as indicator organisms for chemical stress in terrestrial ecosystems. Ecotoxicol Environ Safe 50:25–43

Ding J, Drewes CD, Hsu WH (2001) Behavioral effects of ivermectin in a freshwater oligochaete, *Lumbriculus variegatus*. Environ Toxicol Chem 20:1584–1590

Drewes CD (1997) Sublethal effects of environmental toxicants on oligochaete escape reflexes. Am Zool 37:346–353

Dubilier N, Giere O, Grieshaber MK (1994) Concomitant effects of sulfide and hypoxia on the aerobic metabolism of the marine oligochaete *Tubificoides benedii*. J Exp Zool 269:287–297

Dubilier N, Giere O, Grieshaber MK (1995) Morphological and ecophysiological adaptations of the marine oligochaete *Tubificoides benedii* to sulfidic sediments. Am Zool 35:163–173

Dutta TK, Kaviraj A (1996) Effects of lime acclimation on the susceptibility of two freshwater teleosts and one oligochaete worm to metallic pollutant cadmium. Folia Biol Krakow 44:143–148

Erickson RJ, Ankley GT, DeFoe DL, Kosian PA, Makynen EA (1999) Additive toxicity of binary mixtures of phototoxic polycyclic aromatic hydrocarbons to the oligochaete *Lumbriculus variegatus*. Toxicol Appl Pharm 154:97–105

Erséus C (1982) Taxonomic revision of the marine genus *Limnodriloides* (Oligochaeta, Tubificidae). Verh Naturwiss Ver Hamburg 25:207–277

Erséus C (2005) Phylogeny of oligochaetous Clitellata. Hydrobiologia 535(536):357–372

Ewell WS, Gorsuch JW, Kringle RO, Robillard KA, Spiegle RC (1986) Simultaneous evaluation of the acute effects of chemicals on seven aquatic species. Environ Toxicol Chem 5:831–840

Fargasová A (1994) Toxicity of metals on *Daphnia magna* and *Tubifex tubifex*. Ecotoxicol Environ Safe 27:210–213

Fargasová A (1997) Determination of toxicity of plant growth regulators in selected aquatic organisms *Daphnia magna* and *Tubifex tubifex*. Biologia 52:405–408

Fargasová A (1998) Comparison of effects of Tributyl-, triphenyl-, and Tribenzyltin compounds on freshwater benthos and alga *Scenedesmus quadricauda*. Bull Environ Contam Toxicol 60:9–15

Fargasová A, Kizlink J (1996) Acute toxic effects of organotin compounds on benthic organisms: *Tubifex tubifex* and *Chironomus plumosus*. Biologia 51:677–681
Fischer E, Horvath I (1979) The effect of environmental oxygen concentration on the chloragocytes of *Tubifex tubifex* Müll. Zool Anz 203:283–288
Fischer E, Molnár L (1992) Environmental aspects of the chloragogenous tissue of earthworms. Soil Biol Biochem 24:1723–1727
Fischer E, Filip J, Molnár L (1980a) The effects of catabolic inhibitors on the oxygen-dependent nuclear volume alterations of the chloragocytes in *Tubifex tubifex* Müll. Zool Anz 205:269–271
Fischer E, Filip J, Molnár L (1980b) The effect of bivalent heavy metals on the oxygen-dependent nuclear volume alterations of the chloragocytes in *Tubifex tubifex* Müll. Environ Pollut A 23:261–265
Fisher JA, Beeton AM (1975) The effect of dissolved oxygen on the burrowing behaviour of *Limnodrilus hoffmeisteri* (Oligochaeta). Hydrobiologia 47:273–290
Forbes VE (2000) Is hormesis an evolutionary expectation? Funct Ecol 14:12–24
Fowler DJ, Goodnight CJ (1965) The effect of environmental factors on the respiration of *Tubifex*. Am Midl Nat 74:418–428
Gamenick I, Jahn A, Vopel K, Giere O (1996) Hypoxia and sulphide as structuring factors in a macrozoobenthic community on the Baltic Sea shore: colonisation studies and tolerance experiments. Mar Ecol Prog Ser 144:73–85
Generlich O, Giere O (1996) Osmoregulation in two aquatic oligochaetes from habitat with different salinity and comparison to other annelids. Hydrobiologia 334:251–261
Giere O (1980a) Tolerance and preference reactions of marine Oligochaeta in relation to their distribution. In: Brinkhurst RO, Cook DG (eds) Aquatic Oligochaeta biology. Plenum Press, New York, pp 385–405
Giere O (1980b) The impact of crude oil and oil dispersants on the marine oligochaete *Marionina subterranea*. Cah Biol Mar 21:51–60
Giere O (2006) Ecology and biology of marine Oligochaeta – an inventory rather than another review. Hydrobiologia 564:103–116
Giere O, Pfannkuche O (1982) Biology and ecology of marine oligochaeta, a review. In: Barnes M (ed) Oceanography and marine biology: an annual review, vol 20. Aberdeen Univ Press, Aberdeen, pp 173–308
Giere O, Rhode B, Dubilier N (1988) Structural peculiarities of the body wall of *Tubificoides benedii* (Oligochaeta) and possible relations to its life in sulphidic sediments. Zoomorphology 108:29–40
Giere O, Preusse JH, Dubilier N (1999) *Tubificoides benedii* (Tubificidae, Oligochaeta) - a pioneer in hypoxic and sulfidic environments. An overview of adaptive pathways. Hydrobiologia 406:235–241
Giesy J, Hoke R (1990) Freshwater sediment quality criteria: toxicity bioassessment. In: Baudo R, Giesy J, Muntau H (eds) Sediments: chemistry and toxicity of in-place pollutants. Lewis Publ, Boca Raton, pp 265–348
Gillis PL, Diener LC, Reynoldson TB, Dixon DG (2002) Cadmium-induced production of a metallothioneinlike protein in *Tubifex tubifex* (Oligochaeta) and *Chironomus riparius* (Diptera): correlation with a reproduction and growth. Environ Toxicol Chem 21:1836–1844
Gillis PL, Reynoldson TB, Dixon DG (2004a) Natural variation in a metallothionein-like protein in *Tubifex tubifex* in the absence of metal exposure. Ecotoxicol Environ Safe 58:22–28
Gillis PL, Dixon DG, Borgmann U, Reynoldson TB (2004b) Uptake and depuration of cadmium, nickel, and lead in laboratory-exposed *Tubifex tubifex* and corresponding changes in the concentration of a metallothionein-like protein. Environ Toxicol Chem 23:76–85
Green DWJ, Williams KA, Pascoe D (1985) Studies on the acute toxicity of pollutants to freshwater macroinvertebrates. 2. Phenol. Arch Hydrobiol 103:75–82
Greenberg MS, Burton GA Jr, Rowland CD (2002) Optimizing interpretation of in situ effects in riverine pollutants: impact of upwelling and downwelling. Environ Toxicol Chem 21:289–297

Guérin C, Giani N (1996) Analytical study of the locomotor and respiration movements of tubificid worms by means of video recording. Hydrobiologia 333:63–69

Guérin C, Giani N, Sire M (1994) Étude de la sensibilité *d'Enchytraeus variatus* (Oligochaeta, Enchytraeidae) à certains sels de métaux lourds en vue de son utilisation comme organisme test. Annls Limnol 30:167–178

Hansen DJ, Berry WJ, Mahony JD, Boothman WS, Di Toro DM, Robson DL, Ankley GT, Ma D, Yan Q, Pesch CE (1996) Predicting the toxicity of metal-contaminated field sediments using interstitial concentration of metals and acid-volatile sulfide normalizations. Environ Toxicol Chem 15:2080–2094

Hickey C, Martin M (1995) Relative sensitivity of five benthic invertebrate species to reference toxicants and resin-acid contaminated sediments. Environ Toxicol Chem 14:1401–1409

Hoffman DJ, Rattner BA, Burton GA, Cairns J Jr (1995) Handbook of ecotoxicology. Lewis Publ, Boca Raton, 755 pp

Högger CH, Ammon HU (1994) Testing the toxicity of pesticides to earthworms in laboratory and field tests. IOBC wprs Bull 17:157–178

Hauschildt-Lillge D (1982) Long-term effects of petroleum hydrocarbons on the life cycle and productivity of the littoral oligochaete *Lumbricillus lineatus*. Neth J Sea Res 16:502–510

IJC, International Join Commission (1988) Procedures for the assessment of contaminated sediment problems in the Great Lakes. A report by the Sediment Subcommittee, Windsor, Ontario, 140 pp

Ingersoll CG (1995) Sediment tests. In: Rand GM (ed) Fundamentals of aquatic toxicology: effects, environmental fate and risk assessment, Chapter 8, 2nd edn. Taylor & Francis, Washington, pp 231–255

Inoue T, Kondo K (1962) Susceptibility of *Branchiura sowerbyi, Limnodrilus socialis* and *L. willeyi* for several agricultural chemicals. Bout Kagaku 27:97–99

Ireland MP (1983) Heavy metal uptake and tissue distribution in earthworms. In: Satchell JE (ed) Earthworm ecology from darwin to vermiculture, Chapter 21. Chapman & Hall, London, pp 247–265

Jones JRE (1939) Antagonism between two heavy metals in their toxic action on freshwater animals. Proc Zool Soc Lond A 108:481–499

Kabir SMH, Khatoon N (1980) Toxicity of some common insecticides to *Limnodrilus* spp. (Oligochaeta: Tubificidae). Bangladesh J Zool 8:61–67

Karickhoff SW, Morris SW (1985) Impact of tubificid oligochaetes on pollutant transport in bottom sediments. Environ Sci Technol 19:51–56

Kaster JL, Klump JV, Meyer J, Krezoski J, Smith ME (1984) Comparison of defecation rates of *Limnodrilus hoffmeisteri* Claparède (Tubificidae) using two different methods. Hydrobiologia 111:181–184

Keilty TJ, Landrum PF (1990) Population specific toxicity responses by the freshwater oligochaete *Stylodrilus heringianus* in natural Lake Michigan sediments. Environ Toxicol Chem 9:1147–1154

Keilty TJ, White DS, Landrum PF (1988a) Short-term lethality and sediment avoidance assays with endrin-contaminated sediment and two oligochaetes from Lake Michigan. Arch Environ Contam Toxicol 17:95–102

Keilty TJ, White DS, Landrum PF (1988b) Sublethal responses to endrin in sediment by *Limnodrilus hoffmeisteri* (Tubificidae), and in mixed-culture with *Stylodrilus heringianus* (Lumbriculidae). Aquat Toxicol 13:227–250

Keilty TJ, White DS, Landrum PF (1988c) Sublethal responses to endrin in sediment by *Stylodrilus heringianus* (Lumbriculidae) as measured by a ^{137}Cesium marker layer technique. Aquat Toxicol 13:251–270

Khangarot BS (1991) Toxicity of metals to a freshwater tubificid worm, *Tubifex tubifex* (Müller). Bull Environ Contam Toxicol 46:906–912

Khangarot BS, Rathore RS (2004) Protective action of 24 amino acids on the toxicity of copper to a freshwater tubificid worm *Tubifex tubifex* Müller. Water Air Soil Pollut 157:53–63

Klerks PL, Bartholomew PR (1991) Cadmium accumulation and detoxification in a Cd-resistant population of the oligochaete *Limnodrilus hoffmeisteri*. Aquat Toxicol 19:97–112

Klerks PL, Levinton JS (1989) Rapid evolution of metal resistance in a benthic oligochaete inhabiting a metal polluted site. Biol Bull 176:135–141

Klerks PL, Levinton JS (1993) Evolution of resistance in community composition in metal-polluted environments: a case study on Foundry Cove. In: Dallinger R, Rainbow PS (eds) Ecotoxicology of metals in invertebrates, SETAC Special Publication Series. Lewis Publ, Boca Raton, pp 223–241

Krezoski JR, Robbins JA, White DS (1984) Dual radiotracer measurement of zoobenthos mediated solute and particle transport in freshwater sediments. J Geophysic Res 89:7937–7947

Kukkonen J, Landrum PF (1994) Toxicokinetics and toxicity of sediment-associated pyrene to *Lumbriculus variegatus* (Oligochaeta). Environ Toxicol Chem 13:1457–1468

Lasserre P (1971) Données écophysiologiques sur la repartition des oligochètes marins meiobenthiques. Incidence des parametres salinité, temperature, sur le metabolisme respiratoire de deux espèces euryhalines du genre *Marionina* Michelsen, 1889 (Enchytraeidae). Vie et Milieu (Suppl) 22:523–540

Laughlin RB Jr, Ng J, Guard NE (1981) Hormesis: a response to low environmental concentrations of petroleum hydrocarbons. Science 211:705–707

Learner MA, Edwards RW (1963) The toxicity of some substances to *Nais* (Oligochaeta). Proc Soc Wat Treat Exam 12:161–168

Learner MA, Lochhead G, Hughes BD (1978) A review of the biology of British Naididae (Oligochaeta) with emphasis on the lotic environment. Freshw Biol 8:357–377

Levinton JS, Suatoni E, Wallace W, Junkins R, Kelaher B, Allen BJ (2003) Rapid loss of genetically based resistance to metals after the cleanup of a Superfund site. PNAS 100:9889–9891

Lotufo GR, Fleeger JW (1996) Toxicity of sediment-associated pyrene and phenanthrene to *Limnodrilus hoffmeisteri* (Oligochaeta: Tubificidae). Environ Toxicol Chem 15:1508–1516

Lucan-Bouché ML, Arsac F, Biagianti-Risbourg S, Habets F, Vernet G (1997) Étude expérimentale des effects létaux induits par le cuivre et/ou le plomb chez *Tubifex tubifex*. Bull Soc Zool Fr 122:389–392

Lucan-Bouché ML, Biagianti-Risbourg S, Arsac F, Vernet G (1999a) Autotomy as a mechanism of decontamination used by the oligochaete *Tubifex tubifex*. Bull Soc Zool Fr 124:383–387 (in French)

Lucan-Bouché ML, Biagianti-Risbourg S, Arsac F, Vernet G (1999b) An original decontamination process developed by the aquatic oligochaete *Tubifex tubifex* exposed to copper and lead. Aquat Toxicol 45:9–17

Luoma SM, Carter JL (1993) Understanding the toxicity of contaminants in sediments: beyond the bioassay-based paradigm. Environ Toxicol Chem 12:793–796

Maciorowski HD, Clarke RMcV (1980) Advantages and disadvantages of using invertebrates in toxicity testing. In: Buikema AL Jr, Cairns J Jr (eds) Aquatic invertebrate bioassays. ASTM Special Technical Publication 715, Philadelphia, pp 36–47

Maestre Z, Martinez-Madrid M, Rodriguez P (2009) Monitoring the sensitivity of the oligochaete *Tubifex tubifex* in laboratory cultures using three toxicants. Ecotoxicol Environ Safe 72:2083–2089

Marchese MR, Brinkhurst RO (1996) A comparison of two tubificid oligochaete species as candidates for sublethal bioassay tests relevant to subtropical and tropical regions. Hydrobiologia 334:163–168

Martinez DE, Levinton JS (1996) Adaptation to heavy metals in the aquatic oligochaete *Limnodrilus hoffmeisteri*: evidence for control by one gene. Evolution 50:1339–1343

Martinez-Madrid M, Rodriguez P, Perez-Iglesias JI, Navarro E (1999) Sediment toxicity bioassays for assessment of polluted sites in the Nervion River (northern Spain). 2. *Tubifex tubifex* (Oligochaeta) reproduction sediment bioassay. Ecotoxicology 8:111–124

Martínez-Tabché L, Mora BR, Olivan LG, Faz CG, Ortega MD (1999) Toxic effect of nickel on hemoglobin concentration of *Limnodrilus hoffmeisteri* in spiked sediments of trout farms. Ecotoxicol Environ Safe 42:143–149

Martínez-Tabché L, Ortega MD, Mora BR, Faz CG, Lopez EL, Martinez MG (2001) Hemoglobin concentration and acetylcholinesterase activity of oligochaetes in relation to lead concentration in spiked sediments from Ignacio Ramirez reservoir. Ecotoxicol Environ Safe 49:76–83
Matisoff G, Wang XS, McCall PL (1999) Biological redistribution of lake sediments by tubificid oligochaetes: *Branchiura sowerbyi* and *Limnodrilus hoffmeisteri/Tubifex tubifex*. J Great Lakes Res 25:205–219
McCrary JE, Heagler MG (1997) The use of a simultaneous multiple species acute toxicity test to compare the relative sensitivities of aquatic organisms to mercury. Environ Sci Health A Environ Sci Eng Tox Hazard Subst Control 32:73–81
McMurtry MJ (1984) Avoidance of sublethal doses of copper and zinc by tubificid oligochaetes. J Great Lakes Res 10:267–272
Mekenyan OG, Ankley GT, Veith GD, Call DJ (1994) QSARs for photoinduced toxicity: I. acute lethality of polycyclic aromatic hydrocarbons to *Daphnia magna*. Chemosphere 28:567–582
Meller M, Egeler P, Rombke J, Schallnas H, Nagel R, Streit B (1998) Short term toxicity of lindane, hexachlorobenzene, and copper sulfate to tubificid sludgeworms (Oligochaeta) in artificial media. Ecotoxicol Environ Safe 39:10–20
Milani D, Reynoldson TB, Borgmann U, Kolasa J (2003) The relative sensitivity of four benthic invertebrates to metals in spiked-sediments exposures and application to contaminated field sediment. Environ Toxicol Chem 22:845–854
Milbrink G (1983) Characteristic deformities in tubificid oligochaetes inhabiting polluted bays of Lake Vänein, southern Sweden. Hydrobiologia 106:169–185
Milbrink G (1987) Biological characterization of sediments by standardized tubificid bioassays. Hydrobiologia 155:267–275
Mischke CC, Terhune JS, Wise DJ (2001) Acute toxicity of several chemicals to the oligochaete *Dero digitata*. J World Aquac Soc 32:184–188
Monson PD, Ankley GT, Kosian PA (1995) Phototoxic response of *Lumbriculus variegatus* to sediments contaminated by polycyclic aromatic hydrocarbons. Environ Toxicol Chem 14:891–894
Mosleh YY, Paris-Palacios S, Couderchet M, Biagianti-Risbourg S, Vernet G (2005) Effects of the herbicide isoproturon on metallothioneins, growth, and antioxidative defenses in the aquatic worm *Tubifex tubifex* (Oligochaeta, Tubificidae). Ecotoxicology 14:559–571
Mosleh YY, Paris-Palacios S, Biagianti-Risbourg S (2006) Metallothioneins induction and antioxidative response in aquatic worms *Tubifex tubifex* (Oligochaeta, Tubificidae) exposed to copper. Chemosphere 64:121–128
Mudroch A, Azcue J, Mudroch P (1999) Manual of bioassessment of aquatic sediments quality. Lewis Publisher, Boca Raton, 236 pp
Naqvi SMZ (1973) Toxicity of twenty-three insecticides to a tubificid worm *Branchiura sowerbyi* from the Mississippi Delta. J Econ Entom 66:70–74
OECD (2007) Sediment-water lumbriculus toxicity test using spiked sediment. OECD Guideline for the Testing of Chemicals No 225
Pasteris A, Vecchi M, Reynoldson TB, Bonomi G (2003) Toxicity of copper-spiked sediments to *Tubifex tubifex* (Oligochaeta, Tubificidae): a comparison of the 28-day reproductive bioassay with a 6-month cohort experiment. Aquat Toxicol 65:253–265
Penttinen OP, Kukkonen JVK, Pellinen J (1996) Preliminary study to compare body residues and sublethal energetic responses in benthic invertebrates exposed to sediment-bound 2,4,5-trichlorophenol. Environ Toxicol Chem 15:160–166
Persoone G, Janssen CR (1993) Freshwater invertebrate toxicity tests. In: Calow P (ed) Handbook of ecotoxicology, Chapter 4, vol 1. Blackwell, London, pp 51–65
Peterson GS, Ankley GT, Leopard EN (1996) Effect of bioturbation on metal-sulfide oxidation in surficial freshwater sediments. Environ Toxicol Chem 15:2147–2155
Phipps GL, Ankley GT, Benoit DA, Mattson VR (1993) Use of the aquatic oligochaete *Lumbriculus variegatus* for assessing the toxicity and bioaccumulation of sediment-associated contaminants. Environ Toxicol Chem 12:269–279
Phipps GL, Mattson VR, Ankley GT (1995) Relative sensitivity of three freshwater benthic macroinvertebrates to ten contaminants. Arch Environ Contam Toxicol 28:281–286

Prater BL, Anderson MA (1977) A 96-hour sediment bioassay of Duluth and Superior Harbor Basins (Minnesota) using *Hexagenia limbata, Asellus communis, Daphnia magna* and *Pimephales promelas* as test organisms. Bull Environ Contam Toxicol 18:159–169

Purschke G, Hagens M, Westheide W (1991) Ultrahistopathology of enchytraeid oligochaetes (Annelids) after exposure to pesticides- a means of identification of sublethal effects? Comp Biochem Physiol 100C:119–122

Rand GM (1995) Fundamentals of aquatic toxicology: effects, environmental fate and risk assessment, 2nd edn. Taylor & Francis, Washington, 1125 pp

Rand GM, Wells PG, Mccarty LS (1995) Introduction to aquatic toxicology. In: Rand GM (ed) Fundamentals of aquatic toxicology: effects, environmental fate and risk assessment, 2nd edn. Taylor & Francis, Washington, pp 3–67

Rathore RS, Khangarot BS (2002) Effect of temperature on the sensitivity of sludge worm *Tubifex tubifex* (Müller) to selected heavy metals. Ecotoxicol Environ Safe 53:27–36

Rathore RS, Khangarot BS (2003) Effects of water hardness and metal concentration on a freshwater *Tubifex tubifex* Müller. Water Air Soil Pollut 142:341–356

Redeker E, Blust R (2004) Accumulation and toxicity of cadmium in the aquatic oligochaete *Tubifex tubifex*: a kinetic modelling approach. Environ Sci Technol 38:537–543

Redeker ES, van Campenhout K, Bervoets L, Reijnders H, Blust R (2007) Subcellular distribution of Cd in the aquatic oligochaete *Tubifex tubifex*, implications for trophic availability and toxicity. Environ Pollut 148:166–175

Rehwoldt R, Wirhowsky E, Lakso L, Shaw S (1973) The acute toxicity of some heavy metal ions toward benthic organisms. Bull Environ Contam Toxicol 10:291–294

Reible DD, Popov V, Valsaraj KT, Thibodeaux LJ, Lin F, Dikshit M, Tudaro MA, Fleeger JW (1996) Contaminant fluxes from sediment due to tubificid oligochaete bioturbation. Wat Res 30:704–714

Reish DJ, Oshida PS (1986) *Manual of Methods in Aquatic Environmental Research*. Part 10. Short-term static bioassays. FAO Fish. Tech. Paper 247, 62 pp

Reynoldson TB (1994) A field test of a sediment bioassay with the oligochaete worm *Tubifex tubifex* (Müller, 1774). Hydrobiologia 278:223–230

Reynoldson TB, Day KE (1993) Freshwater sediments. In: Callow P (ed) Handbook of ecotoxicology, Chapter 6, vol 1. Blackwell, London, pp 83–100

Reynoldson TB, Day KE (1998) Biological guidelines for the assessment of sediment quality in the Laurentian Great Lakes. National Water Research Institute (Canada), Report, pp 498–232

Reynoldson TB, Thompson SP, Bamsey JL (1991) A sediment bioassay using the tubificid oligochaete worm *Tubifex tubifex*. Environ Toxicol Chem 10:1061–1072

Reynoldson TB, Rosa F, Mudroch A, Day KE (1995) Investigation of the Spanish River area of the North Channel of Lake Huron. II. Benthic invertebrate community structure and sediment toxicity, with reference to biological sediment guidelines. NWRI Report No 95–14

Reynoldson TB, Rodriguez P, Martinez-Madrid M (1996) A comparison of reproduction, growth and acute toxicity in two populations of *Tubifex tubifex* (Müller, 1774) from the North America Great Lakes and northern Spain. Hydrobiologia 334:199–206

Rodriguez P, Reynoldson TB (1999) Laboratory methods and criteria for sediment bioassessment. In: Mudroch A, Azcue J, Mudroch P (eds) Manual of bioassessment of aquatic sediments quality, Chapter 3. Lewis Publisher, Boca Raton, pp 83–133

Rodriguez P, Arrate J, Martinez-Madrid M, Reynoldson TB, Schumacher V, Viguri J (2006) Toxicity of Santander Bay sediments to the euryhaline freshwater oligochaete *Limnodrilus hoffmeisteri*. Hydrobiologia 564:157–169

Rodriguez P, Maestre Z, Martinez-Madrid M, Reynoldson TB (2011) Evaluating the type II error rate in a sediment toxicity classification using the reference condition approach. Aquat Toxicol 101:207–213

Rogge RW, Drewes CD (1993) Assessing sublethal neurotoxicity effects in the freshwater oligochaete, *Lumbriculus variegatus*. Aquat Toxicol 26:73–90

Römbke J, Knacker TH (1989) Aquatic toxicity test for enchytraeids. Hydrobiologia 180:235–242

Schubauer-Berigan MK, Ankley GT (1991) The contribution of ammonia, metals and nonpolar organic compounds to the toxicity of sediment interstitial water from an Illinois River tributary. Environ Toxicol Chem 10:925–939

Scott-Fordsmand JJ, Weeks JM (2000) Biomarkers in earthworms. Rev Environ Contam Toxicol 165:117–159

Sibley PK, Benoit DA, Balcer MD, Phipps GL, West CW, Hoke RA, Ankley GT (1999) *In situ* bioassay chamber for assessment of sediment toxicity and bioaccumulation using benthic invertebrates. Environ Toxicol Chem 18:2325–2336

Slooff W (1983) Benthic macroinvertebrates and water quality assessment: some toxicological considerations. Aquat Toxicol 4:73–82

Smith DP, Kennedy JH, Dickson KL (1991) An evaluation of a naidid oligochaete as a toxicity test organism. Environ Toxicol Chem 10:1459–1465

Stebbing ARD (2000) Hormesis: interpreting the β-curve using control theory. J Appl Toxicol 20:93–101

Sturmbauer C, Opadiya GB, Niederstrátter H, Riedmann A, Dallinger R (1999) Mitochondrial DNA reveals cryptic oligochaete species differing in cadmium resistance. Mol Biol Evol 16:967–974

Suedel BC, Rodgers JHJ (1996) Toxicity of fluoranthene to *Daphnia magna*, *Hyalella azteca*, *Chironomus tentans* and *Stylaria lacustris* in water-only and whole sediment exposures. Bull Environ Contam Toxicol 57:132–138

Thompson KA, Brown DA, Chapman PM, Brinkhurst RO (1982) Histopathological effects and cadmium-binding protein synthesis in the marine oligochaete *Monopylephorus cuticulatus* following cadmium exposure. Trans Am Microsc Soc 101:10–26

Traunspurger W, Drews C (1996) Toxicity analysis of freshwater and marine sediments with meio and macrobenthic organisms: a review. Hydrobiologia 328:215–261

USEPA (1988) An overview of sediment quality in the United States. EPA 905/9-88-002

van Wijngaarden RPA, Crum SJH, Decraene K, Hattink J, van Kammen A (1998) Toxicity of derosal (active ingredient carbendazim) to aquatic invertebrates. Chemosphere 37:673–683

Vecchi M, Reynoldson TB, Pasteris A, Bonomi G (1999) Toxicity of copper-spiked sediments to *Tubifex tubifex* (Oligochaeta, Tubificidae): a comparison of the 28-day reproductive bioassay with an early life stage bioassay. Environ Toxicol Chem 18:1173–1179

Veltz I, Arsac F, Biagianti-Risbourg S, Habets F, Lechenault H, Vernet G (1996) Effects of platinum (Pt^{4+}) on *Lumbriculus variegatus* Müller (Annelida, Oligochaeta): acute toxicity and bioaccumulation. Arch Environ Contam Toxicol 31:63–67

Veltz-Balatre I (2000) Toxicity studies of some pollutants from the Champagne-Ardenne area on a freshwater invertebrate: *Lumbriculus variegatus* (Annelida, Oligochaeta). Bull Soc Zool Fr 125:333–341

Verdonschot PFM (1981) Some notes on the ecology of aquatic oligochaetes in the Delta Region of the Netherlands. Arch Hydrobiol 92:53–70

Verdonschot PFM, Ter Braak CJF (1994) An experimental manipulation of oligochaete communities in mesocosms treated with chlorphyrifos or nutrient additions: multivariate analyses with MonteCarlo permutation tests. Hydrobiologia 278:251–266

Verdonschot PFM, Smies M, Spers ABJ (1982) Distribution of aquatic oligochaetes in brackish inland waters in the SW Netherlands. Hydrobiologia 89:29–38

Vidal DE, Horne AJ (2003a) Mercury toxicity in the aquatic oligochaete *Sparganophilus pearsei*: I. Variation of resistance among populations. Arch Environ Contam Toxicol 45:184–189

Vidal DE, Horne AJ (2003b) Inheritance of mercury tolerance in the aquatic oligochaete *Tubifex tubifex*. Environ Toxicol Chem 22:2130–2135

Wallace WG, Lopez GR (1996) Relationship between subcellular cadmium distribution in prey and cadmium trophic transfer to a predator. Estuaries 19:923–930

Wallace WG, Lopez GR (1997) Bioavailability of biologically sequestered cadmium and the implications of metal detoxification. Mar Ecol Prog Ser 147:149–157

Wang XS, Matisoff G (1997) Solute transport in sediments by a large freshwater oligochaete, *Branchiura sowerbyi*. Environ Sci Technol 31:1926–1933

Ward GS (1995) Saltwater tests. In: Rand GM (ed) Fundamentals of aquatic toxicology: effects, environmental fate and risk assessment, Chapter 3, 2nd edn. Taylor & Francis, Washington, pp 103–133

Wentsel R, McIntosh A, Atcinson G (1977) Sediment contamination and benthic macroinvertebrate distribution in a metal-impacted lake. Environ Pollut 14:187–193

West CW, Ankley GT (1998) A laboratory assay to assess avoidance of contaminated sediments by the freshwater oligochaete *Lumbriculus variegatus*. Arch Environ Contam Toxicol 35:20–24

West CW, Mattso VR, Leonard EN, Phipps GL, Ankley GT (1993) Comparison of the relative sensitivity of 3 benthic invertebrates to copper – contaminated sediments from the Keweenaw Waterway. Hydrobiologia 262:57–63

Westeheide W, Bethke-Beilfuss D (1991) The sublethal enchytraeid test system: guidelines and some results. In: Esser G, Overdieck D (eds) Modern ecology: basic and applied aspects, Chapter 23. Elsevier, Amsterdam, pp 497–508

Westheide W, Bethke-Beilfuß D, Gebbe J (1991) Effects of Benomyl on reproduction and population structure of enchytraeid oligochaetes (Annelida) – Sublethal test on agar and soil. Comp Biochem Physiol 100C:221–224

White DS, Keilty TJ (1988) Burrowing avoidance assays of contaminated Detroit River sediments, using the freshwater oligochaete *Stylodrilus heringianus* (Lumbriculidae). Arch Environ Contam Toxicol 17:673–681

Whitley LS (1963) Studies on the biology of Tubificidae. PhD Dissertation, Purdue University, Lafayette

Whitley LS (1968) The resistance of tubificid worms to three common pollutants. Hydrobiologia 32:193–205

Whitley LS, Sikora RA (1970) The effect of three common pollutants on the respiration rate of tubificid worms. J Water Pollut Control Fed 42:R57–R66

Whitten BK, Goodnight CJ (1966) Toxicity of some common insecticides to tubificids. J Water Pollut Control Fed 38:227–235

Widdows J (1993) Marine and estuarine systems. In: Callow P (ed) Handbook of ecotoxicology, vol 1. Blackwell, London, pp 145–166

Wiederholm T, Dave G (1989) Toxicity of metal polluted sediments to *Daphnia magna* and *Tubifex tubifex*. Hydrobiologia 176(177):411–417

Wiederholm T, Wiederholm AM, Milbrink G (1987) Bulk sediment bioassays with five species of freshwater oligochaetes. Water Air Soil Pollut 36:131–154

Willmer P, Stone G, Johnston I (2005) Environmental physiology of animals, vol 2. Blackwell, Oxford, 754 pp

Zipperle A, Reise K (2005) Freshwater springs on intertidal sand flats cause a switch in dominance among polychaete worms. J Sea Res 54:143–150

Chapter 5
Bioaccumulation and Trophic Transfer

Abstract Sediments are not simply a passive sink for contaminants, on the contrary, benthic animals can both act as accumulators with subsequent transfer of materials to their predators, and through their physiological processes can transport contaminants either in solution or adsorbed to sediment particles. Worms can accumulate metals and have metabolic routes that are able to eliminate them via excretion or store them in subcellular fractions that are not toxic to the organism. Body burdens of essential metals are physiologically regulated, and show other uptake-elimination models than non-essential metals. Bioavailability of metals through adsorption on organic particles or due to the formation of sulphides, pH, or dissolved oxygen levels has been studied and may explain the frequent lack of correlation between body burden and sediment or pore water toxicant concentration. Surface adsorption of some metals into the mucous layer covering the body and contaminants in food that passes through the gut without being absorbed are confounding factors in bioaccumulation assessment that must be considered to the extent possible. Worms exposed to pesticides, PCBs, HCB, or metals, proved to be toxic to crayfish, leech and fish feeding on them, thus worms play a role in the transfer of toxicants through the food chain. Worms are also important bioturbators in the aquatic systems through their burrowing, feeding and respiratory activities, making toxicants more available to animals at higher trophic levels, even if they are not prey for those species. These results are important for environmental risk assessment when the potential for a chemical to bioaccumulate or biomagnify through the food chain is suspected. There is a considerable amount of work required to develop a full understanding of the toxicokinetics, uptake routes, assimilation efficiencies, detoxification processes, elimination rates, and organs responsible for bioaccumulation and depuration in aquatic oligochaetes.

P. Rodriguez and T.B. Reynoldson, *The Pollution Biology of Aquatic Oligochaetes*,
DOI 10.1007/978-94-007-1718-3_5,

5.1 Introduction

Sediments are not simply a passive sink for contaminants, on the contrary, benthic animals can act both as accumulators, with subsequent transfer of materials to their predators, and through their physiological processes (*e.g.* defecation) can transport contaminants either in solution or adsorbed to sediment particles. Endobenthic organisms and aquatic oligochaetes in particular, influence the bioavailability of chemicals by irrigating the sediment through active burrowing, respiratory activity, as well as their ingestion and defecation processes. These processes can result in the transfer of sediment-bound substances towards the mud/water interface (McMurtry et al. 1983; Ciutat et al. 2005a). Such behaviour and physiological functions lead aquatic organisms to accumulate substances in their tissues both from water and their diet. This process known as bioaccumulation is of potential concern in hazard identification of chemicals both because of the possibility of chronic toxicity to the organisms accumulating substances and to the predators eating those organisms (Schlekat et al. 2007). Thus, bioaccumulation is not an effect in itself, although it may be the mechanism through which toxic effects can be induced.

The measurement of bioaccumulation in benthic organisms can be expressed in an absolute way, *i.e.* as the concentration of a substance in whole animal tissue (body concentration or body burden) or as a ratio of the body burden to the concentration in the pore water, in the sediment or in the diet. The term bioconcentration is specifically used to refer to the net amount of water-borne chemicals accumulated by the organism through non-dietary routes, while bioaccumulation describes the net amount of a substance taken up by the organism from water, sediment and through diet (Chapman et al. 1996). The body concentration is usually measured at several time intervals until steady state, that is, when incorporation and elimination of the toxicant reach equilibrium. A bioaccumulation process occurs not only when the bioaccumulation factor (BAF) is >1, but also when the body concentration of a substance increases over time while the sediment concentration (whole sediment or pore water) remains the same, although bioaccumulation risk is usually regarded as negligible if the BAF is lower than 1 (Beek et al. 2000).

The body concentration can be expressed based on total wet weight, total dry weight and when required (*i.e.* for organic compounds) on the basis of the lipid content of the organism. Similarly, the content of the chemical in the sediment can be normalized for sediment total organic carbon (TOC), and used to calculate the Biota-Sediment Accumulation Factor (BSAF) as the ratio of the body concentration ($\mu g\ g^{-1}$) at steady state normalised for lipid content and the concentration in the sediment normalised for TOC ($\mu g\ g^{-1}$ dw). This expression was proposed by Ankley et al. (1992) to facilitate interspecies comparisons and to standardise the expression of bioaccumulation data. Uptake and elimination constants have been measured for several substances in both aqueous and sediment bioaccumulation bioassays, and are useful descriptors in the study of toxicokinetics. Additionally, BAFs can be calculated as the ratio of the uptake rate constant (k_s) and the elimination rate constant (k_e), assuming first-order kinetics (OECD 2008). This method is particularly useful when a steady-state is not reached. The time required for metal body concentration to reach to steady state can be exposure-concentration dependent. For cadmium Gillis et al. (2004) found that worms (*Tubifex tubifex*) exposed to 0.14 μmol Cd g^{-1} dw did not reach equilibrium after a 6-week exposure period, while those worms exposed to 3.66 μmol Cd g^{-1} dw did not significantly increase tissue concentration after 4 weeks of exposure.

Most bioaccumulation data were initially generated in field studies and more recently in standardised laboratory tests, using either field or formulated sediments spiked with chemicals. Several aquatic oligochaete species have been considered to be suitable for bioaccumulation studies and are included in standard guidelines. They are easily handled and cultured, provide reasonable biomass for residue analysis, are tolerant of varying sediment physical and chemical characteristics, and are exposed to contaminants via pore water and ingested sediment (criteria from USEPA 2000), although they usually require feeding in long-term exposure tests. In a review of the toxicity and bioaccumulation of sediment contaminants, Ingersoll (1995) identified two oligochaete species as potential test-organisms, *Lumbriculus variegatus* and *T. tubifex* based on eight selection criteria (Table 5.1). Among them, *L. variegatus* is probably the species which has been most frequently used in bioaccumulation research and several guidelines have included bioassay methods for this species (USEPA 2000; ASTM 2005). Egeler et al. (1997, 1999) have developed a bioassay for measuring bioaccumulation and toxicokinetics of organic compounds using *T. tubifex*. Based on these and several other contributions, including an international ring-test, a highly standardised guideline has recently been adopted by OECD (2008) for *T. tubifex* and *L. variegatus*. The OECD (2008) test guideline for bioaccumulation studies using benthic oligochaetes includes detailed definitions of all the variables involved in the calculation of bioaccumulation factors (*e.g.* uptake and elimination rates, steady state). This guideline has been implemented for the classification of new substances (neutral organic chemicals, metallo-organic compounds), using radiolabelled chemicals with artificial sediment, but it is not applicable to metals without technical modifications. Appendix 2 (Chap. 7) provides a summary of published bioaccumulation factors, toxicokinetic constants and other data related to bioaccumulation by aquatic oligochaetes for a variety of substances.

The purpose of this chapter is to review and summarize the studies available to the authors on bioconcentration and bioaccumulation using aquatic oligochaetes, and to provide an insight into the use of these benthic, detritivorous organisms as test species. It is important to note that many publications may have been overlooked due, in part, to the fact that bioaccumulation measurements often appear in contexts unrelated to the fields of ecotoxicology or pollution biology. We have focussed on three primary objectives: to describe the bioaccumulation process of contaminants (uptake and elimination rates and pathways), to relate body burden with concentration of the substance in the bulk sediment or in the pore water, and to assess the risk associated with transfer of contaminants from sediment through aquatic oligochaetes to higher trophic levels. Additionally, there is a short discussion

Table 5.1 Criteria used for the selection of aquatic oligochaete species as test organisms in bioaccumulation studies

Criteria	Species	Source
– Sensitivity to contaminant behaviour in the sediment	*Lumbriculus variegatus*	Ingersoll et al. (1995)
– Feeding sediment – Ecological relevance – Geographic distribution – Taxonomic relation to indigenous animals – Standardised methods for toxicity assessment – Availability – Tolerance to natural geochemical sediment characteristics	*Tubifex tubifex*	
– Availability of organisms through the year – Known chemical exposure history – Adequate tissue biomass for chemical analysis – Tolerance to a wide range of sediment physico-chemical characteristics – Low sensitivity to chemicals associated with sediments – Amenability to long-term exposures without adding food – Ability to accurately reflect concentrations of chemicals in field-exposed organisms	*Lumbriculus variegatus*	USEPA (2000)
Refers to criteria in USEPA (2000) (see above)	*Lumbriculus variegatus* *Tubifex tubifex*	ASTM (2005)
– Endobenthic and sediment ingesters	*Lumbriculus variegatus*	OECD (2008)
– May represent the most abundant species in adverse environmental conditions	*Tubifex tubifex*	
– Influence bioavailability through bioturbation and as prey of other animals – Exposure to chemicals by pore water, overlying water and ingested particles	*Branchiura sowerbyi*	

of the relationships between body concentration and lethal or sublethal effects in the organism, an area that needs further exploration for a better understanding of the environmental risk of chemical bioaccumulation.

5.2 Uptake, Storage and Elimination

Bioaccumulation and elimination of surplus materials are normal physiological processes that control the maintenance of essential nutritional chemicals in the body at optimal concentrations, and mechanisms for the reduction or removal of toxic compounds. The requirements of different species for essential elements vary substantially, but optimal concentration ranges in organisms are often narrow and frequently under homeostatic control. Beyond this range, excess material must be actively excreted, stored in cells and tissues, or metabolically immobilised, otherwise toxic effects may occur (Chapman et al. 1996). Except for a small group of publications, bioaccumulation measurements usually ignore the fact that chemicals can be accumulated in separate pools, either as chemically inert forms (such as metals in inorganic granules or bound to metallothionein-like proteins), or as chemicals with the direct potential to bind at sites that provoke a toxic effect (Schlekat et al. 2007).

5.2.1 Uptake

The body wall is an important absorption route for water-borne contaminants, thus for the bioconcentration of chemicals in aquatic organisms. In oligochaete worms, ventilation by the tail, blood flow and cuticle permeability can be important factors which regulate the absorption of chemicals from the water column. Most metal compounds formed in aquatic solutions are hydrophilic and their uptake requires specific transport systems, which is in contrast to non-polar organics which are taken up across the cell membranes of body epithelia (*e.g.* digestive tract) by passive diffusion (Schlekat et al. 2007). It may also be possible that active tegumentary uptake of some organic substances occurs as some studies have demonstrated that freshwater oligochaetes use the integument for the uptake of dissolved organic matter (glycine in *Quistadrilus multisetosus*: Brinkhurst and Chua 1969), long-chain fatty acids (in *Tubifex tubifex*: Testerman 1972), or short-chain carboxylic acids (in *T. tubifex*, *Limnodrilus hoffmeisteri* and *Ilyodrilus templetoni*: Hoffmann and Wulf 1993).

Fischer and Horváth (1977) had already suggested that the worm cuticle and epidermis may function as traps for heavy metals. In a later paper, Fleming and Richards (1982) described Zn uptake from dilution water by *T. tubifex* to not only be a result of absorption but also to surface binding of the element. These authors estimated that this process accounted for 86–90% of the total Zn accumulated in the first 5 min of exposure, with a decreasing proportion of the Zn associated with the body surface to approximately only 4% of the total accumulated after 11–17 days (Fig. 5.1). Electron microscopy showed that in worms preincubated in 10 mg Zn l^{-1} for 5 min, deposits predominantly occurred in the supracuticular mucoid layer and to a lesser extent at the base of the cuticle, close to the junction with the epidermis. In conditions of trace element scarcity, the mucous layer enabled rapid adsorption for the first few minutes, followed by saturation of the mucous layer. The role of the surface mucous was also interpreted as being a buffer against unlimited entry of heavy metals when levels are temporarily high, and in short-term exposures (less than 6 h) this factor may be relevant for accumulation assessment. In 1990, Back investigated the epidermal uptake of the metals Zn, Cd and Pb from the water by *Limnodrilus udekemianus* through electron microscopy, using the sulphide-silver method. He showed that uptake of these metals took place mainly via the integument, especially in the caudal part of the body, and to a lesser extent by the hind part of the

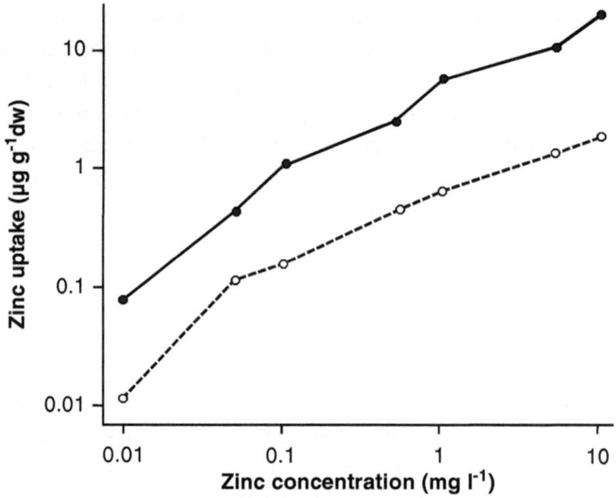

Fig. 5.1 Total body uptake (*solid line*) and adsorption-free uptake (*dashed line*) of zinc (in µg g^{-1} dw) during 5-min incubations over the concentration range of 0.01–10.0 mg Zn l^{-1}. Axes plotted logarithmically (From Fleming and Richards 1982, Fig. 5, redrawn and reproduced by permission of Elsevier Ltd, © conveyed by Copyright Clearance Centre, Inc.)

intestine (but animals were starving). The metals were stored in the double-membrane lysosomal structures of epidermal cells, as shown by electron microscopy, X-ray microanalysis and laser-induced mass analysis. This study also demonstrated the dependence of metal uptake on temperature, and the regulation of uptake by tubificids for the essential metal Zn.

In detritivorous organisms, an important fraction of chemicals is incorporated from contaminated sediments via sediment ingestion. The diet is a major source of metals essential for the organism's metabolism, and therefore regulation of uptake, storage and elimination processes are necessarily associated with digestive function. The intestine is also the site of uptake of small organic molecules, and metals can bind to them and enter the internal milieu using these molecules as transport (Schlekat et al. 2007). Therefore, both tegumentary and digestive routes are commonly used in the uptake of chemical compounds by aquatic oligochaetes, and several studies evaluating the importance of each of these are described below.

Most research in this field has been done using the oligochaete *Lumbriculus variegatus*, a species that reproduces asexually by fragmentation (architomy) and fragments cannot ingest sediment until the mouth or anus has been regenerated. This fact has been used by Leppänen and Kukkonen (1998) and Conrad et al. (2000) to compare the significance of the uptake route via pore-water-only compared with uptake via pore water and ingested sediment, using *L. variegatus* with and without a mouth. After architomy takes place, a new mouth develops in about 1 week or more if it divides again. Leppänen and Kukkonen (1998) exposed three groups of worms, the first did not fragment and ingested pyrene-polluted sediment for the duration of the test; the second group started feeding on day 6 and a third group on

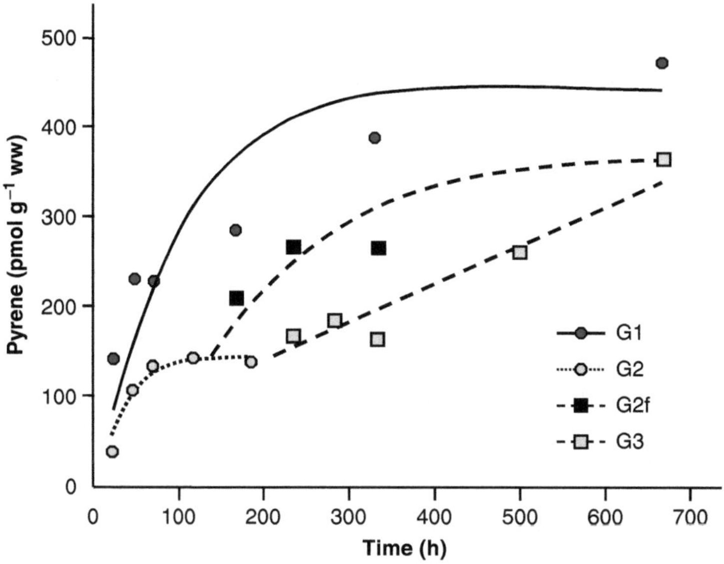

Fig. 5.2 Accumulation of pyrene in complete (*G1* sediment ingesting) and prostomium removed (*G2* non ingesting) *Lumbriculus variegatus* in a 28-days exposure. (*G2f*: prostomium regenerated and feeding recommenced after day 6, *G3*: prostomium regenerated and feeding recommenced after day 9) (From Leppänen and Kukkonen 1998, Fig. 2, redrawn and reproduced by permission of ACS, © conveyed by Copyright Clearance Centre, Inc.)

day 9 (Fig. 5.2). The study demonstrated that the main route for pyrene accumulation in *L. variegatus* was via ingestion although the amount of pyrene accumulated via the integument was estimated to be as high as 39% of the total, after 8 days. The results reported by Conrad et al. (2000) showed that the relevance of uptake routes depend on the chemical compound, with about 20% of the body burden due to ingestion for pyrene, while this route was negligible for dichlorophenol. Loonen et al. (1997) also investigated the relative importance of pore water and sediment ingestion as uptake pathways for two polychlorinated dibenzo-p-dioxins. They exposed *L. variegatus* to radio-labelled dioxins via sediment or via sand and overlying water (assuming that uptake in this case was only possible through water, as sand was not ingested). The Biota-Sediment Accumulation Factor (BSAF) after 28-day exposure was 1.6 ± 0.27 for TCDD ([^{3}H]2,3,7,8-tetrachorodibenzo-p-dioxin) and 0.07 ± 0.02 for OCDD ([^{14}C]octachlorodibenzo-p-dioxin). Comparing the predicted accumulation from pore water concentration (through the Equilibrium Partitioning model: Di Toro et al. 1991) with the observed accumulation in sediment-exposed oligochaetes, Loonen et al. showed that it was approximately 1.4 times greater than expected, due to the additional uptake from particulate sediment.

Feeding behaviour, which includes selection, ingestion, digestion and assimilation, in benthic organisms and its role in bioaccumulation process has been reviewed by Leppänen (1995). Selection of food particles has been demonstrated in some aquatic oligochaete species (*e.g.* Wagner 1968; Brinkhurst and Austin 1979) and this can affect bioaccumulation. It is important to know if particle selection operates over a specific size fraction of the sediment so that correlation of chemical compounds is estimated for this fraction only. In this regard, feeding selection by *T. tubifex* was investigated by Klump et al. (1987). They examined particle size selection in a mixed assemblage of 90% *L. hoffmeisteri*, with 10% *T. tubifex* and *Q. multisetosus*, using a dual tracer technique $^{51}Cr^{3+}$- labelled sediment and ^{14}C-labelled contaminant. The authors reported selection of the finest sediment particles, those of 5–20 μm diameter being the major size fraction (39.6%) in faeces. Faecal particle size fractions of 2–5 μm and 0.2–2 μm increased (32.4% and 23.6%, respectively) compared to the sediment. The use of dual tracer techniques has allowed very efficient measurement of uptake rates of organic substances by tubificid worms. Assimilation efficiencies for [^{14}C]HCBP from sediment labelled with $^{51}Cr^{3+}$ (a conservative tracer for ingestion since it is not assimilated by worms) decreased from 36% to 15% over the initial 10 days of active feeding, and were inversely related to the average defecation rate. From average tracer retention rates calculated over 356 h, assimilation of sediment-associated HCBP was determined to be relatively rapid (4.7 pmol HCBP mg^{-1} worm h^{-1}). Using a different technique, Rodriguez et al. (2001) measured the particle size distribution in the faecal pellets with a Coulter® Multisizer. In all cases, faeces from organisms fed on different sediments in the laboratory were composed of particles with mean diameter <63 μm, i.e. lime and clay particles, and 75% of the total faecal volume was formed of particles with a mean diameter <25 μm, with the volume of the clay fraction (<4 μm) either similar or up to 2 times higher in the worm faeces than in the associated sediment (Table 5.2). The authors hypothesised the existence of two levels of selectivity in this feeding behaviour of *T. tubifex*, primarily based on particle size, avoiding the ingestion of sand particles, and a secondary selection based on organic matter and the bacterial communities associated with the fine sediment particles. A more recent contribution by Ciutat et al. (2006) using X-ray and laser diffraction grain-size analysis revealed similar results, reporting that 78% of the ingested articles by tubificids were <63 μm.

The feeding selectivity index (SI) calculated for *L. variegatus* by Kukkonen and Landrum (1995a), using organic carbon as a tracer, indicated that the selective feeding by the worm is based on size (fine) of the particles, not upon their organic-content. In the same study, these authors also measured assimilation efficiencies (AE) in *L. variegatus* for sediment-sorbed benzo(a)pyrene over a 5-day exposure, using two different methods. The first method compared relative concentration in sediment to that in the faecal material on a carbon-normalised basis, and the second was based on a dual-tracer method used by Klump et al. (1987). The AE ranged 0–26% with the first method, while the second method measured AEs of 23–26% during the 1st day and then decreased to 11–13%, after 1 to 5-day exposure.

Small size particles adsorb the highest proportion of contaminants, therefore, worms might reduce bioaccumulation of toxic substances through selection of particles of higher diameter that are less nutritious but with less contaminant load. Millward et al. (2001) measured pyrene bioaccumulation in *L. hoffmeisteri* as it related to particle selection, showing that BSAF decreased with increasing exposure at pyrene con-

Table 5.2 Particle size characteristics (diameter, μm) of the <63 μm sediment fraction and worm faecal pellets produced after a 4-week period, measured with Coulter® Multisizer (From Rodriguez et al. 2001, Table 3, with permission of Springer Publ., © conveyed by Copyright Clearance Centre, Inc.)

Sediment	Particle size characteristics	Experiment	Sediment (n=3)	Worm faeces (n=3)
Natural	Modal range of particle size (μm)	1	35.9–36.7	5.0–7.6
		2	6.2–7.0	8.7–9.7
		3	2.0–4.0	<2.0
		4	2.0–4.0	2.0–4.0
Natural	Clay (<4 μm)% volume (mean)	1	11.5	22.4
		2	19.1	20.9
		3	27.3	42.5
		4	22.5	22.6
Natural	Particle size (μm) comprising 75% total volume (mean)	1	27.5	14.2
		2	11.8	13.9
		3	15.0	10.3
		4	17.6	21.1
Artificial[a]	Modal range of particle size (μm)	1	12.0–12.1	6.0–6.2
		2	8.0–8.7	6.2–7.0
Artificial[a]	Clay (<4 μm)% volume (mean)	1	14.8	23.1
		2	18.6	21.9
Artificial[a]	Particle size (μm) comprising 75% total volume (mean)	1	18.3	13.5
		2	14.1	14.0

[a]Sediment formulated following OECD Guideline 207

centrations >199 nmol g^{-1} dw sediment, corresponding to a shift in the selection to larger diameter particles. Thus, at the end of the exposure period (10 days), worms exposed at highest concentrations (199 and 1,196 nmol g^{-1} dw) ingested significantly fewer small-size particles (<3.5 μm) and more medium size particles (4.6–33 μm for 199 nmol g^{-1} dw, and 3.5–43 μm for 1,196 nmol g^{-1} dw), in contrast to a preference for the smaller size particles (0.7–2.1 μm) in the control sediment. Selection of the larger sediment particles and a reduction of feeding rates are mechanisms hypothesised by these authors to minimise exposure to hydrophobic contaminants.

The release of metals under conditions (pH, redox potential and enzymes present) associated with the digestive processes, together with gut passage time and the metabolism of intestinal bacterial can affect the bioavailability of diet-borne chemicals. Metal uptake and its relationship to digestive physiological processes in aquatic organisms have been reviewed by Campbell et al. (2005). In the review, much of the data is for benthic marine polychaetes, but it is likely that the general conclusions apply to the analysis of metal uptake by aquatic oligochaete worms.

5.2.2 Storage

Once chemicals have entered the body, either by passive diffusion or active transport, they can be stored or metabolised and excreted. The main storage organ in oligochaetes is the chloragogenous tissue. This consists of modified peritoneal cells (the chloragocytes) arranged in a thick layer around the gut and the dorsal blood vessel. Worm chloragosomes, which are cytoplasmic organelles of chloragocytes consisting of concentric layers of electrodense lipids around a lipid core, may function as a cation-exchange system capable of capturing and storing heavy metals, thus reducing their toxic effects (Ireland 1983). In earthworms, chloragosomes contain mainly lipids (61% of the dry weight) and a varying amount of protein (up to 16% of the dry weight), with only about 20% of inorganic material (reviewed by Ireland 1983). Chloragosomes are presumed to play various functions including storage of trophic material, accumulation and immobilization of heavy metals, and detoxification of xenobiotics (Fischer 1976; Fischer and Molnár 1992). Coelomic eleocytes (free chloragocytes) are able to enter various tissues and organs of the body and are particularly active in nutritive and excretory functions, the maintenance of osmotic balance and the binding of xenobiotics, and they can also invade the vascular system and release granules and chloragosomes for later elimination (Vetvicka et al. 1994). It is, however, noteworthy that most studies on chloragocytes are based on lumbricid taxa, therefore, the results might not be valid for aquatic families.

One of the first contributions on the role of chloragogenous tissue in freshwater organisms was by Say and Giani (1981) who studied the accumulation of zinc in different tissues of aquatic worms collected from river sediments contaminated with metals (River Mort, France). The authors examined 20-μm histological sections using an X-ray diffraction microsonde and electron microscope, and concluded that Zn was primarily located in the alimentary tract and in the chloragocytes. In this part of the body, the quantity of Zn represented between 0.5% and 1% of the animal's dry weight and could be up to 4% in some locations. Back and Prosi (1985) described metal granules that had accumulated in chloragocytes and the intestinal epithelium of the tubificine *L. udekemianus* exposed to both water and sediment spiked with heavy metals. These authors found most of the essential metals, including Zn, but not Cd, concentrated in the chloragosomes. Lead was transported to chloragocytes, stored and detoxified in chloragosomes.

While there is relatively little information on the role of chloragosomes in storage and detoxification for the aquatic oligochaetes, there are background data for lumbricine earthworms on metal granules accumulated in various tissues, which is likely relevant to research on microdriles. The formation of metal-rich granules is described in many invertebrate phyla. Brown (1982) reviewed the major granule types, their formation and composition, and highlighted the specificity of granule elimination mechanisms and also the lack of knowledge on granule interactions within cells and tissues. In oligochaetes, both epidermal and intestinal cells have been described as harbouring an unusually rich variety of lysosomes and these cells store precipitates, often as large aggregates of particles. Klerks and Bartholomew (1991) described two distinct types of granules (about 1-μm diameter) in *L. hoffmeisteri* exposed to cadmium. Backscattered electron images showed high-density granules in most body tissues, but especially in chloragocytes, the body wall and the gut, while low-density granules were only found in the chloragocytes. In the same paper, the analysis of various body tissues avoiding the high density granules showed that only the chloragogenous tissue contained significant concentrations of Cd (10–80 μmol g^{-1}). Occasionally, the granules formed aggregations, and their analysis by scanning electron microscopy-electroprobe microanalysis (SEM-EPMA) revealed major amounts of Cd and S, in a 1:1 ratio, probably in the form of cadmium sulphide. Higher accumulation of Cd in the particulate fraction (granules) was demonstrated in *L. hoffmeisteri* from a metal-polluted site compared with worms from a control site, and proposed granule formation as a sequestering mechanism for Cd that contributes to detoxification and increases worm resistance. The presence of granules is important in bioaccumulation studies of metals since this storage mechanism often results in little or no toxicity to the organism or reduced bioavailability to its predator (Schlekat et al. 2007) (see Sect. 5.5).

Storage of organic hydrophobic substances occurs mainly in lipid-rich tissues. In oligochaetes, the chloragogenous layer of the gut is the main tissue for storing lipids and it has an important role in binding organic xenobiotics (Jamieson 1981). The accumulation of PAHs was reported by Giere and Pfannkuche (1982) for the littoral enchytraeid *Lumbricillus lineatus*. Animals maintained in oiled *Fucus* fed on the algae and incorporated oil through the gut into the chloragogenous tissue, chloragocytes became blackened, and apparently transformed the oil through metabolic processes.

5.2.3 Elimination

Granule formation in the chloragogenous tissue and the gut wall and their deposition into the alimentary canal was proposed as a feasible elimination mechanism for *L. hoffmeisteri* exposed to cadmium (Klerks and Bartholomew 1991). The chloragocytes in oligochaetes are efficient in detoxification processes since they first accumulate and immobilize toxic metals and then eliminate them by stimulation of chloragosome exocytosis or even extrusion of whole chloragocytes (Cancio et al. 1995). In earthworms exposed to pesticides chloragocyte-eleocyte transformation is stimulated and chloragogenous tissue is depleted (Fischer 1976), processes that may be related to toxicant elimination through the gut. The release of chloragosomes into extracellular milieu increased in *T. tubifex* when exposed to carbofuran and dyes (azin, thiazine and xanthene) where they were further phagocytised by amoebocytes (Fischer and Horváth 1976). Autometallography, TUNEL-test and histological studies have been used to measure loss of chloragogenous and intestinal tissues due to cell death (apoptosis) in lumbricid worms exposed to metals (Amaral and Rodrigues 2005), and these methods could be applied to aquatic species.

There is evidence suggesting that some type of elimination can occur through the worm integument, via epidermal vesicles (Giere et al. 1988; Gustavsson 2001). In the marine species *Tubificoides benedii* mucus secretion occurs through a dense papillated epidermis. This mucous layer accumulates a dark debris and encloses a rich variety of bacteria, and Giere et al. (1988) suggested that the periodic removal

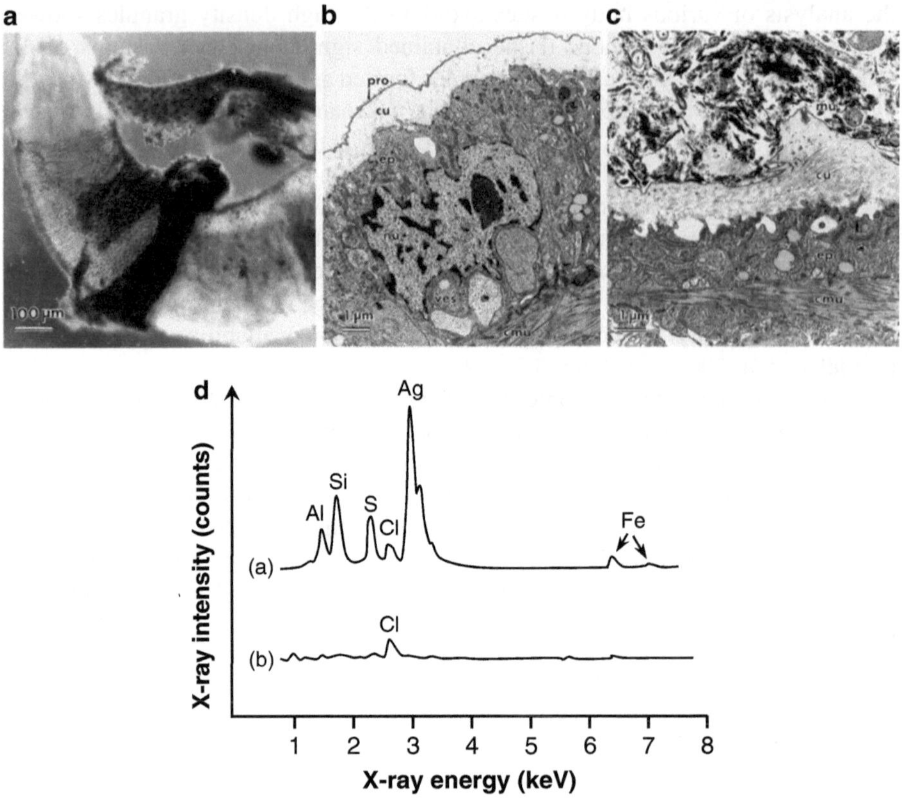

Fig. 5.3 *Tubificoides benedii:* (**a**) live individual shedding its dark papillate epicuticle (anterior end); (**b**) cross section through the body wall of a freshly moulted body region; (**c**) *ibid.* from a non-moulted region; (**d**) two micro-X-ray spectrograms (simplified from original) of specimens from sulphidic sediment after incubation in $AgNO_3$, showing (*a*) elemental contents of precipitates in mucus cap above papillae, and (*b*) elemental content of cuticular matrix underlying this papilla (*cmu* circular musculature, *cu* cuticle, *ep* epidermis, *mu* cover of mucus, *nu*; nucleus, *pro* peg-like projection, *ves* mucus vesicles) (From Giere et al. 1988, Fig. 18, 23–25, redrawn (d) and reproduced by permission of Springer Publ., © conveyed by Copyright Clearance Centre, Inc.)

of the black mucous layer containing xenobiotics by "moulting" of the integumentary epicuticle could represent a drastic response to highly sulphidic biotopes (Fig. 5.3). Through this process, anthropogenic toxicants accumulated in the integument may also be eliminated.

Oligochaetes do not accumulate metals homogeneously in the body rather metals predominantly accumulate within the posterior segments (Morgan and Morgan 1990). Thus, the elimination of toxicants by autotomy (fragmentation) of the posterior region of the body might be a "cheap" alternative for worms. Lucan-Bouché et al. (1999a) described *T. tubifex* exposed to Cd losing the posterior segments, from only 17% of the individuals when exposed for 96 h to 0.01 mg Cd l^{-1} to 100% when

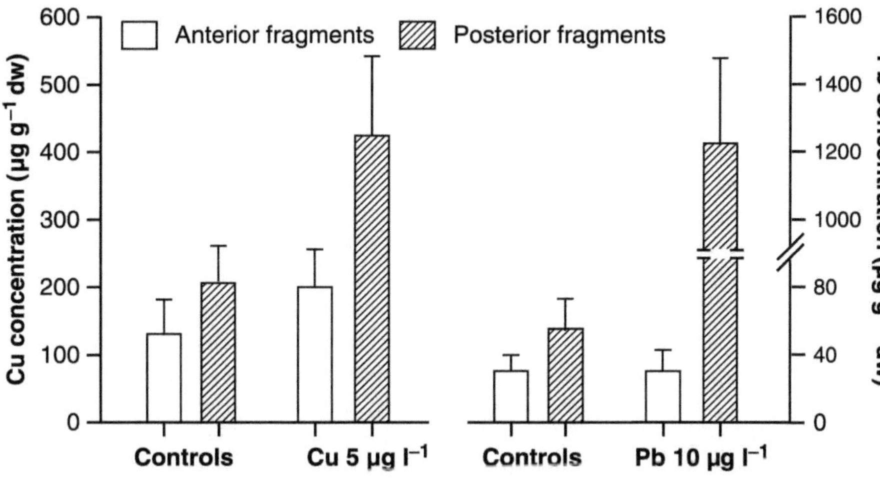

Fig. 5.4 Copper and lead body concentrations in the anterior and posterior regions of *Tubifex tubifex* in 4-days exposure of 5 µg Cu l^{-1} and 10 µg Pb l^{-1} (From Lucan-Bouché et al. 1999b, Figs. 11, 12, redrawn and reproduced by permission of Elsevier Ltd, conveyed by Copyright Clearance Centre, Inc.)

exposed to 0.05 mg Cd l^{-1}. The concentration of 0.10 mg Cd l^{-1} which was lethal to all the worms in a 72-h exposure, caused autotomy on 13±2% worms in 24 h, and on 100% in 48 h. However, analysis of the anterior and posterior regions in worms exposed to Cd levels lower than those that induce autotomy did not reveal any Cd-concentration gradient. Bouché et al. (2000) also found autotomy of the posterior region (either a beaded appearance or loss of segments) in 50% of *T. tubifex* exposed to 0.015 mg Cd l^{-1} for 96 h, and the number of worms undergoing this process increased proportionally with the duration of the exposure and the metal concentration. A similar experiment was conducted with Cu and Pb (Lucan-Bouché et al. 1999b), accumulation was measured in the posterior third of the body as well as autotomy, and loss of this caudal region was interpreted as a detoxification mechanism. These authors reported 79% autotomised worms when exposed for 96 h to 0.01 mg Cu l^{-1}, and 100% at 0.02 mg Cu l^{-1}, 31% when exposed to 0.05 mg Pb l^{-1}, and 60% at 0.2 mg Pb l^{-1}. The rear portion of the worms exposed to 0.005 mg Cu l^{-1} accumulated almost double that of the anterior segments, and when exposed to 0.01 mg Pb l^{-1} accumulation was up to 40 times higher in the posterior segments (Fig. 5.4). In the field, worms from a basin polluted with vineyard wastes (up to 400 mg Cu kg^{-1} and 40 mg Pb kg^{-1} in sediment) showed a high proportion of worms with missing or regenerating caudal segments, and the posterior region was also shown to accumulate twice as much Cu than the anterior region, however, this gradient was not observed for Pb. These studies suggested that the ability to accumulate metals in the posterior segments might represent a potential risk for predator fishes that graze on worm tails protruding from the sediment. Another interesting observation by these authors is that *T. tubifex* worms that loose their caudal region by autotomy spend 7 days regenerating a functional anus (Bouché et al. 2003).

Thus, detoxification through the loss of the posterior region of the body appears to involve two processes: first, the elimination of accumulated metals in this region, and second, a probable reduction of the toxicant uptake during the regeneration process as the worms presumably stop feeding.

Biotransformation of parent substances also affects measurement of elimination rates and should be considered when assessing the bioaccumulation hazard in polluted sediments. Oxidation of aromatic xenobiotics may occur in the chloragocytes, as has been suggested for leeches (Fischer 1993). The apparent elimination of some chemicals (*e.g.* PAHs) due to their metabolic transformation or degradation may be of concern since their metabolites can be dangerous for the organisms and may also be transferred to fishes through predation. Some chemicals, such as trans-chlordane, benzo(a)pyrene and pyrene, were not metabolised by *L. variegatus* after 4–7 days exposure (Harkey et al. 1994). However, using different solvent extractions and measuring the relative proportions of parent substances in this species, Leppänen and Kukkonen (2000) suggested slow biotransformation processes of pyrene and benzo(a)pyrene in the worms which decreased steadily during the 504-h exposure period. However, oligochaetes have an apparently limited ability to metabolise and excrete most PCB homologue groups, as suggested by Ankley et al. (1992) from BSAFs values close to 1 for *L. variegatus* (laboratory experiments, 30 days duration) and *Limnodrilus* sp. (field worms). Radiolabelled test substances can facilitate the analysis of water, sediment and biological samples, and may be used to determine whether identification and quantification of degradation products should be undertaken. The method described by OECD (2008) for aquatic oligochaetes (*L. variegatus* and *T. tubifex*) recommends the use of ^{14}C-labelled compounds. The concentration of total radioactivity can be used to calculate the bioaccumulation factor (BAF) based on the concentration of the parent compound including any labelled degradation product. The total radioactive residues in samples taken at steady state or at the end of the uptake phase should be analysed for the % radioactivity associated with the parent compound, and allows the BAF of the parent compound to be calculated.

5.3 Body Burden and Sediment Concentration

The ratio of body burden of a chemical to its concentration in the environment is the basis for the calculation of bioaccumulation factors, and in particular the BSAF (Biota-Sediment Accumulation Factor, Ankley et al. 1992). The use of benthic detritivorous organisms such as aquatic oligochaetes is preferred when the sediment, either whole sediment or a particular compartment (*e.g.* pore water or the organic carbon fraction), represents the main pathway for bioaccumulation. Research addressing the study of body burdens in oligochaete worms compared to sediment concentration is summarised below for metals and organics. Additional information on tissue concentration and bioaccumulation factors of several chemical compounds can be found in Appendix 2 (Chap. 7).

5.3.1 *Metals*

The correlation between body burden and the chemical sediment concentration depends on the existence of physiological regulation of uptake and elimination processes with consideration of whether the metals are essential or non essential. Likewise, the various factors mediating metal bioavailability in sediments are extremely complex and relate to variables such as physical structure of the sediment, mineralogy, metal speciation, and the redox regime. The assessment of the potential hazard associated with accumulated metals requires to distinguish between nutritional accumulation (*i.e.* essential metals), benign accumulation (sequestration in granules and binding proteins), and accumulation that causes adverse chronic effects (Schlekat et al. 2007).

Free-metal ions are the most bioavailable chemical species, but there are different metal-binding elements in the sediments that have a high affinity for metals and reduce the bioavailability of metals to organisms, *e.g.* particulate organic carbon (POC) and acid volatile sulphides (AVS) (Chapman et al. 1998). Thus, normalisation of metal concentrations for these factors is a common approach taken when comparing body burden in organisms exposed to metal contaminated sediments. Gunn et al. (1989) studied the bioavailability of heavy metals in sediment to freshwater tubificid worms and compared metal accumulation to measures of chemical extractability using a sequential extraction procedure. Results indicated little correlation between metal body burden and total metal concentration in sediment, although good correlations were reported with the exchangeable fraction (extracted with ammonium acetate/calcium chloride, at pH 6) for Cd, Cu and Pb. Nickel uptake was very low, and Zn levels in the worms were found to be constant, suggesting regulation. In general, these authors found that metals spiked to the sediment or adsorbed to the clay were more bioavailable than those bound to carbonate, hydrous ferric oxide or sewage sludge phases. Likewise, Cd uptake rate constants showed an exponential reduction in *L. variegatus* with increasing water hardness and -log ($[Cd^{2+}]/[Ca^{2+}]$) (Xie et al. 2008).

The role of sediment AVS in metal bioavailability was investigated by Carlson et al. (1991) who spiked three natural sediments with different levels of AVS (3.6, 8.8 and 42 μmol g^{-1}), at five $CdCl_2$ concentrations (0.3–16 mg Cd g^{-1} dw). After 10-day exposure, whole-body Cd concentration in *L. variegatus* was analysed (a measurement compromised by the inclusion of the gut content). Results indicated that mortality occurred when the Cd sediment concentration to AVS ratios were >1. At ratios ≤1, no mortality was observed and maximum whole-body Cd tissue concentrations ranged between 240 and 690 μg g^{-1} dw, the higher value being approximately equal to the maximum body concentration obtained in survivors of the Cd exposure in water-only tests (mean maximum body burden due to bioconcentration was 670 μg g^{-1}). The difficulty of measuring mortality in *L. variegatus* is related to the asexual reproduction of the species, where mortality is only assumed to have occurred when the number of final fragments in the sediment test was less than the initial number, even when the number of fragments depends on both mortality and

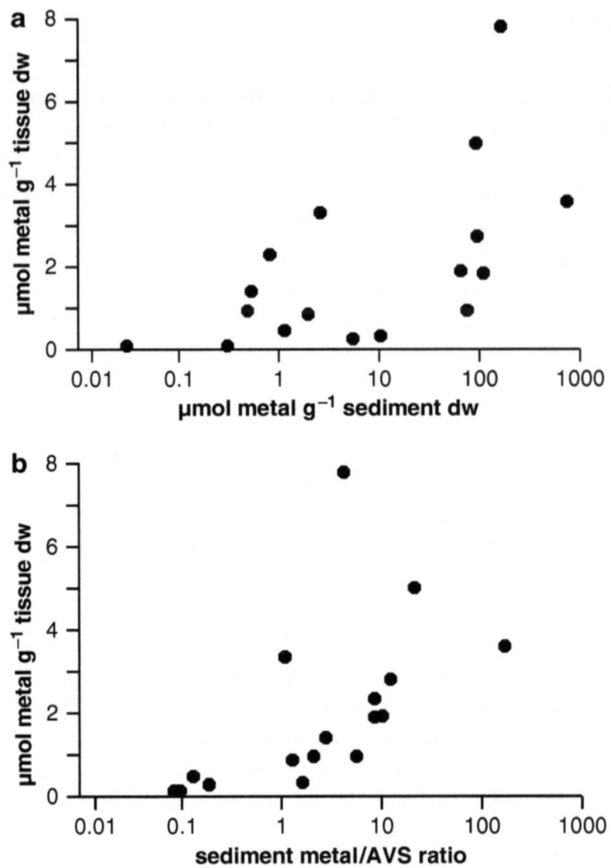

Fig. 5.5 Bioaccumulation of cadmium plus nickel by *Lumbriculus variegatus* relative to (**a**) total metal concentration in sediments, and (**b**) sediment metal (Cd and Ni, simultaneously extracted with AVS) concentration normalised to acid-volatile sulphide (AVS) (From Ankley et al. 1991, Fig. 5, redrawn and reproduced with permission of John Wiley & Sons Ltd., © conveyed by Copyright Clearance Centre, Inc.)

the frequency of autotomy. In a concurrent study, Ankley et al. (1991) also evaluated the role of AVS in the bioaccumulation of Cd and Ni by *L. variegatus*, in a short-term (10 days) bioassay, showing that the worms (after 24-h gut purging) exhibited bioaccumulation only at metal to AVS molar concentration ratios greater than one (Fig. 5.5). This model has been shown to improve the prediction of bioaccumulation in long-term tests of 30-day exposure with the oligochaete *L. variegatus* (Ankley et al. 1994). The authors also confirmed that certain cationic metals (Cd, Ni, Pb, Zn, Cu) in sediments are not bioavailable when AVS molar concentrations are sufficient to bind these metals (SEM/AVS <1, where SEM is the molar sum of divalent metals). However, at concentrations where autotomy occurs in *L. variegatus*, the inability to feed could also explain a lower than expected chemical uptake, mainly in short-term exposures until the regeneration of mouth and anus has been completed.

The sources of metal contamination for earthworms were reviewed by Ireland (1983) and Morgan et al. (1993) and much of the information and suggestions derived from these studies may be applied or tested in aquatic oligochaetes. There are also a few studies on metal bioaccumulation in enchytraeids which are useful, as some species occur in both terrestrial and aquatic habitats. Rüther and Greven (1990) exposed *Enchytraeus buchholzi* to several metals in agar, for 12 days, and determined bioconcentration factors of 262.4 (Cd), 11.4 (Cu), 9.7 (Pb) and 20.0 (Zn). Willuhn et al. (1994a) studied *E. buchholzi* exposed for 13 days to several concentrations of $CdCl_2$ in a 1% agar substrate and demonstrated a rapid, almost linear accumulation of the metal by the worms at a concentration of 1 mg Cd l^{-1}, although they reported no acute effects below 4 mg Cd l^{-1}. The exposure to 3 mg Cd l^{-1} in an aquatic test system induced the synthesis of a new mRNA encoding a non-metallothionein 33-kDa protein, suggesting a possible indicator for Cd intoxication in enchytraeids. Willuhn et al. (1994b) have also investigated the gene expression of *E. buchholzi* at sublethal Cd concentrations (3 mg Cd l^{-1}) in an aquatic medium for 6 days, at 20°C, showing that the CRP gene is a candidate for monitoring bioavailable Cd at sublethal levels.

Another set of studies have examined the body burden of worms sampled in the field and compared this with metal concentrations in sediment or pore water. One of the first studies that investigated the relationship of body burden and concentration in field sediments was made by Say and Giani (1981). They found a good correlation for Zn in both water column and sediment concentration, for the species *Limnodrilus hoffmeisteri, L. udekemianus, Tubifex tubifex* and *Lumbricillus rivalis*. Cocoons collected from two sites were also analysed for metal content and results indicated that a certain amount of accumulation also took place in cocoons. This seems to have potential significance, although we are unaware of any work that has investigated the consequences of bioaccumulation in cocoons on developmental alterations or the success in hatching of young worms.

Sager and Pucsko (1991) studied sediments from a reservoir in the River Danube (Austria) and examined bioaccumulation of As, Cu, Cd, Pb and Zn in relation to particle size, organic content, pH and other characteristics of the sediment, using a multivariate approach. Increase of the clay mineral content appeared to be associated with increasing Cd, Zn and Cu levels in oligochaete tissues. Arsenic body burden correlated positively with total As concentration in sediment. Copper availability seemed to be governed by the presence of Fe/Mn oxides, while Pb in tissues was closely related to Pb pore water concentration. From factor analysis of trace element concentrations in worms with physical and chemical data of sediment and pore water, the authors concluded that the influence of pore water was small.

Measurements of metal bioaccumulation in worms from laboratory bioassays and in worms sampled from the field usually give different results. The fact that the species used in the laboratory and those sampled in the field are not generally the same, the differences in bioavailability due to metal speciation resulting from manipulation of sediment for laboratory bioassays, and the different periods of

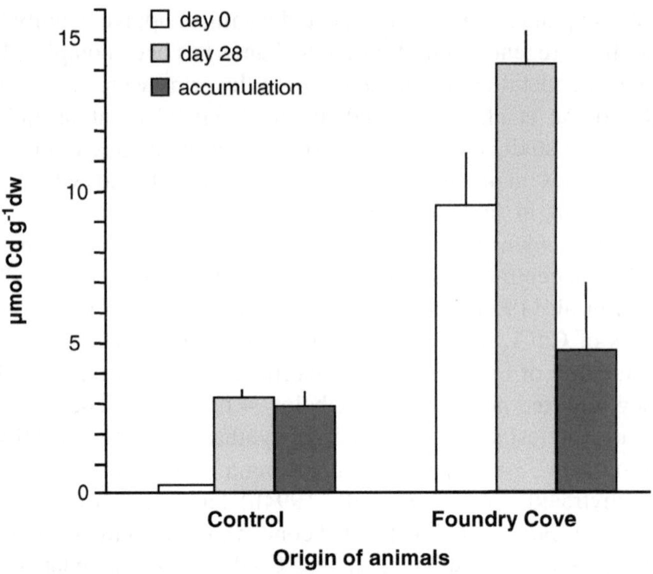

Fig. 5.6 Cadmium accumulation in *Limnodrilus hoffmeisteri* after 28-days exposure to control sediment and Foundry Cove Cd-rich sediment. Accumulation is calculated as the difference between mean of day 0 and day 28 (From Klerks and Bartholomew 1991, Fig. 2, redrawn and reproduced by permission of Elsevier Ltd, conveyed by Copyright Clearance Centre, Inc.)

exposure, all contribute to affecting the results based on the experimental approach selected. Metal body burden measured in fathead-minnows from laboratory bioassays and in aquatic oligochaetes sampled in the field from contaminated sediments of Hamilton Harbour (Lake Ontario, Canada) was compared by Krantzberg (1994). In most cases, metal body burden in oligochaetes was higher than in fish, except for Zn, Ni and Hg, but at some sites the results were contradictory. The author attributed the differences in body burden obtained by the two approaches to interspecific differences and disparity in exposure conditions and bioavailability. However, one cannot discount the fact that resident field populations may be adapted (by physiological acclimation or natural selection) to high metal concentrations and may bioaccumulate higher levels of metals (Klerks and Bartholomew 1991; Vidal and Horne 2003) (Fig. 5.6). A later study (Bervoets et al. 1997) compared the body burden in different sediment-dwellers such as tubificids and chironomids from field sediments. These authors compared Cu, Zn, Cd and Pb body burden in organisms sampled in natural sediments polluted by metals at four different river basins in Belgium. They found that in most cases metal concentrations in organisms were positively related to the reducible metal fraction (Mn oxides and Fe oxides), as well as negatively related to the TOC and Fe sediment content. However, the low determination coefficients from non-linear regression models relating body burden in tubificids and chironomid larvae ($r^2=0.16$ for Zn, 0.28 for Cd, 0.52 for Cu) indicate that the understanding of availability cannot be restricted to chemical characteristics of the sediment alone, but the biology of the species must also be considered.

5.3.2 Organic Chemicals

The Equilibrium Partitioning theory (EqP) was developed in the 1990s as a means of predicting toxicity of hydrophobic substances to sediment-dwelling organisms (Di Toro et al. 1991). The approach utilises the fact that the partition coefficient between an organism and water (K_b) is directly related to the octanol:water partition coefficient (K_{ow}), provided that the chemical is not reactive nor metabolized by the organism (Spacie et al. 1995). The logarithm of K_{ow} of a substance can be used as an indication of the potential for bioaccumulation by aquatic organisms, and a large number of correlations of both partition coefficients have been found for hydrophobic substances (log $K_{ow} > 1$). This implies that sediment bioaccumulation of organic compounds in benthic organisms can be estimated on the basis of the chemical concentration in pore water and a bioconcentration factor. In tubificids, this approach is suggested to operate for lipid-normalized BCF data and organic compounds up to log K_{ow} values of 7.5 (Kraaij et al. 2003). It is important to note, however, that benthic organisms are exposed to chemicals via many uptake routes, including direct contact and ingestion of contaminated sediment particles, not only via pore water (OECD 2008), and the expected body burdens estimated from the log K_{ow} are only satisfied by direct empirical measurement for some substances, but not for others (*e.g.* compounds with high partition coefficients, log $K_{ow}>6$, such as chlorinated dioxins, in Loonen et al. 1997).

Most commonly, the bioaccumulation factors of organic chemicals are reported as a normalised ratio of the chemical concentrations in worms and sediments, that is, the biota-sediment accumulation factor (BSAF, g carbon g^{-1} lipid). There is a growing tendency to use BSAF for benthic organisms, although it is not always possible due to the extended periods required to reach steady state for some substances. Standley (1997) compared a simple bioaccumulation factor (BAF) with the normalised bioaccumulation factor (BSAF) for *L. variegatus*, showing that the latter reduced the range of measured dieldrin accumulation factors from 38-fold to 8-fold.

One of the earliest studies on bioaccumulation of organic substances in aquatic oligochaetes was published by Oliver (1984) who investigated the uptake of chlorinated chemicals by mixed populations of oligochaete worms, mostly *T. tubifex* and *L. hoffmeisteri*, exposed to field polluted sediments from Lake Ontario. The worms accumulated many of the 24 organic chemicals and for most of them the bioaccumulation factors (BAF, corrected for control levels) increased over the 110-day study period. In a later study, large populations of worms, the equivalent of 7,000 ind. m^{-2}, were exposed to contaminated or spiked sediments at both 8°C and 20°C, for up to 79 days (Oliver 1987). The worms were then transferred to tanks with clean

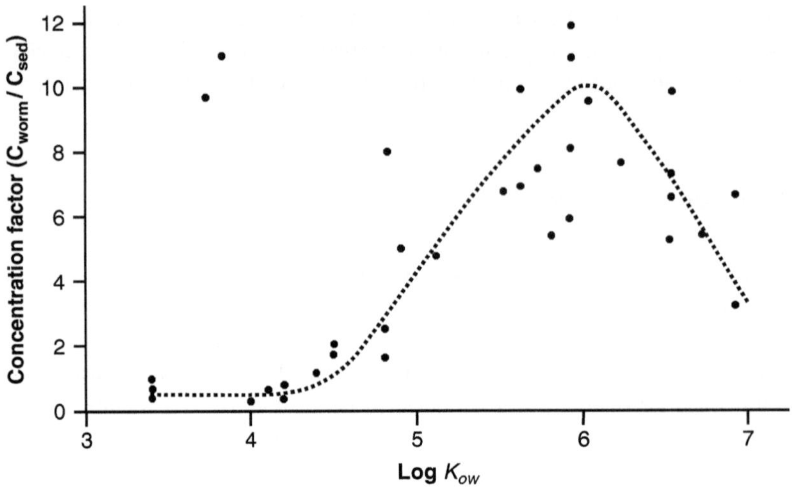

Fig. 5.7 Concentration factors of chlorinated hydrocarbons from laboratory spiked sediments plotted against the partition coefficient octanol-water (log K_{ow}) (From Oliver 1987, Fig. 3, redrawn and reproduced by permission of ACS, © conveyed by Copyright Clearance Centre, Inc.). C_{worm} chemical concentration in worms, ng g^{-1} dw; C_{sed} chemical concentration in sediment, ng g^{-1} sediment dw

Lake Superior sediment and allowed to depurate for up to 84 days to estimate the substances half-lives. Thirty-seven chlorinated hydrocarbons were tested, although unfortunately only a small proportion of the data was published for brevity. There was a general agreement between worm and fish bioconcentration factors, expressed for the worms as chemical concentration (ng g^{-1}, dw) divided by pore water concentration (ng l^{-1}) (Oliver 1987). Worm vs. sediment concentrations from field samples revealed that the lowest body residues were associated with sediments having the highest organic content, indicating the lower availability of chemicals in these sediments. Worm lipid content was about 8% dry weight, similar to that in fish, which may explain the similarity in bioconcentration factors in these taxa. Both studies showed that BAF increased until log K_{ow} reached a value of about 6, followed by a marked decline for larger molecules with log K_{ow} over 6–7 (Fig. 5.7). This behaviour was explained by the author as a reduced availability or a resistance to chemical transport across worm membranes due to the larger molecular size (Oliver 1984).

Interspecific differences in bioaccumulation factors can be attributed to different specific feeding strategies. Thus, endrin bioaccumulation in *Limnodrilus hoffmeisteri* and *Stylodrilus heringianus* in single and mixed species microcosms, after 980–1,312-h exposure was measured by Keilty et al. (1988a) who found that *Stylodrilus* had bioaccumulation factors that were on average 3–4 times higher than *Limnodrilus* (*S. heringianus* BAF = 2.4–43.8, *L. hoffmeisteri* BAF = 3.1–13.6), suggesting that differences could be related to a more surficial feeding zone for *S. heringianus*, where endrin levels were higher. These authors noted that moderate levels of endrin contamination appeared to stimulate the worms, so that the highest accumulation factors were observed at the intermediate concentrations (6,300 ng g^{-1} dw), probably due to an increase in worm feeding activity. In another set of experiments (1,300-h exposure period), Keilty et al. (1988b) reported higher endrin bioaccumulation factors (BAF, on a dw basis) for *S. heringianus*, in the range of 34–67 times the experimental sediment concentration. The authors found no relationship between sediment concentration and bioaccumulation, but the concentrations were not normalised. Variation in uptake and accumulation were apparent among the oligochaete *L. variegatus* and two arthropod species (*Chironomus riparius* and *Diporeia* sp.) (Harkey et al. 1994). This was attributed in some cases to differences in lipid content which may increase the potential for bioaccumulation of hydrophobic compounds, but also to differences in feeding behaviour (*e.g.* selection of particles, continuous vs. intermittent feeding) or in metabolism. In the same study, the authors also reported a wide variation in contaminant uptake among individuals over time, probably due to differences in age which was not standardised.

Penttinen et al. (1996) measured the uptake of trichlorophenol (TCP) in *L. variegatus, Chironomus riparius* and *Sphaerium cornatum* in uncontaminated sediment spiked to achieve 25–100 μg TCP g^{-1} dw sediment. It was observed that TCP uptake in *L. variegatus* was considerably higher (up to 481 μg g^{-1} ww) than in the other species (4–23 times lower), probably due to their continuous ingestion of sediment, higher lipid content, and inability to metabolize TCP. Similar conclusions were derived from a study on PCB bioaccumulation in oligochaetes and chironomids from a Swedish polluted lake (Bremle and Ewald 1995). In this study, the body burden ranged between 2.6 and 6.2 μg g^{-1} lipid in worms, and was significantly lower than in chironomids, which in this case was explained by the higher lipid content in the insects.

Field and laboratory bioconcentration factors showed a reasonable agreement for most persistent compounds (Oliver 1987). Bioaccumulation of PCBs in fish and worms in both laboratory and field exposures was reviewed by Ankley et al. (1992). Laboratory work was done with fathead minnows and the oligochaete *L. variegatus*, but field work was done with black bullhead and a mixture of worms, mostly *Limnodrilus* sp. Bioaccumulation factors (BSAF) for most PCB homologues and total PCBs were similar for laboratory and field-collected oligochaetes (see Appendix 2, Chap. 7), and were relatively close to 1, suggesting that oligochaete species possess only a limited ability to metabolise (and excrete) this class of compound. The values of body concentrations and BSAFs for the worms were intermediate between those for black bullhead (the highest) and the fathead minnow (the least), leading Ankley et al. (1992) to suggest that the use of *L. variegatus* would provide useful quantitative estimates of accumulation. Similar conclusions were obtained by Brunson et al. (1998) comparing PAH bioaccumulation factors, indicating that laboratory results can be reasonably extrapolated to the field. However, as the number of comparison studies of laboratory bioassays and field measurement is very low, it is important to stress the relevance of field validation studies for any test species/exposure regime when attempting quantitative estimates of exposure for the purposes of ecological risk assessment.

Harkey et al. (1995) examined bioaccumulation by *L. variegatus* from sediment core sections from different depths in the sediment, and accordingly of different periods (from 1899 to 1993). They demonstrated that bioaccumulation measured after a chronic sediment test (4 weeks) was maximum at the highest PAH concentration and was unrelated to the age of the sediment. Kinetic curves for low molecular weight PAHs peaked around 96 h, while most higher molecular weight PAHs peaked at about 2 weeks. Difference in bioavailability between surficial and profundal sediments with comparable sediment concentrations was discussed by the authors who suggested different binding of PAHs to colloids, micro-particles, or various forms of humic or fulvic acids at different sediment depths. Organic matter in sediment acts as a mechanism for sequestering organic chemicals in a non-bioavailable pool, and accordingly relates inversely to bioaccumulation. Dieldrin bioaccumulation in *L. variegatus* exposed (2 and 7 days) to four sediments with different organic content was studied by Standley (1997). Sediment characteristics which correlated positively with log K_{oc} (the partition coefficient to organic carbon content of sediments) correlated inversely with BSAF, as would be expected since increased sorption should reduce the bioavailable pool of contaminants. In addition, sediment characteristics correlated with log K_{doc} (the partition coefficient to dissolved and colloidal organic carbon) and BSAF were both either positive or negative, a trend that supports the hypothesis, at least for some chemicals, that contaminant binding by DOC increased the bioavailable pool by solubilising residues into faster desorbing compartments. Several other studies have shown that bioaccumulation is inversely related to sediment organic matter (*e.g.* MeHg: Nuutinen and Kukkonen 1998; LAS and 4-NP: Mäenpää and Kukkonen 2006). Moreover, in some cases increasing sediment chemical concentration can result in declining uptake rates, as reported for PAHs (Landrum et al. 2002), and is explained by differences in desorption rates of the substances in the sediments.

Field bioaccumulation of a substance can be influenced by the presence of other substances. For example, the presence of Selenium has been demonstrated to influence the bioaccumulation of MeHg by *L. variegatus* exposed to [^{14}C]methylmercuric iodide-spiked sediment (Nuutinen and Kukkonen 1998). The dose of 2.5 mg Se kg^{-1} dw resulted in a 25% reduction of the MeHg body burden after 2-week exposure. When 15 and 50 mg Se kg^{-1} were added to the sediment, the accumulation of MeHg was decreased by 75% and 86%, respectively, compared to the reference. However, it is not clear if this is the result of a chemical reaction in the sediment or an alteration of MeHg toxicokinetics in the organisms. Salinity has also been proved to be inversely related to bioaccumulation of fluoranthene in the estuarine oligochaete *Monopylephorus rubroniveus* (Weinstein 2003).

The effect of animal density on bioaccumulation could also account for differences between field and laboratory data. Kukkonen and Landrum (1994) examined the effect of worm density on pyrene bioaccumulation in laboratory tests and were surprised to observe that accumulation declined at lower densities (Fig. 5.8). The authors attributed this result to either higher sediment porosity at high worm density resulting in greater pore water volumes and higher uptake by this route, or to increased feeding rates at high worm densities where there was competition for

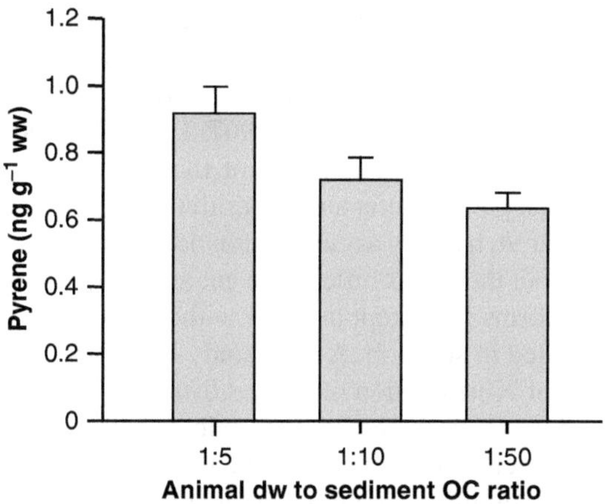

Fig. 5.8 Body burden in *Lumbriculus variegatus* exposed to 0.4 ng pyrene g^{-1} dw sediment in a 7-day exposure at 3 density ratios (From Kukkonen and Landrum 1994, Fig. 5, redrawn and reproduced with permission of John Wiley & Sons Ltd., © conveyed by Copyright Clearance Centre, Inc.)

limited food resources. However, it is also possible that feeding avoidance behaviour of the pyrene contaminated particles is more efficient at low worm densities than at high densities.

Egeler et al. (1997, 1999) described a highly standardised bioassay with artificial sediment for measuring uptake and depuration of radiolabelled lindane, hexachlorobenze (HCB), and 3,4 dichloroaniline (DCA) in the tubificines *T. tubifex* and *L. hoffmeisteri.* They reported wet weight based bioaccumulation factors (BAF) of 4 for lindane, 6–7 for HCB and 13.2 for DCA (Egeler et al. 1999). Although no major metabolites were detected, the bioaccumulation factors calculated from total radioactivity were later corrected with respect to the actual content of the parent compound. Both lindane and HCB were taken up rapidly during the first hour of the uptake phase and reached equilibrium after 5.9–12.8 h and after 109–152 h, respectively (see bioaccumulation factors and toxikokinetic coefficients in Appendix 2, Chap. 7). Sediment-based tubificine bioaccumulation factors were lower than the corresponding fish bioconcentration factors, although the body burden in the tubificines was notably higher. This demonstrated that the extrapolation of fish BCF to other organisms in other aquatic compartments (*i.e.* sediment) is inappropriate for bioaccumulation risk assessment (Egeler et al. 1999). Organic substances with high affinity to sediments and low partitioning in the overlying water (HCB: 0.1% and DCA: 3.1% of sediment concentration) may cause higher body concentrations in tubificines, than substances with higher partitioning into the overlying water (lindane: 4.3%). Therefore, lipophilicity is not the only factor to be considered in the bioaccumulation of organic substances, since DCA (log K_{ow} = 2.69) was reported to have BAF mean values of 13.2, while lindane (log K_{ow} = 3.63) and HCB (log K_{ow} = 5.72) had much lower BAFs (4.4 and 6.5, respectively).

There has been little interest in the role of worms in accumulation and sediment release of radioactive materials in relation to fallout from atomic power generation. The first studies examined accumulation of ^{89}Sr and ^{45}Ca and transfer of ^{32}P in an aquatic ecosystem (Whitten and Goodnight 1967, 1969). In experiments with strontium and calcium, histological examination of the worms revealed that after 3-h exposure the epidermis, gut epithelia and nephridial tissue contained some radioactive elements, and after 96 h heavy accumulation occurred in the nephridial ampulla and in the epidermis, but there was little in the gut and chloragogen cells. This is not surprising since the worms were kept in water without sediment and the radioactive elements were presented in solution. A later study by Block and Goodnight (1976) evaluated the effects of X-irradiation on *Limnodrilus* sp. When exposed to 3,000 R the worms migrated away from the source, chloragogen cells became black and granulated, and sperm cells were killed, but the worms survived.

Anoxic or hypoxic conditions are quite frequent in polluted sediments, and oligochaetes are often the dominant organisms in these habitats. However, there has been little investigation of bioaccumulation under these conditions. Penttinen and Kukkonen (2000) did use a direct calorimetric approach to measure the effect of the chemical pentachlorophenol (PCP) in energy metabolism, under conditions of normoxia and anoxia. No fluctuation occurred in metabolic heat dissipation by unexposed *L. variegatus,* under different oxygen conditions. However, after 24-h normoxic exposure, the rate of PCP metabolism at body residues above 0.2 μmol PCP g^{-1} ww was enhanced and there was a significant linear relation between PCP tissue residue (μmol g^{-1}) and the metabolic response (heat output, μW mg^{-1}), up to 0.6 μmol PCP g^{-1}. An anoxic PCP exposure did not produce any notable change in heat output, but the amount of chemical accumulated in tissues was only 18% of that accumulated under normoxia. This laboratory evidence helps with the understanding of the deviation from expected values in the field, where anoxic conditions may periodically occur. Field and laboratory differences are usually explained as a result of changes in bioavailability associated with sediment chemistry, but may be a result of changes in the physiology of the worms under anoxic conditions. Heat dissipation and its relationship with uptake of trichlorophenol (TCP) had also been measured by Penttinen et al. (1996) using a microcalorimeter, but did not yield useful results. Under air-saturated water, no significant effects of TCP on worm heat dissipation were detected and the metabolic response was similar to the control animals, although body residues varied from 1.4 to 2.43 μmol g^{-1} ww.

There is a considerable amount of work required to develop a full understanding of the toxicokinetics, uptake routes, assimilation efficiencies, detoxification processes, elimination rates, and organs responsible for bioaccumulation and depuration in aquatic oligochaetes. Many studies on bioaccumulation have simply measured tissue concentration which provides little information in terms of environmental risk assessment, unless it can be associated with a biological effect or can be used to predict tissue concentrations at higher trophic levels. Both these issues are examined below. There are several biological factors that control bioaccumulation processes, such as organism age, lipid content, reproductive stage, density or feeding behaviour which also require further research. Compared to toxicity studies, bioaccumulation methods for laboratory tests still need further standardisation, although among invertebrates, aquatic oligochaetes are at present one of the few faunistic groups with standardised guidelines for measuring bioaccumulation.

5.4 Bioaccumulation and Toxicity

There are basically two different approaches for determining cause/effect relationships in ecotoxicology. Both approaches provide a dose variable for assessing either lethal or sublethal toxic effects: the traditional approach measures effects based on chemical concentrations in the environment (water or sediment), and a more recent body burden approach. McCarty (1986, 1991) established that internal toxicant concentration producing defined acute or chronic responses provides a means of evaluating the toxicological risk of the body burden of organic chemicals. McCarty defined the Critical Body Residue (CBR) as the molar tissue concentration of a chemical that consistently produces a defined toxic effect (*e.g.* the 50% mortality). The CBR concept relates the body burden with a predetermined effect level, and assumes that there is a concentration of toxicant within the organism that produces an effect (*e.g.* mortality, inhibition of growth or reproduction). This approach has the theoretical advantage of separating bioavailability issues from the toxic response. The consistency of CBRs and the applicability of the CBR approach have been reviewed by Barron et al. (2002) for eight chemical classes.

Behavioural and other sublethal effects related to chemical body burdens were investigated by Fisher et al. (1999) in *Lumbriculus variegatus* exposed to PCBs. These authors described narcotic effects, survival, and impairments in growth and reproduction (fragmentation) in water-exposures by adding spiked algae every 2 days. Those chemicals with higher elimination rates (MCBP and DCBP) required much higher exposure concentrations to achieve body residues that resulted in changes in biomass, reproduction or mortality. PCB body concentrations in dead organisms after 35-day exposure were 0.88–1.35 μmol g^{-1} ww. The lowest body residues at which significant differences in growth and reproduction relative to controls were detected ranged between 0.34 and 0.56 μmol g^{-1} ww. As expected, the results indicated that sublethal effects occur at a significantly lower body residue levels than those required to cause death, and the sublethal effects associated with sublethal body concentrations varied from 2.7% to 37.7% reduction of biomass and 22.8–38.4% reduction in young, relative to controls. The CBR approach was also useful with *L. variegatus* exposed to chlorophenols, since lethal body residues (LBR_{50}, critical body concentration corresponding to LC_{50} values) were independent of exposure conditions (aqueous or sediment exposure), or bioavailability due to differences in toxicant binding (*e.g.* sediment organic content) (Nikkilä et al. 2003). The estimated body burdens corresponding to LC_{50} in *L. variegatus* exposed to herbicides showed similar values to those corresponding to oral LD_{50} obtained for avian species (Mäenpää et al. 2003). In other instances, differences between species for the CBR values seem to be linked to differences in herbicide degradation, although the authors did not analyse the metabolites.

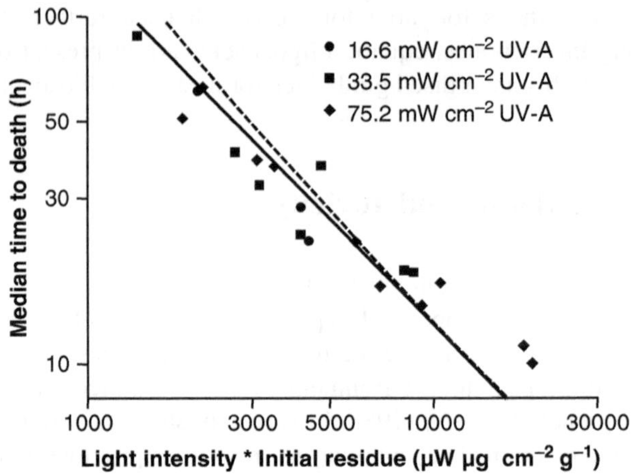

Fig. 5.9 Time-dependent mortality of *Lumbriculus variegatus* in relation to light intensity × initial tissue concentration of fluoranthene (From Ankley et al. 1995, Fig. 5, modified and reproduced by permission of ACS, © conveyed by Copyright Clearance Centre, Inc.). Expected response for zero elimination rate and no damage repair (*solid line*), expected response for 0.1 day^{-1} elimination rate and no damage repair (*dash line*)

The CBR approach for metals was first used by Redeker and Blust (2004) who determined the CBR for *Tubifex tubifex* using two methods: first directly, measuring body concentration and the associated mortality at different times in water-only exposure, and second indirectly, as the body concentration corresponding to LC_{50} values (CBR_{50}) estimated from a pharmacokinetic model. The estimated value of CBR_{50} for cadmium was 0.32 μmol g^{-1} ww, in good agreement with directly measured concentration associated with 50% mortality (0.37 μmol g^{-1} ww). Penttinen et al. (2008) have estimated CBR values for Cu, Cd, Pb and Cr related to mortality and feeding activity in experiments with *L. variegatus*, in water-only tests and in the presence of sediment. The cause-effect link for body burden and mortality could only be established in aqueous tests, while in sediment experiments the observed body concentrations of different metals were far below a dose that would be expected to result in death in acute tests, even at the highest sediment concentrations. LC_{50} values varied by a factor of 22, whereas LBR_{50} varied only by a factor of 4.7. Interestingly, the relative toxicity of the metals expressed in LC_{50} or LBR_{50} changed, with Cu being the most toxic metal with the former approach and Cd with the latter.

Using the body residue approach, the relationship between chemical body burden or bioaccumulation factors and lethal or sublethal endpoints has been demonstrated. One of the first studies that modelled worm toxicity including tissue residue was Ankley et al. (1995), where the median lethal time for fluoranthene could be predicted through the product of light intensity and PAH body burden, after 96-h exposure (Fig. 5.9). In a later study, the oligochaete *L. variegatus* was exposed to multiple concentrations of anthracene, pyrene and fluorene for 96 h, followed by another 96-h holding period in clean water under three different UV light intensity (Ankley et al. 1997). Time-dependent lethality of the three PAHs was modelled as previously for fluoranthene, by plotting median lethal time as a function of the product of initial body burden and UV light intensity. Behavioural symptoms associated with photoinduced toxicity were quite characteristic: violent twitching, followed by tight spiralling, relaxation and finally death. The toxicity of anthracene and pyrene was proportional to both tissue residue and UV intensity, but fluorene did not exhibit any phototoxicity at tissue residues of about 190 μg g^{-1} ww.

Chapman et al. (1999) described the negative relationships of Cu and Cd accumulated by *L. variegatus* and reproduction (body burden could not be calculated for *T. tubifex* due to insufficient biomass for analyses), in 14-day chronic bioassays. In a 28-day chronic exposure of *Tubifex tubifex* to heavy metals, Gillis et al. (2002) studied the relationship of Cd body burden and reproduction (number of young per adult) and described a threshold concentration for reproduction impairment (number of young per adult) of 2.7 μmol Cd g^{-1} sediment dw, where worms accumulated 30.38 μmol Cd g^{-1} dw (see Fig. 4.13). Using data in Table 1 from this publication in a logistic regression model, a CBR value of 45.1 μmol Cd g^{-1} dw corresponding to 50% reproductive output inhibition has been estimated, which is more than tenfold the body concentration calculated for 50% mortality in other studies performed in water-only tests (*e.g.* 2.3–2.7 μmol Cd g^{-1} dw, in Penttinen et al. 2008). However, it is likely that the absence of sediment in experiments with benthic organisms results in unrealistic values with the CBR approach.

The potential for the use of body concentration for risk assessment has also been investigated by Landrum et al. (2004a) who estimated the body burden in *L. variegatus* at 50% reduction of sediment reworking rates (using ^{137}Cs marker in sediment). After 24-day (at 22°C) and 26-day (at 10°C) exposure to tetrachlorobiphenyl (TCBP) spiked in a reference sediment, biological burial rates (W_b) declined with increasing TCBP concentrations and also with temperature. Interestingly, body residue associated with a 50% reduction of W_b was temperature-independent (96 and 124 nmol g^{-1} ww at 10°C and 22°C, respectively). However, using microspheres (<1 μm diameter) and luminophores (63–315 μm) to measure reworking rates by tubificids, Ciutat et al. (2005b) did not find significant effects with worms exposed to cadmium (20 μg Cd l^{-1} in the overlying water, for 56 days), despite the fact that worm body concentrations attained values close to 50 μg Cd g^{-1} dw.

There is a relatively small amount of research dealing with Critical Body Residue approach or the relationships between toxicity and body burden in aquatic oligochaetes. Several authors have criticized the lack of standardization of the CBR approach (Barron et al. 2002; Landrum and Meador 2002). These authors advocate the consideration of temporal scale, since different exposure times can cause diverse modes of toxic action, as well as for a standardization of the environmental variables (temperature, pH, salinity), since they not only affect bioavailability but also physiological processes and the sensitivity of organisms, and therefore bioaccumulation. However, this seems to be a promising field for further bioaccumulation studies, as the approach may offer a more accurate and precise measurement of dose for chemical exposure than chemical concentration in the environment, as demonstrated for surfactants and *L. variegatus* by Mäenpää and Kukkonen (2006). When comparing the various studies, there should be also some consideration of differences between genetically different populations of the same species that may result in higher tolerances to higher body residues (Klerks and Bartholomew 1991), and thus in higher CBR values. Therefore, the importance of a controlled culture of the test species is equally important when using the CBR approach.

5.5 Transfer Through Food-Chain and Biomagnification

In the absence of toxic effects, the ability of benthic invertebrates to accumulate relatively high concentrations of chemicals may be relevant to the transfer of contaminants through trophic chains to potentially sensitive vertebrate species.

Oligochaetes feed selectively on small size particles of sediment and organic matter, as well as bacteria and microscopic algae. Wavre and Brinkhurst (1971) demonstrated a reduction of about 72% of the heterotrophic aerobic bacteria present in the sediment on passage through the gut, suggesting that these bacteria represent an important food source for tubificids. Bacteria accumulate metals, acting as intermediates for the transfer of heavy metals from solution to deposit feeders, including oligochaete species, that are prey for fish (Patrick and Loutit 1976, 1978). These authors showed that Cr, Cu, Mn, Fe, Pb and Zn were concentrated by worms fed for 4 days on metal-enriched bacteria and then these metals were transferred to fish (Fig. 5.10). Food selection of sediment-dwellers varies among species that show preferences for different particle sizes or microorganisms, and selective feeding can therefore influence the transfer of toxicants, bioaccumulation and biomagnification through the food chain. For instance, Bott and Standley (2000) investigated the increase of bioaccumulation of two organics from radiolabelled benzo(a)pyrene (BaP) and 2,2′,5,5′-tetrachlorobiphenyl (PCB) in *Lumbriculus variegatus* and chironomid larvae. Results indicate that worms accumulated more when the labelled source was in sediment or bacteria than in algae.

Oligochaetes are well known to be important prey organisms for other benthic invertebrates and fish (Kennedy 1969; Loden 1974; Bouguenec 1992) and, therefore, they can act as passive agents for the transfer of pollutants associated with the sediment to other organisms through the food chain. In laboratory studies, they have also been demonstrated to be an excellent model for toxicant exposure to fish via diet (Egeler et al. 2001; Mount et al. 2006).

Whitten and Goodnight (1969) investigated uptake from water, bacteria and sediment by *Limnodrilus* sp. over a 7-day period, in an aqueous solution containing a mixture of bacteria, autoclaved plant material, and sediment labelled with radioactive phosphorus. Worms labelled with ^{32}P were also fed to fish (bluegills and bluntnose minnows). After 96 h, radioactive elements were detectable in the epidermis, gut epithelium and chloragogen cells of the worms. Uptake from bacteria and sediment was estimated, and higher uptake rates were obtained using bacteria than other materials. Fish accumulated P from worms presented as food, and it appeared that

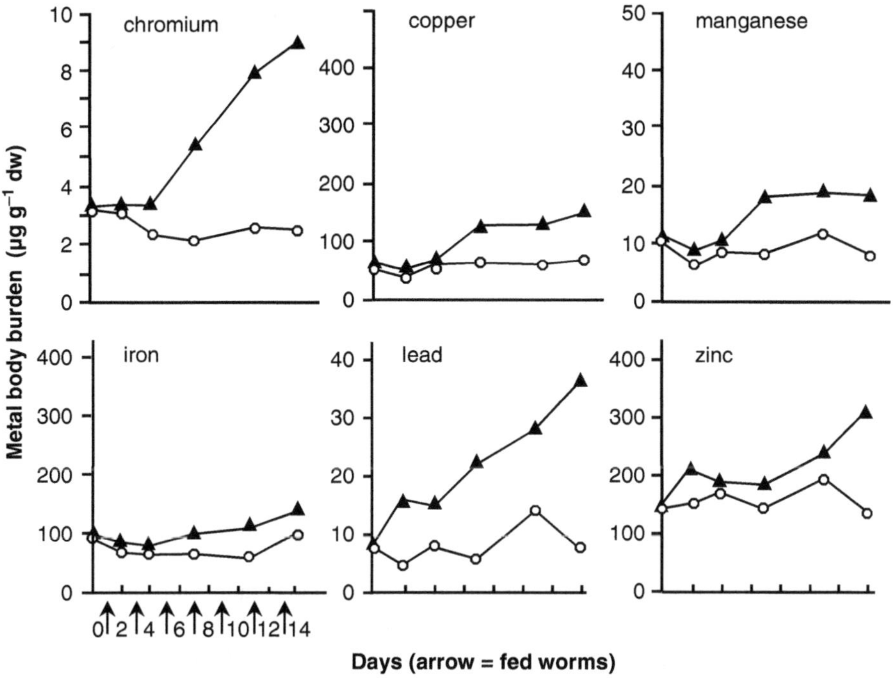

Fig. 5.10 Estimates of metal levels (mg kg^{-1} dw) in fish fed on tubificid worms containing various metals. Tubificids fed on bacteria grown without metals (○), and with 1 mg l^{-1} of each metal (▲) (From Patrick and Loutit 1978, Fig. 2, redrawn and reproduced by permission of Elsevier Ltd, conveyed by Copyright Clearance Centre, Inc.)

the response was species specific. In a study by Naqvi (1973), the crayfish *Procambarus clarkii* was fed with insecticide-exposed *Branchiura sowerbyi*, resulting in mortality of the crayfish that was inversely proportional to the treatment time of the worms and directly proportional to the insecticide concentration in the worms, suggesting degradation of pesticides by worm metabolism. Thus, the average lethal time for crayfish was 1,440 min when fed worms treated in 1 ppm DDT for 90 days, while it was only 43 min when fed worms treated in 2.5 ppm DDT for about 3 h. Poisoning symptoms of the crayfish included tremors, loss of balance and hyperactivity until death, and the reactions were also specific to the toxicant used. Working with pesticides, Kabir and Khatoon (1980) demonstrated the toxicant specificity of toxic transfer and its dependence on dose. They fed guppies (*Lebistes reticulatus*) with *Limnodrilus* sp. previously exposed to pesticides and reported 5–10% mortality of worm-fed fish when worms had been exposed to the maximum concentration (10 ppm) of Cabicron and Nogos. However, they observed no response in fish fed with worms exposed to lower concentrations of those pesticides or to any concentration of Diazinon, Malathion and Dimecron.

The ability of some substances to biomagnify through aquatic food webs is problematic for regulatory agencies and government when establishing water and sediment quality guidelines for the protection of aquatic life, however, there are relatively few studies that have addressed this problem. Nuutinen and Kukkonen (1998) emphasised the role of sediment-ingesting organisms, such as oligochaetes, in the transfer of sediment-bound methyl mercury to food webs, where concentration of this compound has been shown to increase through trophic levels. In the field, the increase of metals in fish is not always attributable to food transfer from prey, since worm burrowing activity can also make metals more available. In their review of heavy metal toxicity, Chapman et al. (1980) reported increasing bioavailability for uptake by fish of inorganic and MeHg from sediments via resuspension of fine material into the water column, caused by *L. hoffmeisteri* burrowing activity. Jernelöv (1970) also recorded the accumulation of MeHg in fish (*Lebistes reticulatus*) caused by worm burrowing activity in the top 3-cm sediment layer.

The position of organisms in the food web affects their accumulation of PCBs so that chemical biomagnification occurs (Zaranko et al. 1997). In a study in Pottersburg Creek during 1989–1993, these authors regarded trophic transfer as the primary mechanism determining PCB levels in the biota rather than bioconcentration (i.e. uptake from water phase), and aquatic oligochaetes and chironomids which accumulate chemicals by ingesting sediment were a major food source for fish and leeches and, thus, a source of PCBs. Biomagnification of PCBs and chlorinated pesticides in oligochaetes from sediments in the lower reaches of the River Po (Italy) was studied by Galassi et al. (1994). They calculated bioconcentration factors (ml pore water g^{-1} dw) for 13 species (fish, molluscs and oligochaetes) which were compared to theoretical values obtained by equations based on the Equilibrium Partitioning model (Di Toro et al. 1991). Body concentration values predicted for oligochaetes from the model agreed with values measured in field for PCBs and chlorinated pesticides, except for HCB whose estimated body concentration (102 μg kg^{-1} dw) was much higher than the measured concentration (2 μg kg^{-1} dw). The biomagnification of the organic compounds in fish species via food transfer was also demonstrated in this study using field fugacity ratios, although bioaccumulation was less than that predicted for the top predators using the model (four trophic levels) proposed by Connolly and Pedersen (1988).

In a different approach to the biomagnification, Wallace and Lopez (1996) tested the hypothesis that exposure-related alterations in the subcellular cadmium distribution of prey determined changes in cadmium absorption by a predator. The oligochaete *Limnodrilus hoffmeisteri* was used as prey in an experiment of Cd bioavailability to the shrimp *Palaemonetes pugio*. The first contribution demonstrated that exposure conditions (Cd concentration and duration) significantly influenced the percentage of Cd in different subcellular fractions, and there was an increase in Cd in the cytosol (metallothionein-like and other proteins) from worms exposed to 0.5 and 47 μg l^{-1} after 6 weeks of exposure. There was also a positive correlation between the amount of Cd in the worm cytosol and the Cd transferred to the predator, showing that the absorption efficiency of shrimps was influenced by Cd concentration and exposure duration to oligochaetes (Table 5.3). In the second

Table 5.3 Cadmium transfer and absorption efficiency in the shrimp (*Palaemonetes pugio*) (mean ± SD) fed on Cd-contaminated worms (*Limnodrilus hoffmeisteri*). The second column shows the Cd concentration to which worms where exposed in the feeding experiments (From Wallace and Lopez 1996, Table 2, with permission of Springer Publ., © conveyed by Copyright Clearance Centre, Inc.)

Treatment time	Cd concentration ($\mu g\ l^{-1}$)	Absorption efficiency (%)	Cd transferred to shrimp (μg Cd/$g_{worm\ ingested}$ ww)
1 week	0.5	43.6 ± 3.8	0.12 ± 0.02
	47	43.4 ± 3.2	11.08 ± 0.10
	140	65.9 ± 3.0	12.65 ± 3.01
6 weeks	0.5	79.4 ± 2.5	0.78 ± 0.04
	47	66.4 ± 4.1	60.39 ± 15.33
	140	–[a]	–

[a]100% mortality of worms

paper (Wallace and Lopez 1997), worms were radiolabelled with ^{109}Cd for 6 days and shrimps were fed with whole and homogenised worms, in order to investigate the availability to the shrimp of Cd sequestered by worms and to allow predictions on Cd trophic transfer. The subcellular fractionation showed that the cytosol accounted for the largest proportion (42.3%) of the ^{109}Cd, the debris fraction (metal-rich granules and tissue fragments) was the second largest pool (39.3%) and the intracellular fraction had only a minor role in Cd storage, containing only 9.3% of the ^{109}Cd in *L. hoffmeisteri*. This study showed that ^{109}Cd bound to metallothionein-like proteins was transferred with high efficiency, representing 35.8% of the total Cd absorbed by shrimp.

More recently, Redeker et al. (2007) have analysed the implication of the subcellular distribution of cadmium in *T. tubifex* in trophic availability. They divided the body concentration into two distinct fractions, one a pool available for metabolic use, and the other a detoxified fraction in which metals are captured by heat stable proteins (*e.g.* metallothionein) or granules. Compartmentalization of total metal changes over time and Cd accumulates linearly in the trophically available fraction (TAM, all metal fractions except granules) over a 12-day exposure period. However, when expressed as percentage of body burden, TAM stays constant, due to detoxification processes, probably because in this study metals did not exceed certain internal levels (Fig. 5.11). Cadmium accumulates in the metabolically available pool over time but is at the same time detoxified and stored and the detoxified pool levels off. At the start of the experiment, 33% and 32% Cd was found in the granular and heat-sensitive protein fractions, respectively, and the smallest fraction was in the organelles (10%). After 12-day exposure, the organelle fraction was the largest Cd pool (26%). Up to 72% of the total accumulated Cd was in a potentially available form (non-granular), however, carp fed on radiolabelled contaminated tubificids only assimilated 9.8% of the total Cd, suggesting that metal transfer efficiency can change with the metal exposure conditions of the prey as well as with the predator assimilation efficiency. Ng and Wood (2008) have measured lower dietary transfer

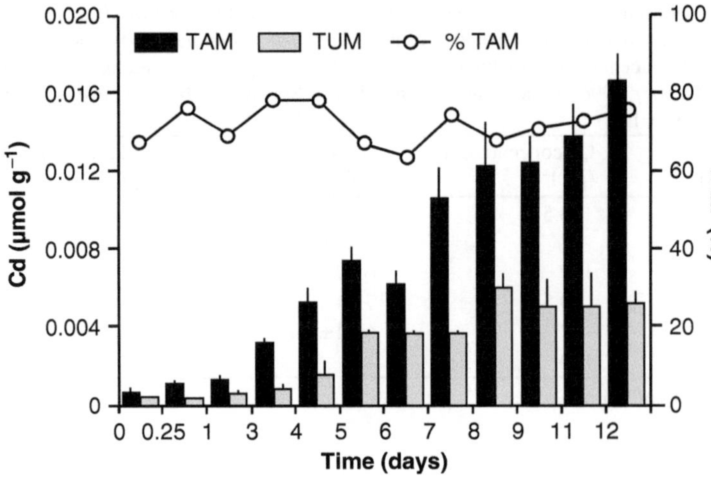

Fig. 5.11 Subcellular partitioning in *T. tubifex* of cadmium in a pool of trophically available metals (*TAM*, all metal fractions except granules) and trophically unavailable metals (*TUM*, metal-rich granules fraction) (From Redeker et al. 2007, Fig. 4, redrawn and reproduced by permission of Elsevier Ltd, conveyed by Copyright Clearance Centre, Inc.)

efficiencies (0.9–6.4%) of Cd from *L. variegatus* to rainbow trout (*Onchorhynchus mykiss*), even if the amount Cd in the TAM fraction was comparable (72–80%) to that previously reported by Redeker et al. (2007). Trophic transfer was only weakly correlated with worm Cd body burden, suggesting that reduced assimilation and increasing excretion could have occurred over the exposure period (1 month). The increase in Cd concentration in fish kidney and liver and in the MTL-protein fraction was also interpreted as a more efficient detoxification mechanism in fish. Nevertheless, growth reductions of 50% relative to the control were found up to the 3rd and 4th week for fish fed with worms exposed to the highest Cd dose (200 µg l^{-1}), suggesting an environmentally relevant Cd body burden through trophic transfer.

In a highly standardised experimental design, Egeler et al. (2001) investigated the role of oligochaetes in the transfer of hexachlorobenzene (HCB) to fish (*Gasterosteus aculeatus*). They spiked artificial sediment with [^{14}C]hexachlorobenzene and exposed *T. tubifex* for 60- and 63-day periods. In a set of experiments, fish were exposed to spiked water, spiked sediment, pre-exposed oligochaete worms and a combination of these exposure routes. In the first set of dietary experiments, the fish were kept in uncontaminated water and fed pre-exposed contaminated live worms for 60 days (dietary exposure), producing a linear increase in the [^{14}C]HCB accumulation factor ($AF_{fish/worm}$) over time. In a second set of experiments, fish were kept in a system containing spiked water and sediment and were also fed with pre-exposed worms (combined exposure) for 63 days (Fig. 5.12). Steady state was reached after 30 days, after which the concentration in the fish remained constant.

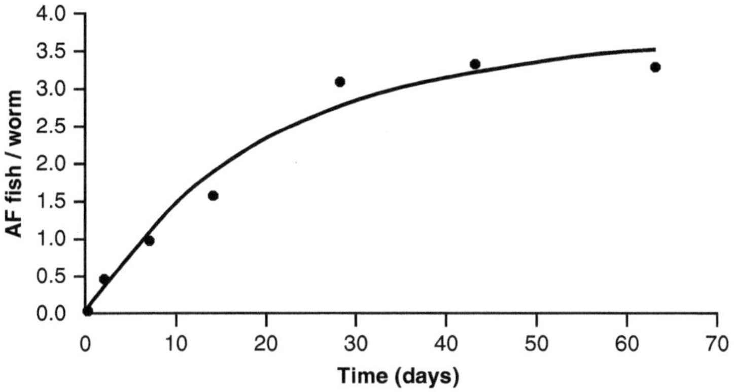

Fig. 5.12 Bioaccumulation kinetics of [^{14}C]HCB in the fish *Gasterosteus aculeatus* fed on pre-exposed *Tubifex tubifex* worms and exposed to spiked water and sediment, in a combined exposure scenario. Accumulation factor (*AF*) measured as the ratio of fish concentration and worm concentration (ww/ww) (From Egeler et al. 2001, Fig. 4, modified and reproduced by permission of Springer Publ., conveyed by Copyright Clearance Centre, Inc.)

Table 5.4 Accumulation of [^{14}C]HCB in fish (*Gasterosteus aculeatus*), fed on pre-exposed worms (*Tubifex tubifex*), in a combined exposure to the pollutant in water, sediment, and food and showing the relative contribution of different exposure routes (SE in parentheses) (From Egeler et al. 2001, Table 8, with permission of Springer Publ., © conveyed by Copyright Clearance Centre, Inc.)

	C_{fish}	C_{water}	$C_{sediment}$	C_{worm}
C_{source} µg kg^{-1} (SE)	–	0.215 (0.09)	238 (3)	3,586 (120)
		$C_{fish\ (water)}$	$C_{fish\ (sediment)}$	$C_{fish\ (worm)}$
C_{fish} µg g^{-1}	11.1	5.0	1.6	4.6
C_{fish} (in% total C_{fish})	100	45	14	41
$AF=C_{fish}/C_{source}$	–	23,100 (on a ww basis)	6.6 (lipid/OC sediment)	1.3 (lipid/lipid)

$C_{fish\ (water)}$ concentration in the fish that results from uptake via contaminated water; $C_{fish\ (sediment)}$ concentration in the fish that results from uptake via contaminated sediment; $C_{fish\ (worm)}$ concentration in the fish that results from uptake via contaminated worms

Regression analysis indicated a BAF $_{fish/worm}$ of 3.5 ± 0.3(SE). Each potential uptake route (bioconcentration and dietary exposure, as single factors and combined) was studied and the accumulation factors are shown in Table 5.4. Biomagnification is judged to occur if the ratio $AF_{fish/worm}$ >1. When using the lipid-normalised concentrations in the fish and worms, the accumulation factor for the dietary exposure ($AF_{fish/worm}$) was 0.53, therefore, biomagnification could not be demonstrated. Nevertheless, the $AF_{fish/worm}$ calculated in the combined exposure was 1.3, which implies biomagnification. Results suggested that the major uptake routes for HCB in a benthivorous fish are the overlying water (45%) and the food (41%), in this case represented by the oligochaete worm *T. tubifex*. Therefore, the assessment of bioaccumulation of HCB in fish must consider the dietary uptake route.

5.6 Conclusions

Worms can accumulate metals and have metabolic routes that are able to eliminate them via excretion or store them in subcellular fractions that are not toxic to the organism. Body burdens of essential metals are physiologically regulated, and show other uptake-elimination models than non-essential metals. Bioavailability of metals through adsorption on organic particles or due to the formation of sulphides, pH, or dissolved oxygen levels has been studied and may explain the frequent lack of correlation between body burden and sediment or pore water toxicant concentration. Surface adsorption of some metals into the mucus layer which covers the body and the level of contaminants in the food passing through the gut without being absorbed are confounding factors in bioaccumulation assessment that must be considered to the extent possible. Worms exposed to pesticides, PCBs, HCB, or metals, proved to be toxic to crayfish, leech and fish feeding on them, thus worms play a role in the transfer of toxicants through the food chain. Worms are also important bioturbators in aquatic systems through their burrowing, feeding and respiratory activities, making toxicants more available to animals at higher trophic levels, even if they are not prey for other species. These results are important for environmental risk assessment when the potential for a chemical to bioaccumulate or biomagnify through the food chain is suspected.

References

Amaral AFS, Rodrigues AS (2005) Metal accumulation and apoptosis in the alimentary canal of *Lumbricus terrestris* as a metal biomarker. Biometals 18:199–206

Ankley GT, Phipps GL, Leonard EN, Benoit DA, Mattson VR (1991) Acid-volatile sulfide as a factor mediating cadmium and nickel bioavailability in contaminated sediments. Environ Toxicol Chem 10:1299–1307

Ankley GT, Cook PM, Carlson AR, Call DJ, Swenson JA, Corcoran HF, Hoke RA (1992) Bioaccumulation of PCBs from sediments by oligochaetes and fishes: comparison of laboratory and field studies. Can J Fish Aquat Sci 49:2080–2085

Ankley GT, Leonard EN, Mattson VR (1994) Prediction of bioaccumulation of metals from contaminated sediments by the oligochaete *Lumbriculus variegatus*. Water Res 28:1071–1076

Ankley GT, Erickson RJ, Phipps GL, Mattson VR, Kosian PA, Sheedy BR, Cox JS (1995) Effects of light intensity on the phototoxicity of fluoranthene to a benthic macroinvertebrate. Environ Sci Technol 29:2828–2833

Ankley GT, Erickson RJ, Sheedy BR, Kosian PA, Mattson VR, Cox JS (1997) Evaluation of models for predicting the phototoxic potency of polycyclic aromatic hydrocarbons. Aquat Toxicol 37:37–50

ASTM (2005) Standard guide for conducting sediment toxicity tests with freshwater invertebrates. ASTM E1706-05, Philadelphia

Back H, Prosi F (1985) Distribution of inorganic cations in *Limnodrilus udekemianus* (Oligochaeta, Tubificidae) using laser induced microprobe mass analysis, with special emphasis on heavy metals. Micron Microsc Acta 16:145–150

Barron MG, Hansen JA, Lipton J (2002) Association between contaminant tissue residues and effects in aquatic organisms. Rev Environ Contam Toxicol 173:1–37

Beek B, Böhling S, Bruckmann U, Franke C, Jöhncke U, Studinger G (2000) The assessment of bioaccumulation. In: Beek B (ed) The handbook of environmental chemistry, vol 2, part J, bioaccumulation. Springer, Berlin, pp 235–276

Belden JB, Lotufo GR, Lydy MJ (2005) Accumulation of hexahydro-1,3,5-trinitro-1,3,5-triazinone in channel catfish (*Ictalurus punctatus*) and aquatic oligochaetes (*Lumbriculus variegatus*). Environ Toxicol Chem 24:1962–1967

Bervoets L, Blust R, De-Wit M, Verheyen R (1997) Relationships between river sediment characteristics and trace metal concentration in tubificid worms and chironomid larvae. Environ Pollut 95:345–356

Block EM, Goodnight CJ (1976) The effect of X-irradiation on the coelomic cells of the tubificid *Limnodrilus hoffmeisteri*. Trans Am Microsc Soc 95:23–34

Bott TL, Standley LJ (2000) Transfer of benzo[a]pyrene and 2,2′,5,5′-tetrachlorobiphenyl from bacteria and algae to sediment-associated freshwater invertebrates. Environ Sci Technol 34:4936–4942

Bouché ML, Habets F, Biagianti-Risbourg S, Vernet G (2000) Toxic effects and bioaccumulation of cadmium in the aquatic oligochaete *Tubifex tubifex*. Ecotoxicol Environ Safe 46:246–251

Bouché ML, Arnoult F, Vernet G (2003) Caudal regeneration in *Tubifex tubifex* (Oligochaeta, Tubificidae) following copper exposure. Invertebr Biol 122:42–51

Bouguenec V (1992) Oligochaetes (Tubificidae and Enchytraeidae) as food in fish rearing: a review and preliminary tests. Aquaculture 102:201–217

Bremle G, Ewald G (1995) Bioconcentration of polychlorinated biphenyls (PCBs) in chironomid larvae, oligochaete worms and fish from contaminated lake sediment. Mar Freshwater Res 46:267–273

Brinkhurst RO, Austin MJ (1979) Assimilation by aquatic Oligochaeta. Int Rev Ges Hydrobiol 64:863–868

Brinkhurst RO, Chua KE (1969) Preliminary investigation of the exploitation of some potential nutritional resources by three sympatric tubificid oligochaetes. J Fish Res Bd Can 26:2659–2668

Brown BE (1982) The form and function of metal-containing "granules" in invertebrate tissues. Biol Rev 57:621–667

Brunson EL, Canfield TJ, Dwyer FJ, Ingersoll CG, Kemble NE (1998) Assessing the bioaccumulation of contaminants from sediments of the Upper Mississippi River using field-collected oligochaetes and laboratory-exposed *Lumbriculus variegatus*. Arch Environ Contam Toxicol 35:191–201

Campbell PGC, Clearwater SJ, Brown PB, Fisher NS, Hogstrand C, Lopez GR, Mayer LM, Meyer JS (2005) Digestive physiology, chemistry and nutrition. In: Meyer JS, Adams WJ, Brix KV, Luoma SN, Mount DR, Stubblefield WA, Wood CM (eds) Toxicity of dietborne metals to aquatic organisms, Chapter 2. SETAC Press, Pensacola, pp 13–57

Cancio I, Gwynn I, Ireland MP, Cajaraville MP (1995) The effect of sublethal lead exposure on the ultrastructure and on the distribution of acid phosphatase activity in chloragocytes of earthworms (Annelida, Oligochaeta). Histochem J 27:965–973

Carlson AR, Phipps GL, Mattson VR, Kosian PA, Cotter AM (1991) The role of acid-volatile sulfide in determining cadmium bioavailability and toxicity in freshwater sediments. Environ Toxicol Chem 10:1309–1320

Chapman PM, Churchland LM, Thomson PA, Michnowsky E (1980) Heavy metal studies with oligochaetes. In: Brinkhurst RO, Cook DG (eds) Aquatic oligochaete biology. Plenum Press, New York, pp 477–502

Chapman PM, Allen HE, Godfredsen K, Z'graggen MN (1996) Evaluation of bioaccumulation factors in regulating metals. Environ Sci Technol 30:448A–452A

Chapman PM, Wang F, Janssen C, Persoone G, Allen HE (1998) Ecotoxicology of metals in aquatic sediments: binding and release, bioavailability, risk assessment, and remediation. Can J Fish Aquat Sci 55:2221–2243

Chapman KK, Benton MJ, Brinkhurst RO, Scheuerman PR (1999) Use of the aquatic oligochaetes *Lumbriculus variegatus* and *Tubifex tubifex* for assessing the toxicity of copper and cadmium in a spiked-artificial sediment toxicity test. Environ Toxicol 14:271–278

Ciutat A, Anschutz P, Gerino M, Boudou A (2005a) Effects of bioturbation on cadmium transfer and distribution into freshwater sediment. Environ Toxicol Chem 24:1048–1058

Ciutat A, Gerino M, Mesmer-Dudons N, Anschutz P, Boudou A (2005b) Cadmium bioaccumulation in Tubificidae from overlying water source and effects on bioturbation. Ecotoxicol Environ Safe 60:237–246

Ciutat A, Weber O, Gerino M, Boudou A (2006) Stratigraphic effects of tubificids in freshwater sediments a kinetic study based on X-ray images and grain-size analysis. Acta Oecol 30:228–237

Conder JM, Lapoint TW, Bowen AT (2004) Preliminary kinetics and metabolism of 2,4,6-trinitrotoluene and its reduced metabolites in an aquatic oligochaete. Aquat Toxicol 69:199–213

Connell DW, Bowman M, Hawker DW (1988) Bioconcentration of chlorinated hydrocarbons from sediment by oligochaetes. Ecotoxicol Environ Safe 16:293–302

Connolly JP, Pedersen CJ (1988) A thermodynamic-based evaluation of organic chemical accumulation in aquatic organisms. Environ Sci Technol 22:99–103

Conrad AU, Comber SD, Simkiss K (2000) New method for the assessment of contaminants uptake routes in the oligochaete *Lumbriculus variegatus*. Bull Environ Contam Toxicol 65:16–21

Croce V, de Angelis S, Patrolecco L, Polesello S, Valsecchi S (2005) Uptake and accumulation of sediment-associated 4 nonylphenol in a benthic invertebrate (*Lumbriculus variegatus*, freshwater oligochaete). Environ Toxicol Chem 24:1165–1171

Di Toro DM, Zarba CS, Hansen DJ, Berry WJ, Swartz RC, Cowan CE, Pavlou SP, Allen HE, Thomas NA, Paquin RP (1991) Technical basis for establishing sediment quality criteria for non ionic organic chemicals using equilibrium partitioning. Environ Toxicol Chem 10:1–43

Egeler P, Römbke J, Meller M, Knacker Th, Franke C, Studinger NR (1997) Bioaccumulation of lindane and hexachlorobenzene by tubificid sludgeworms (Oligochaeta) under standardized laboratory conditions. Chemosphere 35:835–852

Egeler P, Rombke J, Meller M, Knacker Th, Nagel R (1999) Bioaccumulation test with tubificid sludgeworms in artificial media. Hydrobiologia 406:271–280

Egeler P, Meller M, Roembke J, Spoerlein P, Streit B, Nagel R (2001) *Tubifex tubifex* as a link in food chain transfer of hexachlorobenzene from contaminated sediment to fish. Hydrobiologia 463:171–184

Filipowicz AB, Weinstein JE, Sanger DM (2007) Dietary transfer of fluoranthene from an estuarine oligochaete (*Monopylephorus rubroniveus*) to grass shrimp (*Palaemonetes pugio*): influence of piperonyl butoxide. Mar Environ Res 63:132–145

Fischer E (1976) Chloragogenzelle-eleocyt transformation, induziert mit Benomyl- und Carbofuran- Vergiftung der Lumbriciden (Oligochaeta). Zool Anz 197:225–233

Fischer E (1993) The myelo-erytroid nature of the chloragogenous-like tissues in the annelids. Comp Biochem Physiol 106A:449–453

Fischer E, Horváth I (1976) The effect of carbofuran-toxication on the chloragogen tissue of *Tubifex tubifex* Müll. (Oligochaeta). Z Mikrosk Anat Forsch 4:720–736

Fischer E, Horváth I (1977) Cytochemical studies on the cuticle and epidermis of *Tubifex tubifex* Müll. with special regard to the localization of polysaccharides, heavy metals and the DAB-reactivity. Histochemistry 54:259–271

Fischer E, Molnár L (1992) Environmental aspects of the chloragogenous tissue of earthworms. Soil Biol Biochem 24:1723–1727

Fisher SW, Chordas SW III, Landrum PF (1999) Lethal and sublethal body residues for PCB intoxication in the oligochaete, *Lumbriculus variegatus*. Aquat Toxicol 45:115–126

Fleming TP, Richards KS (1982) Uptake and surface adsorption of zinc by the freshwater tubificid oligochaete *T. tubifex*. Comp Biochem Physiol C Comp Pharmacol 71:69–75

Galassi S, Guzzella L, Battegazzore M, Carrieri A (1994) Biomagnification of PCBs, p, p′-DDE, and HCB in the River Po ecosystem (northern Italy). Ecotoxicol Environ Safe 29:174–186

Giere O, Pfannkuche O (1982) Biology and ecology of marine Oligochaeta, a review. In: Barnes M (ed) Oceanography and marine biology: an annual review, 20. Aberdeen University Press, Aberdeen, pp 173–308

Giere O, Rhode B, Dubilier N (1988) Structural peculiarities of the body wall of *Tubificoides benedii* (Oligochaeta) and possible relations to its life in sulphidic sediments. Zoomorphology 108:29–40

Gillis PL, Diener LC, Reynoldson TB, Dixon DG (2002) Cadmium-induced production of a metallothioneinlike protein in *Tubifex tubifex* (Oligochaeta) and *Chironomus riparius* (Diptera): correlation with a reproduction and growth. Environ Toxicol Chem 21:1836–1844

Gillis PL, Dixon DG, Borgmann U, Reynoldson TB (2004) Uptake and depuration of cadmium, nickel, and lead in laboratory-exposed *Tubifex tubifex* and corresponding changes in the concentration of a metallothionein-like protein. Environ Toxicol Chem 23:76–85

Gunn AM, Hunt DTE, Winnard DA (1989) The effect of heavy metal speciation in sediment on bioavailability to tubificid worms. Hydrobiologia 188(189):487–496

Gustavsson LM (2001) Comparative study of the cuticle in some aquatic oligochaetes (Annelida: Clitellata). J Morphol 248:185–195

Harkey GA, Landrum PF, Klaine SJ (1994) Comparison of whole-sediment, elutriate and pore-water exposures for use in assessing sediment associated organic. Environ Toxicol Chem 13:1315–1329

Harkey GA, van Hoof PL, Landrum PF (1995) Bioavailability of polycyclic hydrocarbons from a historically contaminated sediment core. Environ Toxicol Chem 14:1551–1560

Hernandez M, Rovira JV, Antonio MT (1988) Relationship between Cu and Pb levels in sediments and dwelling tubificidae in the Jarama River (Madrid, Spain). In: Astruc M, Lester JN (eds) Heavy metals in the hydrological cycle. Selper Ltd, London, pp 531–538

Higgins CP, McLeod PB, MacManus-Spencer LA, Luthy RG (2007) Bioaccumulation of perfluorochemicals in sediments by the aquatic oligochaete *Lumbriculus variegatus*. Environ Sci Technol 41:4600–4606

Hirsch MP (1998) Bioaccumulation of silver from laboratory-spiked sediments in the oligochaete (*Lumbriculus variegatus*). Environ Toxicol Chem 17:605–609

Hoffmann K, Wulf A (1993) Comparative studies of the integumentary uptake of short chain carboxylic acids by freshwater oligochaetes. Comp Biochem Physiol 104:169–174

Hyötiläinen T, Oikari A (2004) Bioaccumulation of PAHs from creosote-contaminated sediment in a laboratory-exposed freshwater oligochaete, *Lumbriculus variegatus*. Chemosphere 57:159–164

Ingersoll CG (1995) Sediment tests. In: Rand GM (ed) Fundamentals of aquatic toxicology: effects, environmental fate and risk assessment, 2nd edn. Taylor & Francis, Washington, D.C., pp 231–255

Ingersoll CG, Ankley GT, Benoit DA, Bronson EL, Burton GA, Dwyer FJ, Hoke R, Landrum PF, Norberg-King TJ, Winger PV (1995) Toxicity and bioaccumulation of sediment-associated contaminants using freshwater invertebrates: a review of methods and applications. Environ Toxicol Chem 14:1885–1894

Ingersoll CG, Buson EL, Wang N, Dwyer FJ, Ankley GT, Mount DR, Huckins J, Petty J, Landrum PF (2003) Uptake and depuration of nonionic organic contaminants from sediment by the oligochaete, *Lumbriculus variegatus*. Environ Toxicol Chem 22:872–885

Ireland MP (1983) Heavy metal uptake and tissue distribution in earthworms. In: Satchell JE (ed) Earthworm ecology from Darwin to vermiculture, Chapter 21. Chapman & Hall, London, pp 247–265

Jamieson BGM (1981) The ultrastructure of the Oligochaeta. Academic, London, p 462

Jernelöv A (1970) Release of methyl mercury from sediments with layers containing inorganic mercury at different depths. Limnol Oceanogr 15:958–960

Kabir SMH, Khatoon N (1980) Toxicity of some common insecticides to *Limnodrilus* spp (Oligochaeta: Tubificidae). Bangladesh J Zool 8:61–67

Kaiser M, Irmer U, Weiler K (1989) Monitoring of water quality: seasonal variation of heavy metals in sediments, suspended particulate water and tubificids of the Elbe River. Environ Technol Lett 10:845–855

Keilty TJ, White DS, Landrum PF (1988a) Sublethal responses to endrin in sediment by *Limnodrilus hoffmeisteri* (Tubificidae), and in mixed-culture with *Stylodrilus heringianus* (Lumbriculidae). Aquat Toxicol 13:227–250

Keilty TJ, White DS, Landrum PF (1988b) Sublethal responses to endrin in sediment by *Stylodrilus heringianus* (Lumbriculidae) as measured by a ^{137}Cesium marker layer technique. Aquat Toxicol 13:251–270

Kennedy CR (1969) Tubificid oligochaetes as food of dace *Leuciscus leuciscus* (L.). J Fish Biol 1:11–15

Klerks PL, Bartholomew PR (1991) Cadmium accumulation and detoxification in a Cd-resistant population of the oligochaete *Limnodrilus hoffmeisteri*. Aquat Toxicol 19:97–112

Klump JV, Krezoski JR, Smith ME, Kaster JL (1987) Dual tracer studies of the assimilation of an organic contaminant from sediments by deposit feeding oligochaetes. Can J Fish Aquat Sci 44:1574–1583

Kraaij R, Mayer P, Busser FJM, van het Bolscher M, Seinen W, Tolls J (2003) Measured pore-water concentrations make equilibrium partitioning work – a data analysis. Environ Sci Technol 37:268–274

Krantzberg G (1994) Spatial and temporal variability in metal bioavailability and toxicity of sediment from Hamilton Harbour, Lake Ontario. Environ Toxicol Chem 13:1685–1698

Kukkonen J, Landrum PF (1994) Toxicokinetics and toxicity of sediment-associated pyrene to *Lumbriculus variegatus* (Oligochaeta). Environ Toxicol Chem 13:1457–1468

Kukkonen J, Landrum PF (1995a) Measuring assimilation efficiencies for sediment-bound PAH and PCB congeners by benthic organisms. Aquat Toxicol 32:75–92

Kukkonen J, Landrum PF (1995b) Effects of sediment bound polymethylsiloxane on the bioavailability and distribution of benzo(a)pyrene in lake sediment to *Lumbriculus variegatus*. Environ Toxicol Chem 14:523–531

Landrum PF, Meador JP (2002) Is the body residue a useful dose metric for assessing toxicity? SETAC Globe 3:32–34

Landrum PF, Gedeon ML, Burton GA, Greenberg MS, Rowland CD (2002) Biological responses of *Lumbriculus variegatus* exposed to fluorathene-spiked sediment. Arch Environ Contam Toxicol 42:292–302

Landrum PF, Leppänen M, Robinson SD, Gossiaux DC, Burton GA, Greenberg M, Kukkonen JVK, Eadie BJ, Lansing MB (2004a) Effect of 3,4,3′,4′-Tetrachlorobiphenyl on the reworking behavior of *Lumbriculus variegatus* exposed to contaminated sediment. Environ Toxicol Chem 23:178–186

Landrum PF, Leppänen M, Robinson SD, Gossiaux DC, Burton GA, Greenberg M, Kukkonen JVK, Eadie BJ, Lansing MB (2004b) Comparing behavioural and chronic endpoints to evaluate the response of *Lumbriculus variegatus* to 3,4,3′,4′-tetrachlorobiphenyl sediment exposures. Environ Toxicol Chem 23:187–194

Leibig M, Egeler P, Römbke J, Oehlmann J, Knacker T (2005) Bioaccumulation of ^{14}C-17α-ethinylestradiol by the aquatic oligochaete *Lumbriculus variegatus* in spiked artificial sediment. Chemosphere 59:271–280

Leppänen M (1995) The role of feeding behaviour in bioaccumulation of organic chemicals in benthic organisms. Ann Zool Fennici 32:247–255

Leppänen MT, Kukkonen JVK (1998) Relative importance of ingested sediment and pore water as bioaccumulation routes for pyrene to oligochaete (*Lumbriculus variegatus*, Müller). Environ Sci Technol 32:1503–1508

Leppänen MT, Kukkonen JVK (2000) Fate of sediment-associated pyrene and benzo[a]pyrene in the freshwater oligochaete *Lumbriculus variegatus* (Muller). Aquat Toxicol 49:199–212

Leppänen MT, Kukkonen JVK (2004) Toxicokinetics of sediment-associated polybrominated diphenylethers (flame retardants) in benthic invertebrates (*Lumbriculus variegatus*, Oligochaeta). Environ Toxicol Chem 23:166–172

Loden MS (1974) Predation by chironomid (Diptera) larvae on oligochaetes. Limnol Oceanogr 19:156–159

Loonen H, Muir DCG, Parsons JR, Govers HAJ (1997) Bioaccumulation of polychlorinated dibenzo-p-dioxins in sediment by oligochaetes: influence of exposure pathway and contact time. Environ Toxicol Chem 16:1518–1525

Lu X, Reible DD, Fleeger JW (2004) Bioavailability and assimilation of sediment-associated benzo[a]pyrene by *Ilyodrilus templetoni* (Oligochaeta). Environ Toxicol Chem 23:57–64

Lucan-Bouché ML, Biagianti-Risbourg S, Arsac F, Vernet G (1999a) Autotomy as a mechanism of decontamination used by the oligochaete *Tubifex tubifex*. Bull Soc Zool Fr 124:383–387

Lucan-Bouché ML, Biagianti-Risbourg S, Arsac F, Vernet G (1999b) An original decontamination process developed by the aquatic oligochaete *Tubifex tubifex* exposed to copper and lead. Aquat Toxicol 45:9–17

Lyytikäinen M, Sormunen A, Peräniemi S, Kukkonen JVK (2001) Environmental fate and bioavailability of wood preservatives in freshwater sediments near an old sawmill site. Chemosphere 44:341–350

Mäenpää K, Kukkonen JVK (2006) Bioaccumulation and toxicity of 4-nonylphenol (4-NP) and 4-(2-dodecyl)-benzene sulfonate (LAS) in *Lumbriculus variegatus* (Oligochaeta) and *Chironomus riparius* (Insecta). Aquat Toxicol 77:329–338

Mäenpää K, Sormunen AJ, Kukkonen JVK (2003) Bioaccumulation and toxicity of sediment associated herbicides (Ioxynil, Pendimethalin and Bentazone) in *Lumbriculus variegatus* (Oligochaeta) and *Chironomus riparius* (Insecta). Ecotoxicol Environ Safe 56:398–410

McCarty LS (1986) The relationship between aquatic toxicity QSARs and bioconcentration for some organic chemicals. Environ Toxicol Chem 5:1071–1080

McCarty LS (1991) Toxicant body residues: implications for aquatic bioassays with some organic chemicals. In: Mayes MA, Barron MG (eds) Aquatic toxicology and risk assessment, vol 14. ASTM special technical publication 1124, Philadelphia, pp 183–192

McMurtry MJ, Rapport DJ, Chua KE (1983) Substrate selection of tubificid oligochaetes. Can J Fish Aquat Sci 40:1639–1646

Millward RN, Fleeger JW, Reible DD, Keteles KA, Cuningham BP, Zhang L (2001) Pyrene bioaccumulation, effects of pyrene exposure on particle size selection, and fecal pyrene content I the oligochaete *Limnodrilus hoffmeisteri* (Tubificidae, Oligochaeta). Environ Toxicol Chem 20:1359–1366

Morgan JE, Morgan AJ (1990) The distribution of cadmium, copper, lead, zinc and calcium in the tissues of the earthworm *Lumbricus rubellus* sampled from one uncontaminated and four polluted soils. Oecologia 84:559–566

Morgan AJ, Morgan JE, Turner M, Winter C, Yarwood A (1993) Metal relationships of earthworms. In: Dallinger R, Rainbow PS (eds) Ecotoxicology of metals in invertebrates. A special publication of SETAC, Chapter 17. Lewis Publ, Boca Raton, pp 333–358

Mosleh YY, Paris-Palacios S, Couderchet M, Biagianti-Risbourg S, Vernet G (2005) Effects of the herbicide isoproturon on metallothioneins, growth, and antioxidative defenses in the aquatic worm *Tubifex tubifex* (Oligochaeta, Tubificidae). Ecotoxicology 14:559–571

Mount DR, Highland TL, Mattson VR, Dawson TD, Lott KG, Ingersoll CG (2006) Use of the oligochaete, *Lumbriculus variegatus*, as a prey organism for toxicant exposure of fish through the diet. Environ Toxicol Chem 25:2760–2767

Naqvi SMZ (1973) Toxicity of twenty-three insecticides to a tubificid worm *Branchiura sowerbyi* from the Mississippi Delta. J Econ Entom 66:70–74

Ng TY-T, Wood CM (2008) Trophic transfer and dietary toxicity of Cd from the oligochaete to the rainbow trout. Aquat Toxicol 87:47–59

Nikkilä A, Halme A, Kukkonen JVV (2003) Toxicokinetics, toxicity and lethal body residues of two chlorophenols in the oligochaete worm, *Lumbriculus variegatus*, in different sediments. Chemosphere 51:35–46

Nuutinen S, Kukkonen JVK (1998) The effect of selenium and organic material in lake sediments on the bioaccumulation of methylmercury by *Lumbriculus variegatus*. Biogeochemistry 40:267–278

OECD (2008) Bioaccumulation in sediment-dwelling benthic oligochaetes. OECD guideline for the testing of chemicals no 315

Oliver BG (1984) Uptake of chlorinated organics from anthropogenically contaminated sediments by oligochaete worms. J Can Fish Aquat Sci 41:878–883

Oliver BG (1987) Bio-uptake of chlorinated hydrocarbons from laboratory-spiked and field sediments by oligochaete worms. Environ Sci Technol 21:785–790

Patrick EM, Loutit M (1976) Passage of metals in effluents, through bacteria, to higher organisms. Water Res 10:333–335

Patrick EM, Loutit M (1978) Passage of metals to freshwater fish from their food. Water Res 12:395–398

Penttinen OP, Kukkonen JVK (2000) Metabolic response of *Lumbriculus variegatus* to respiratory uncoupler in cold and anoxic water. Environ Toxicol Chem 19:2073–2075

Penttinen OP, Kukkonen JVK, Pellinen J (1996) Preliminary study to compare body residues and sublethal energetic responses in benthic invertebrates exposed to sediment-bound 2,4,5-trichlorophenol. Environ Toxicol Chem 15:160–166

Penttinen OP, Kilpi-Koski J, Jokela M, Toivainen K, Väisänen A (2008) Importance of dose metrics for lethal and sublethal sediment metal toxicity in the oligochaete worm *Lumbriculus variegatus*. J Soils Sediments 8:59–66

Redeker E, Blust R (2004) Accumulation and toxicity of cadmium in the aquatic oligochaete *Tubifex tubifex*: a kinetic modelling approach. Environ Sci Technol 38:537–543

Redeker ES, van Campenhout K, Bervoets L, Reijnders H, Blust R (2007) Subcellular distribution of Cd in the aquatic oligochaete *Tubifex tubifex*, implications for trophic availability and toxicity. Environ Pollut 148:166–175

Rodriguez P, Martinez-Madrid M, Arrate JA, Navarro E (2001) Selective feeding by the aquatic oligochaete *Tubifex tubifex* (Tubificidae, Clitellata). Hydrobiologia 463:133–140

Rüther U, Greven H (1990) The effect of heavy metals on enchytraeids. I. Uptake from an artificial substrate and influence on food preference. Acta Biol Benrodis 2:125–131

Sager M, Pucsko R (1991) Trace element concentrations of oligochaetes and relations to sediment characteristics in the reservoir at Altenwörth/Austria. Hydrobiologia 226:39–49

Say PJ, Giani N (1981) The Riou Mort, a tributary to the river Lot polluted by heavy metals. II. Accumulation of zinc by oligochaetes and chironomids. Acta Oecologica 2:339–355

Schlekat CE, McGeer JC, Blust R, Borgmann U, Brix KV, Bury N, Couilard Y, Dwyer RL, Luoma SN, Robertson S, Sappigton KG, Schoeters I, Sijm DTHM (2007) Bioaccumulation: hazard identification of metals and inorganic metal substances. In: Adams WJ, Chapman PM (eds) Assessing the hazard of metals and inorganic metal substances in aquatic and terrestrial system. SETAC Press, Pensacola, pp 55–87

Sheedy B, Mattson VR, Cox JS, Kosian PA, Phipps GL, Ankley GT (1998) Bioconcentration of polycyclic aromatic hydrocarbons by the freshwater oligochaete *Lumbriculus variegatus*. Chemosphere 36:3061–3070

Singh RK, Chavan SL, Sapkale PH (2007) Heavy metal concentrations in water, sediments and body tissues of red worm (*Tubifex* spp.) collected from natural habitats in Mumbai, India. Environ Monit Assess 129:471–481

Spacie A, McCarty LS, Rand GM (1995) Bioaccumulation and bioavailability in multiphase systems. In: Rand GM (ed) Fundamentals of aquatic toxicology: effects, environmental fate, and risk assessment, 2nd edn. Taylor & Francis, Washington, D.C., pp 493–521

Standley LJ (1997) Effect of sediment organic matter composition on the partitioning and bioavailability of dieldrin to the oligochaete *Lumbriculus variegatus*. Environ Sci Technol 31:2577–2583

Testerman JK (1972) Accumulation of fatty acids from sea water by marine invertebrates. Biol Bull 142:160–177

USEPA (2000) Methods for measuring the toxicity and bioaccumulation of sediment-associated contaminants with freshwater invertebrates, 2nd edn. EPA 600/R-99/064

Vetvicka V, Sima P, Cooper EL, Bilej M, Roch P (1994) Immunology of annelids. CRC Press, Boca Raton, p 299

Vidal DE, Horne AJ (2003) Mercury toxicity in the aquatic oligochaete *Sparganophilus pearsei* II: autotomy as a novel form of protection. Arch Environ Contam Toxicol 45:462–467

Wagner G (1968) Zur Bezeihung zwischen der Besiedlungsdichte von Tubificiden und dem Nahrungsangebot im Sediment. Int Rev Ges Hydrobiol 53:715–721

Wallace WG, Lopez GR (1996) Relationship between subcellular cadmium distribution in prey and cadmium trophic transfer to a predator. Estuaries 19:923–930

Wallace WG, Lopez GR (1997) Bioavailability of biologically sequestered cadmium and the implications of metal detoxification. Mar Ecol Prog Ser 147:149–157

Wavre H, Brinkhurst RO (1971) Interactions between some tubificid oligochaete and bacteria found in the sediments of Toronto Harbour, Ontario. J Fish Res Bd Can 28:335–341

Weinstein JE (2003) Influence of salinity on the bioaccumulation of fluoranthene to an estuarine shrimp and oligochaete. Environ Toxicol Chem 2:2932–2939

Weinstein JE, Sanger DM, Holand AF (2003) Bioaccumulation and toxicity of fluoranthene in the estuarine oligochaete *Monopylephorus rubroniveus*. Ecol Environ Safe 55:278–286

West CW, Ankley GT, Nichols JW, Elonen GE, Nessa DE (1997) Toxicity and bioaccumulation of 2,3,7,8,-tetrachlorodibenzo-p-diaoxin in long-term tests with the freshwater benthic invertebrates *Chironomus tentans* and *Lumbriculus variegatus*. Environ Toxicol Chem 16:1287–1294

Whitten BK, Goodnight CJ (1967) The accumulation of SR-89 and CA-45 by an aquatic oligochaete. Physiol Zool 40:371–385

Whitten BK, Goodnight CJ (1969) The role of tubificid worms in the transfer of radioactive phosphorous in an aquatic ecosystem. In: Nelson DJ, Evans FC (eds) Symposium on radioecology CFSTI National Bureau Studies. US Department Commerce, Springfield, pp 270–277

Willuhn J, Schmitt-Wrede HP, Greven H, Wunderlich F (1994a) Cadmium-induced m-RNA encoding a non-metallothionein 25-kDa protein in *Enchytraeus buchholzi* (Oligochaeta). Ecotoxicol Environ Safe 29:93–100

Willuhn J, Schmitt-Wrede HP, Greven H, Wunderlich F (1994b) cDNA cloning of a cadmium-inducible m-RNA encoding a novel cysteine-rich, non-metallothionein 25-kDa protein in an enchytraeid earthworm. J Biol Chem 269:24688–24691

Winger PV, Lasier PJ, White DH, Seginak JT (2000) Effects of contaminants in dredge material from the lower Savannah River. Arch Environ Contam Toxicol 38:128–136

Xie L, Lambert D, Martin C, Cain DJ, Luoma SN, Buchwalter D (2008) Cadmium biodynamics in the oligochaete *Lumbriculus variegatus* and its implications for trophic transfer. Aquat Toxicol 86:265–271

Zaranko DT, Griffiths RW, Kaushik NK (1997) Biomagnification of polychlorinated biphenyls through a riverine food web. Environ Toxicol Chem 16:1463–1471

Chapter 6
Methodological Issues

Abstract Suggestions are made on sampling devices, sample processing, measurement of biomass, and preparation of material for identification. In addition, issues relating to the use of biomass rather than abundance as a response variable are addressed. Advice is also provided on many aspects of culturing oligochaetes and factors that should be considered or verified, such as genetic strains of the same species that can have different reproductive or growth patterns. The culture of marine and estuarine oligochaetes has not been standardised to the same degree as for freshwater species. Identification of worms inhabiting polluted areas does not generally require specialised training or extensive taxonomical expertise, as typically only a few species occur. However, for reference sites in a field survey the assistance of an oligochaete specialist is strongly recommended. Various factors relevant to toxicity testing are addressed including the use of artificial sediment, removal of indigenous organisms, identification of cocoons, the use of food supplements and confounding factors. Finally, gut purging, duration and density of exposures for bioaccumulation studies and handling of specimens is discussed.

6.1 Introduction

The objectives of this chapter are to provide practical information on various methodological issues on the use of aquatic oligochaetes in pollution biology. While not an exhaustive description of methodologies, it hopefully provides advice to those considering working with aquatic oligochaetes both in the field and in the laboratory. We provide advice that is often not available in the descriptions of methods found in the literature but that we consider important to those working in the field. This is often based on work that we ourselves or colleagues have done and is the type of information we wished was available to us before we

P. Rodriguez and T.B. Reynoldson, *The Pollution Biology of Aquatic Oligochaetes*,
DOI 10.1007/978-94-007-1718-3_6,

undertook this work. The material described below addresses topics that will affect decisions on the development and selection of a sampling design, laboratory methods (including sieving, subsampling and biomass estimation), culturing, and the preparation of specimens for identification. We also have included some specific details relevant to studies in toxicity testing and bioaccumulation.

6.2 Sampling

Decisions related to where to sample and what comprises a sampling site are basic, and there are a number of publications on this subject (*e.g.* Rosenberg and Resh 1993; Bailey et al. 2004). Grabs, cores, artificial substrates, Surber-type samplers and hand or kick nets described for sampling benthic fauna (*e.g.* Downing 1984; Peckarsky 1984) are all suitable devices for sampling oligochaetes. The selection will be made depending on the objectives of the survey, and the characteristics of the habitat (depth, nature of substrate, vegetation, water flow, and so on). Kick nets are suitable for lotic systems with flowing water and stony substrate (*rithron* sections of the rivers). The use of kick nets is typically associated with qualitative or semi-quantitative studies. For rivers, semi-quantitative sampling methods are sufficient for establishing a comprehensive list of oligochaete taxa. Kicking or washing the stones, plants or pieces of submerged wood with the hand, just upstream of the net is appropriate in numerous shallow river habitats (sands, cobbles and gravel, algae, aquatic rooted plants, and so on). In fact, this is generally why hand or kick nets are used in many regional bioassessment programmes (*e.g.* RIVPACS, CABIN and EPA rapid bioassessment). For studies that include oligochaetes, a net mesh size of 500 μm is usual in general field community studies (Nijboer et al. 2004), or smaller when sampling special habitats (*e.g.* 160 μm: Juget and Lafont 1982; 100 μm: Giani et al. 2001; Sambugar et al. 2005 for groundwater oligochaetes) or in life history studies.

In deep river sections, hand-net sampling can be restricted to the banks less than 1-m depth. For deeper river sections or lentic environments (lakes and reservoirs), the use of grabs and cores is required (Brinkhurst 2003). At greater depth, grabs or air-lift samplers (Drake and Elliott 1982) can be used, the latter equipped with a core-like device, that "vacuum up" bottom material by a pump of compressed air (see for instance, Ankley et al. 1992). With any sampling device the primary considerations are: lack of bias toward any particular group of organisms; consistency, in so far as the device works in the same way and with the same efficiency across the sites being sampled; and practicality. Given that the sampler behaves consistently over the range of substrata to be sampled, and sampling effort (time and area) is controlled, taxa richness, relative abundance and density values in several sites will be comparable. In general, most oligochaete species prefer specific habitats (Verdonschot 2001). Thus, sampling sites should be selected to minimize habitat variation, or separate sampling protocols for different habitats must be selected.

6.3 Fixation, Preservation and Preparation of Material

Once samples are acquired, there are a number of issues that can affect the ability to suitably identify specimens. Preservation of the sample may also affect estimates of both abundance and diversity, as some methods are unsuitable for certain taxa. The use of alcohol for field samples fixation is not appropriate as it can result in fragmentation to a greater or lesser degree and disintegration of worms. In field surveys, worm samples with sediment and other animals should be fixed in 5–10% formaldehyde, paying attention to the fact that there is a certain degree of dilution of the fixative when it is added to the sample. Concentrated formaldehyde (commercial formalin = 40% aqueous solution of formaldehyde) can be added to the sample (eliminating as much water as possible) to achieve the desired concentration of formaldehyde (Hangay and Dingley 1985), however, from a safety perspective, pre-diluted 10% buffered formalin is recommended.

After fixation in formalin for 48–72 h, samples can be washed with tap water by elutriation using the same or smaller mesh size than that used for sampling. Typically, sieves of 200 μm are used for population studies where juvenile worms need to be collected and 400–500 μm in community studies for bioassessment. However, if the objective is to acquire a comprehensive list of the oligochaete taxa, sieves of 200 μm or less are required for small-sized species.

Sorting live oligochaete worms from field sediments to identify or measure tissue concentration can be done directly using dental hooks, under a binocular microscope if necessary, after washing the sample by elutriation. The sample sorting strategy depends on the objective of the work. For population studies, samples often require subsampling to calculate abundances per area or relative abundances since samples can contain very large numbers of worms (up to densities of several hundred thousands per square meter). Juget and Lafont (1982) have proposed sorting 125–200 individuals as a representative subsample for each habitat, although series of studies and papers have suggested 300, 500 and more as fixed count subsamples, the decision in these cases often relates to the perceived importance of rare taxa in the study. Studies that rely on richness-based metrics tend to recommend higher fixed count sub-samples, while those using community composition metrics tend to favour lower fixed count sub-samples. Subsampling can be done with large gridded white trays. We would always recommend knowing the proportion of the sample that was sub-sampled so that a back calculation to provide a whole sample estimate of abundance can be made, and saving the remainder sample of the sample until identification is complete is important, both for QAQC procedures and when rare taxa are important. Animals can be sorted by eye when they are large, or under a binocular microscope for small-sized specimens. To help in the sorting of small worms some authors have used staining methods, such as the addition of 0.25–1% alcoholic-eosin to the sample for several minutes (based in Korínková and Sigmund 1968) or aqueous-eosin (2%) (Juget and Lafont 1982). Living oligochaetes can also be sorted from field sediment samples without fixation, either by eye, or by using infrared light that forces tubificids to leave the drying sediment (Kaiser et al. 1989).

Similarly, Krantzberg (1994) placed the sample material retained by a 500-μm sieve in trays fitted with Nytex screen bottoms placed over a container with site water. The trays were held under lighted conditions for 12–24 h and the negative phototactism of the worms (and probably the higher temperature close to the light source, too) resulted in the organisms passing through the top tray into the collecting basin, where they were separated using Pasteur pipettes. However, before any of these methods are incorporated into a programme, an estimate of their efficiency should be made, preferably by sub-sampling and then processing the residue. Again, consistency in any selected method is key. Fixed worms are commonly preserved in 70% ethanol. The addition of a small quantity of glycerine (less than 5%) to the alcohol reduces the risk of drying if the alcohol evaporates or is lost from the vial. All vials should be adequately labelled with date and site information (locality, river or lake).

For microscopic examination, worms should be placed in a few drops of water, glycerine or a clearing agent, and covered by a glass coverslip. When identifying samples with large numbers of worms (typically, from polluted sites) it is helpful to use a clearing mounting media. Brinkhurst (1986) recommended Hydramount® or CMC-10. Amman's lactophenol (1 part phenol + 1 part lactic acid + 2 parts glycerol + 1 part water) is commercially available, but great care should be taken as this is a toxic material. Thinner specimens mounted in lactophenol can be examined immediately after mounting, but normally several hours to 1 day is required for sufficient clearing of tissues in reproductive worms and for making visible chaetal and penial structures. The use of a hot plate or oven (about 40°C) can help to speed up the process, but higher temperatures should be avoided as the lactophenol may boil. Lactophenol is toxic, and careful attention must be paid to prevent inhalation of any toxic fumes, either by using a well-ventilated room or preferably working under a fume hood. Prepared slides can be sealed with fingernail polish or another sealant. The use of Hydramount®, Berlese fluid, Apathy's gum-syrup, and Levulose syrup as mounting media was discussed by Lafont (1983). However, these are non-permanent preparations that give good resolution to chitinous structures, such as chaetae, and to penial sheaths, but internal organs can not readily seen and are partially or completely destroyed. Non-permanent preparations usually have a short shelf life (several months to a few years) due to the formation of bubbles, shrinking or crystallisation.

It is always important in field community studies to keep some specimens in alcohol or in permanent mounts for specialist verification, if required. It is also valuable to prepare a reference collection for any study, so that if future taxonomic issues arise they can be addressed. Permanent mounts can be prepared with whole stained worms, using hematoxylin-eosin, rose bengal or borax carmine, followed by differentiation in acid-alcohol. These should be dehydrated using a series of ethyl alcohol solutions and cleared in methyl salicylate (Brinkhurst 1986, S. Fend pers. commun.). For safety, xylene or toluene should be avoided as clearing agents for mounting specimens, because of their high toxicity if inhaled. After clearing, worms are mounted in Canada balsam or artificial resins, as Permount®. Creosote has been used for both dehydrating and clearing of the worm after staining, rinsing the specimens for one to several hours or even several days (under a fume hood), depending on the size of the worm, before it is mounted in resin. It can take several days or weeks until Canada balsam is sufficiently dried, until then, preparations must be kept flat. Worms mounted in Canada balsam can be recovered for further study or for dissection by re-immersion in methyl salicylate (or toluene or xylene) used in the slide preparation, followed by rehydration in the alcohol series of decreasing concentrations.

6.4 Biomass Estimation

The measurement of the worm biomass can be used as a variable for quality assessment in community studies, as endpoint in toxicity tests, and it is compulsory to calculate bioaccumulation. Abrahamsen (1973) calculated the biomass of enchytraeids and lumbricids from length or width to weight or width to volume regression models. However, worm length is dependent on the degree of contraction of the animal. In relaxed animals, length or width of a fixed segment (*e.g.* segment XI) to weight relationships can be used. Length to weight and width to weight relationships for four species were determined by Reynoldson (1983), resulting a good fit for length to wet weight regression model ($r^2=0.42-0.81$), but very low fit for width to weight ($r^2=0.00-0.06$). Smit et al. (1993) used biovolume data (calculated by the displacement of water by live organisms in a 5-ml burette, and from the diameter of the eleventh segment through a regression function) as a non-destructive technique for estimating biomass for several tubificid species. Factors for conversion of biovolume into wet weight, dry weight and ash-free dry weight, using a pycnometer (5.477 ml) were discussed.

Most authors obtain biomass either through the direct wet weight of the blotted-dry individuals or the dry weight after desiccation of the samples in a heater to constant weight (commonly, about 24 h at 60°C, followed by cooling to room temperature in a desiccator). Measuring wet weight has several technical problems as there is constant dehydration of the worms while weighing. However, in experiments on individual animals through time there is little choice, as obtaining a dry-weight measurement is somewhat final. The use of a silica fluid drop for calculation of wet weight was proposed by Lundkvist (1978) for small enchytraeids to minimize the loss of water in the organisms during the weighing operation. This same method was used by Bouguenec and Giani (1989).

In bioaccumulation studies and toxicity tests, it is important to measure worm biomass after gut purging for several hours. Brooke et al. (1996) measured the gut content of several macroinvertebrates, including *Lumbriculus variegatus*, by low-pressure oxygen plasma ashing, a technique that measures the volatile oxidizable matter at a low temperature (≤60°C). The authors measured the gut clearance rate and found that the worm gut was completely empty at the end of the first 12-h depuration period. Martinez-Madrid et al. (1999) measured the egestion rates in *Tubifex tubifex* after the worms were placed in dechlorinated tap water and reported that the faecal production during the first 4 h represented more than the 95% of the total produced in 24 h. These results were similar to those reported by Gnaiger and Staudigl (1987) who measured defecation rates under aerobic and anaerobic conditions. These authors observed that defecation velocity (mm faeces s^{-1}) was a function of worm length and, under aerobic conditions at 20°C, the gut was empty after 6 h. A more sophisticated system based in $^{51}Cr^{3+}$ labelled sediment was used by Klump et al. (1987) who estimated maximum gut clearing times to average 3.6 ± 3.1 h, in a mixture of aquatic oligochaetes composed largely of *Limnodrilus hoffmeisteri*. Gillis et al. (2004) did a different approximation to the gut-clearance problem using data on wet-weight loss of *T. tubifex* during 96 h. They suggested using a gut-clearance time of 24 h, although some biomass loss was still measured after this period (2.3% per hour), which was imputed to probable biomass losses due to starvation.

6.5 Culture of Animals

In many laboratory animal-housing facilities, live tubificids are kept in sediment in aerated water, using running water channels or dishes arranged in a cascade system with recycled water pumped through a filtering reservoir. They are very easy to culture and benign neglect usually provides large populations, mostly consisting of *Tubifex tubifex* and *Limnodrilus hoffmeisteri*. As soon as attempts are made to systematize culturing for only one species, or for less adaptable species, problems arise. Firstly, because we need to be able to clearly distinguish the presence of individuals of species different to the one that we wish to culture, which requires some basic knowledge of the taxonomy of the species or a temporary contact with a specialist. Another problem is that many species are sensitive to handling and to laboratory conditions which can make their permanent culture difficult. However, acclimation of a sampled population over a certain period of time in the laboratory in healthy conditions (*i.e.* with low mortality and without a significant loss of biomass) is usually an accessible objective for many aquatic oligochaete species.

Several protocols have been proposed over the last 50 years for culturing oligochaete species of different families for different purposes. Many of the methods use different types of substrate, food source and ambient conditions. However, one thing is common to all methods as worms are extremely sensitive to dissolved chlorine in tap water that can cause mortality and disintegration of the body, therefore, the use of chlorinated tap water must be avoided even in relatively short sieving or sorting processes. Active carbon filters connected to the tap water system, or reconstituted water, are possible alternatives to ensure chlorine free water.

Learner and Edwards (1963) and Harper et al. (1981a, b) provided methods for keeping *Nais* in laboratory culture, using hard water with agar as substrate. Population growth rates of three naidid species in culture were documented by Lochhead and Learner (1983), also using agar for growing bacteria for food. The carrying capacity of the cultures increased with temperature up to the maximum used (20°C). *Pristina aequiseta* performed better than two *Nais* species in these conditions, and Agar No. 3 was a better food source than CPS agar. *Nais* was cultured in a semi-natural way by Juget et al. (1989) using jars with gas-porous nylon immersed in a pond. Daily growth rate increased with temperature up to the maximum at 30°C. This method is probably inadequate for the permanent maintenance of large populations because of the seasonal cycles experienced in the pond.

Pristina longiseta (as *P. leidyi*) was cultured by Smith et al. (1991). The specimens were originally obtained from a commercial animal supplier as a culture named *Stylaria lacustris*, which illustrates a common problem with commercial cultures which may well be taxonomically unreliable. These authors used a 16-h light period (100 foot candles) and fed the culture with a rabbit chow suspension, which also provided the substratum as it settled. Most recently, Collado and Schmelz (2001) have successfully cultured *Pristina* species in a mixture of 0.5% Agar-Agar and soil in Petri dishes. This method has allowed not only survival but also reproduction of *P. jenkinae* for several months, and of *P. notopora* for more than 2 years. Culture in agar has also been successfully used for toxicological purposes with *Enchytraeus* spp. (Westheide and Bethke-Beilfuss 1991; Arrate et al. 2002).

Cultures of lumbriculids have been developed largely for toxicological studies. Bailey and Liu (1980) bred *Lumbriculus variegatus* in 19-l aquaria under flow-through conditions (4 turnovers per day), with dechlorinated tap water, a substrate of No. 16 sand, 5 cm deep, at 20°C, in a 16:8 h light:dark photoperiod, and at an initial stocking density of about 100 g per tank. Animals were fed with trout food, sprinkled onto the water surface. Leppänen and Kukkonen (1998) cultured *L. variegatus* in the laboratory in a 5-l aquarium containing reconstituted water (pH = 7, hardness = 1.0 mmol l^{-1} as Ca + Mg), at 20 ± 2°C, in a 16:8 h light:dark cycle. Shredded and pre-soaked paper towels were used as a substrate and the entire water volume was renewed once or twice a week. The worms were fed with approximately 100-mg grounded fish food (Tetramin®) every day. This species has never been observed to reach sexual maturity in cultures maintained for many years in the laboratory, and under stable laboratory conditions reproduction is only done through fragmentation (Ruth Collado, Philipp Egeler, and Tarmo Timm, pers. commun.). Cohort cultures have also successfully been established for the lumbriculid *Bichaeta sanguinea* by Bonacina et al. (1987).

A review of culturing methods by Timm (1980) was published in Russian in 1972. In his own laboratory, Timm has maintained for many years small oligochaete populations in glass jars, in sieved sediment from the lake Vörtsjatv (Estonia), with the sediment being changed 3–4 times per year. From this experience, Timm (1980) suggested that *T. tubifex*, *L. hoffmeisteri* and perhaps *L. variegatus* were suitable candidate species for mass production. In Canada, at the Canada Centre for Inland Waters, cultures of *T. tubifex* and other species have been successfully maintained for use in toxicity testing since 1989. There are several references that support the successful use of tubificids for mass production. Marian et al. (1989) used cow manure in a recycled water system of troughs, improving on their earlier methods of culturing tubificids. The flow rate used flushed out many of the newly hatched worms, but they could still produce 5.6 kg of worms per month, with a harvest rate of 0.8 kg m^{-2} $month^{-1}$. A more efficient culture medium for tubificid worms was developed by Ahamed and Mollah (1992), who used fine sand (20%) with cow dung (25%), wheat bran (35%) and mustard oil (20%) in a culvert system with running water. This culture media produced 2.1 kg m^{-2} $month^{-1}$. But these methods are not commonly suitable for laboratory experimentation.

Many scientific laboratories and commercial operators have cultured oligochaetes for a variety of purposes. Earthworms are used in biomass conversion in Asia, and this activity has led to research on the profitable use of oligochaetes, for example, in animal feed. They are also used as fish food in the wholesale aquarium trade, as they provide good nutrition and stimulate normal feeding behaviour in many species. Bouguenec (1992) provided a review of the oligochaetes cultured as food for fish rearing, and compared several freshwater fish species growth rates fed on different diets of both live (Enchytraeidae) or freeze-dried (*Tubifex*) oligochaetes and commercial fish food. These papers may be helpful to those considering experimental work on trophic transfer of toxicants.

There is an extensive literature on the life history and productivity of a small number of tubificid species (reviewed by Poddubnaya 1980; Reynoldson 1987) that is a valuable source of information for those contemplating starting a new culture. One of the main variants in culturing tubificids is the food source that is used. Decomposed lettuce has been the food source of choice for Timm's and other's cultures. Thus, Kosiorek (1974) described the culture of *T. tubifex* in a 5-cm layer of calcined sand and fed on buried rotten lettuce in the substrate. She maintained cultures at 24°C in the dark, water and food were renewed every 3–4 days, and sand every 16–20 days. Other food tested by the author (balls of yeast, agar cultures of bacteria, mouse food) was less effective. Similar culture procedures have been used for culturing cohorts of the tubificid species *Psammoryctides barbatus, Spirosperma ferox, T. tubifex, L. hoffmeisteri* (Adreani et al. 1984; Bonacina et al. 1987, 1989; Paoletti 1989; Pasteris et al. 1999). Bonacina et al. (1994) have reared cohort cultures of *Branchiura sowerbyi* in sand and used deep frozen spinach as food. Pasteris et al. (1999) reared *L. hoffmeisteri* in groups of about 40–50 individuals in circular glass containers (ø=11 cm, height=6 cm), half filled with sand and then filled with aerated tap water. Cultures were kept in the dark at 15±1°C, frozen lettuce was put under sand as food source, and every week each container was washed and fresh lettuce added. One of the advantages of using a food source such as lettuce or spinach rather than natural sediment is that eliminates the likelihood of culture contamination from wild worms. Activated sludge has also been used as food source when mixed with sand or river mud (1:3) for *B. sowerbyi* (Aston and Milner 1981; Aston et al. 1982) and for *T. tubifex* (Finogenova and Lobasheva 1987). The former obtained population-doubling times of only 1.6 weeks, at the optimum temperature of 25°C. Cellulose has also been shown to be an appropriate substrate for culturing *B. sowerbyi* by Aston (1984). Pasteris et al. (1994) worked with cohort cultures of *T. tubifex* using cellulose mixed with sand in different proportions as a substrate, and demonstrated that worm production increased when feeding on cellulose, however newly-hatched worms could not grow and survivorship was low when cellulose was present in high proportion.

Attempts have been made to create a standard formulated substrate that has similar characteristics to natural sediment, and could be reasonably used in chronic toxicological experiments. Egeler et al. (1997) used a formulated substrate, following OECD guideline No. 207, composed of 76% quartz sand, 22% kaolinite clay and 2% *Sphagnum* peat, in which *T. tubifex* and *L. hoffmeisteri* can be successfully cultured. This has also proved to be a good substrate for standardised bioaccumulation and toxicity approaches with *T. tubifex*, *B. sowerbyi* and *L. variegatus* (Egeler et al. 1997, 1999; OECD 2007, 2008).

Finally, when culturing species one should be aware of the existence of genetic strains of the same species that can have different reproductive or growth patterns (*e.g. T. tubifex*: Anlauf 1994; Anlauf and Neumann 1997) or tolerance to toxics (*e.g. L. hoffmeisteri:* Klerks and Bartholomew 1991, *T. tubifex*: Reynoldson et al. 1996; Sturmbauer et al. 1999). These characteristics should be verified and reported so that appropriate comparisons with results obtained from different populations can be made. The type of round-robin testing conducted by ASTM is a good example of the type of study that needs to be done.

The culture of marine and estuarine oligochaetes has not been standardised to the same degree as in freshwater species. However, several authors have successfully kept worms from field samples under laboratory conditions before experimentation. Chapman and Brinkhurst (1980) kept the estuarine *Monopylephorus irroratus* and *Tubificoides fraseri* (as *T. gabriellae*) in the dark, in Petri dishes with 0.5-ml sieved mud and 150 ml of water with different salinities from 0 to 25‰, at 10°C, and with some aeration to prevent anoxia. Chapman (1987) bred *Monopylephorus cuticulatus* in aquaria with natural substrate and clean water (25‰ salinity), at 10±0.5°C, 12:12 h dark:light cycle, with aeration, and fed with a mixture of ground *Enteromorpha* and commercial Tetramin® *ad libitum*. Dubilier et al. (1994) kept *Tubificoides benedii* for several months in the laboratory, at 12–14°C, in azoic sediment (frozen and thawed twice) from the collection site, and artificial water of 32‰ salinity. The animals reproduced regularly under these conditions. Most recently, Rodriguez et al. (2006) have acclimated in the laboratory the euryhaline freshwater species *Limnodrilus hoffmeisteri* to 7–8‰ salinity, for toxicity testing of estuarine sediments. Sediment was sampled in a clean estuary, sieved in the field through 250 μm to eliminate indigenous fauna and coarse sediment particles and mixed in trays with 2–3 cm of overlying water, built with tap water or seawater to achieve the desired salinity. The acclimation of worms to the required salinity was performed carefully, avoiding daily increments of more than 3‰. Although these authors did not establish a permanent culture of the worms, they showed normal behaviour with no mortality, and reproduction was observed. The method may be useful for culturing brackish waters species to the desired salinity.

6.6 Identification

Identification of worms inhabiting polluted areas does not generally require specialised training or extensive taxonomical expertise, as typically only a few species occur. However, for reference sites in a field survey the assistance of an oligochaete specialist is strongly recommended, since many taxa of different degree of taxonomic difficulty may appear. In these sites the use of the lactophenol or any other non-permanent mounting media is not recommended, since anatomical study of internal organs by dissection or histological sections can be necessary for accurate identification.

Most worms in field surveys can be identified at genus or even species level by examination of whole-mounts. The different form, number and position of chaetae in both the dorsal and ventral bundles are common characters used in aquatic oligochaete systematics, and these can better be examined by differential interference contrast microscopy. In several common taxa, such as many tubificines, the form and size of the cuticular penial sheath and the presence of modified spermathecal or penial chaetae is also required for identification to genus or species level. The identity of immature worms is difficult if not impossible to achieve, except for naidines and a few other taxa with peculiar external features, such as the gill-bearing species *Branchiura sowerbyi.* In the case of the *Limnodrilus* group of species, one of the most abundant taxa in polluted sites where several species coexist in the community, immature animals have a very similar form of chaetae, except for *L. udekemianus* that can be readily distinguished by the form of its anterior chaetae. Some authors have proposed chaetal measurements to separate *Limnodrilus* species when immature (Steinlechner 1987). Other authors assign immatures to the different species of *Limnodrilus* in the same proportion as the identified mature worms with the same type of chaetae. This can be problematic in cases where more than one species occurs with immatures that are similar in appearance, for example, if breeding occurs at different times; then, immatures can be misassigned to the wrong species. Unfortunately, until alternative identification methods (*e.g.* barcoding) become widely available, the selection of study sites for life-history studies should be made very carefully and with a knowledge of what species are present. Another group of species, commonly inhabiting polluted sites, which share hair-like and pectinate dorsal chaetae are *Tubifex tubifex* and some *Potamothrix* species. These species are distinguished when mature by the presence of spermathecal chaetae and the structure of the male ducts and penial cuticular layers when present; however, they are similar when immature.

In Chap. 2, we summarise the present state of oligochaete systematics and provide information on the available worldwide and regional keys. Readers are also asked to consider the information in that chapter to update the taxonomy of the group in order to use the current species names for identification.

6.7 Toxicity Tests

6.7.1 Water-Only Toxicity Tests

Aquatic oligochaetes are for the most part infaunal organisms which are under stress in the absence of sediment. However, the use of short-term (48–96 h) water-only tests can be useful in testing the sensitivity of test organisms, or in preliminary tests to set the range of concentrations to be tested in longer-term tests. Subsequent tests can be performed using concentration ranges below those that cause a lethal response or below the LC_{50} value.

Because of the sensitivity of these organisms to chlorine in water, most authors use dechlorinated tap water for toxicity testing with oligochaete worms, but reconstituted water can be used if a greater degree of standardisation is required. A few authors have used distilled water for acute toxicity tests since aquatic oligochaetes can regulate although the physiological stress of osmoregulation is additional to that associated with the chemical toxicity. In fact, the LC_{50} values reported for *T. tubifex* exposed to Zn, Ni, Cu, Cr, and Cd in distilled water were more than two orders of magnitude lower than in tap water (Brkovic-Popovic and Popovic 1977).

6.7.2 Sediment Toxicity Tests

While there are a number of papers describing various methodologies, there are only a few well documented methodological studies that have carefully addressed the various factors that could affect test results and have developed optimised test protocols. The two best documented protocols for oligochaete toxicity tests are those for *Lumbriculus variegatus* originally proposed by Bailey and Liu (1980) and refined and thoroughly tested for sediment assessment by Phipps et al. (1993), and secondly, the method proposed by Reynoldson et al. (1991) for *T. tubifex.* In both cases, the various factors (*e.g.* temperature, density, food supply) that influence test results have been examined and documented. Other publications are also useful guides in making decisions on sampling, storage and handling of sediment for toxicological testing (ASTM 1990, 2005; OECD 2007; Rodriguez and Reynoldson 1999).

One of the most vexing issues in sediment testing is the control or reference substrate to be used. Since natural unpolluted sediments are not always available, and the characteristics of sediment (*e.g.* distribution of particle size and organic content) are variable, the option exist to use formulated (also called artificial) sediments if a high degree of standardisation is required. Formulated sediments that simulate physical and chemical characteristics of natural sediments constitute a suitable habitat for aquatic oligochaetes and have been used for both culture and toxicity testing of tubificids (Egeler et al. 1997) and lumbriculids (OECD 2007). The OECD guideline details all methodological issues related with preparation, conditioning, spiking of chemicals and test performance using formulated sediments.

The use of any animal, including oligochaetes, in sediment bioassays must involve some manipulation of the natural sediment to remove competing or predaceous resident invertebrates before the test animals are added. Sieving has usually been preferred over heating, freezing or drying. Day et al. (1995) exposed *Tubifex tubifex* and two other invertebrates to sediments that were unsieved, wet-sieved (250 μm), autoclaved, frozen, or irradiated (10 and 30 kGy) and responses in

Table 6.1 Reduction in growth relative to control (zero worms) in three invertebrate species in response to density of the oligochaete worm *Tubifex tubifex* (From Reynoldson et al. 1994, Table 3, with permission of John Wiley & Sons Ltd., © conveyed by Copyright Clearance Centre, Inc.)

		Reduction in growth (%)		
No. worms per beaker	Oligochaete density (worms m^{-2})	*Chironomus riparius*	*Hexagenia limbata*	*Hyalella azteca*
5	1,250–1,500	38.7[a]	11.8	4.9
10	3,000	59.5[a]	17.4	6.0[b]
25	6,000–7,000	76.4[a]	30.3	56.3[b]
50	12,500–15,000	90.0[a]	40.8	69.5[b]
75	20,000–22,000	92.2[a]	48.3	66.6[b]

[a]Significantly less than control ($p<0.001$)
[b]Significantly less than control ($p<0.05$)

sediment bioassays were compared. Production of cocoons and young in the *Tubifex* bioassays were affected in sediments that were highly irradiated, frozen or autoclaved, and increased nutritional sources in sediments was explained as a result of enhancement of organic material from death benthic indigenous organisms in the manipulated sediments. Authors recommended wet sieving of the sediment through a fine mesh prior to testing. This procedure does not result in important losses of organic material, metals or PAHs. Most whole-sediment bioassay protocols recommend restricting handling to a minimum in order not to alter the chemical and physical characteristics as well as the toxicity of the sediments to be tested. However, Day et al. (1995) showed that it was impossible to calculate the number of cocoons or young in a 28-day bioassay with *T tubifex* in unsieved or coarsely sieved sediments (2.0 mm), since other oligochaetes were present in the sediment (*L. hoffmeisteri, T. tubifex* and *Q. multisetosus*) and had also reproduced. When sieved through 500-μm mesh, the presence of indigenous worms was low (mean ≤5 worms) and none were observed when sieved through 250-μm mesh. Thus, although sieving through 250 μm was recommended, the standard protocol for this bioassay requires sieving through at least 500 μm. For instance, when the natural sediment is coarse, which is common in rivers, the larger mesh size is preferred, otherwise very large volumes of sediment must be processed to get sufficient material for testing (see Maestre et al. 2007). In a separate study, Reynoldson et al. (1994) looked at the effect of indigenous benthic fauna on the survival and growth of *Chironomus riparius*, *Hyalella azteca* and *Hexagenia limbata* in sediment bioassays, using different densities of the oligochaete *T. tubifex,* typically occurring in polluted sites. Although their presence in sediments had only marginal effects on the survival ($p>0.05$), they did significantly ($p<0.05$) affect the growth of all three species, and the effect on growth was density dependent (Table 6.1). The effects on growth were different among the species suggesting different interspecific interactions, and the authors (Reynoldson et al. 1994) strongly recommended sieving sediments. The limited experience of one of these authors (P. Rodriguez) also supports the need for sieving estuarine sediments for use in sediment bioassays, since some nereid polychaete worms appear to feed on oligochaetes. In a few batches where a polychaete worm

Fig. 6.1 Cocoons of *T. tubifex*. (**a**) Different size and form of cocoons from a sediment bioassay, showing empty, and full cocoons with eggs and embryos. (**b**) Typical oval cocoons from culture and unpolluted sites, full of eggs. (**c**) Cocoons with one or two eggs or embryos left after hatching. (**d**) *T. tubifex* (Tt) and *L. hoffmeisteri* (Lh) cocoons, may both be present in some bioassays, but the latter are easy to distinguish due to the abundant sediment particles attached to their surface and flat appearance

was present no test organism (*Limnodrilus hoffmeisteri*) was found and predation was the likely reason (Rodriguez et al. 2006).

When working with both *T. tubifex* and *Limnodrilus hoffmeisteri* chronic bioassays, one of the endpoints is a count of the total number of empty and full cocoons. In unpolluted sites, they usually show a typical oval appearance, but when laid by worms exposed to test sediments, it is common to find different sizes and forms in the cocoons (Fig. 6.1). Occasionally, the presence of indigenous fauna (*e.g. L. hoffmeisteri*) can be identified in toxicity tests by the occurrence of characteristically shaped cocoons (Fig. 6.1d).

Some authors have included homogenization of field collected sediment prior to testing. In order to ensure an even mix of sediments in the experimental chambers, Keilty and Landrum (1990) dried the sediment at room temperature and thoroughly mixed it by sieving through a 250-μm mesh size. Sediments were further reconstituted adding water and a few millilitres of natural sediment to provide an active bacterial flora. In the same study, other sediment was study for which the slurry was allowed to settle naturally without subsequent mixing. These authors found no significant differences in the ecotoxicological endpoints measured between mixed and unmixed sediments in long-term experiments (42 days). However, authors noted that mixing of the sediments resulted in an increased availability of contaminants sorbed to the fine sediment fraction. The implications of homogenizing field collected sediment is that no estimate of sample scale spatial variation can be made. Depending on the nature of the study, this may or may not be an important issue.

Another methodological issue that can affect toxicity test results is the use of food supplements. This is a concern for long-term toxicity tests where feeding may reduce the confounding effects of limited food supply on endpoints such as growth and reproduction. Differences of feeding worms in sediment tests have been investigated by some authors. Ankley et al. (1994a) used a water-renewal system to examine various confounding factors associated with the physical and chemical properties of the sediment that may affect the *L. variegatus* test. After 10-day exposure, Ankley et al. noted that the survival and growth (mean dry weight per individual) of *L. variegatus* differed markedly in fed and unfed tests with 50 different uncontaminated sediments. However, their results showed that the species was relatively insensitive to the range of physical and chemical characteristics of uncontaminated sediments, which differed markedly with regard to grain-size distribution, organic carbon content and mineralogical composition encountered. In a further study, Leppänen and Kukkonen (1998) examined the effects of the sediment characteristics and feeding on reproduction in *L. variegatus* and found somewhat different results, reporting effects on reproduction, growth and egestion rates (expressed as mg dry faeces mg^{-1} dry worm h^{-1}) in two fine grained reference sediments, related with differences in feeding regime. They described that individuals purged prior the fragmentation or just after since the gut in all newborn fragments was empty; and it took 6–9 days to regenerate new ends, resulting that anterior ends started feeding earlier than posterior ones. The impossibility of ingestion of sediment by posterior fragments during so a long time may be cause of lower growth in a chronic test and give false positives. In a study in the Great lakes to develop the natural response range for the *T. tubifex* test endpoints, Reynoldson et al. (1995) tested a wide range of natural sediments and showed relatively little variation in survival and reproductive endpoints in *T. tubifex,* or growth and survival in three other species (Reynoldson and Day 1998) and less variation than in a single type of sediment used in multiple negative control tests (Fig. 6.2). Thus, the concern regarding the effect of variability in natural sediment characteristics on test endpoints may be unwarranted.

Ammonia can be present in sediment (elutriates or whole) at toxic concentrations as a result of both natural and human activities and can be a confounding factor.

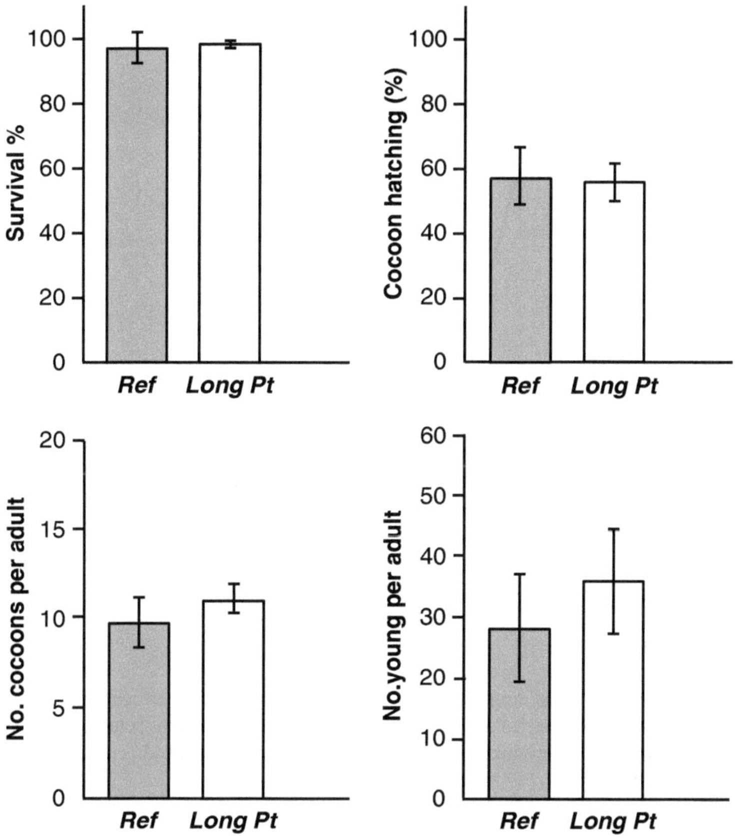

Fig. 6.2 Variation (mean and SD) in four test endpoints in the *T. tubifex* bioassay showing the results from 170 different reference sites (*Ref*) and a single culture sediment (*Long Pt*) used as a negative control in toxicity tests in 46 separate toxicity tests, over 2 year period

Toxicity due to the presence of ammonia is one of the problems that can occur, particularly in longer duration tests. Methods for reducing ammonia are required in some situations for both freshwater and estuarine natural sediments. Usually, aeration of overlying water and keeping the pH below 8 is sufficient to avoid unionised ammonia reaching dangerous levels. However, in some polluted sites removal of overlying water after aeration may be necessary to eliminate dissolved nitrates and reduce the risk of forming unionised ammonia during chronic bioassays. Schubauer-Berigan et al. (1995) examined the influence of pH on the toxicity of ammonia to *L. variegatus* in 10-day flow-through tests and showed that total ammonia was more toxic at elevated pH values, approximately 43 times greater at pH 8.6 than at pH 6.5. They reported 96-h $LC_{50}=21.4$ mg l^{-1} total N-NH_3 at pH = 7.8 (0.77 mg l^{-1} unionised N-NH_3), and 6.60 mg l^{-1} at pH = 8.6 (1.20 mg l^{-1} unionised N-NH_3). Unionised N-NH_3 values were calculated from total ammonia using ammonia speciation

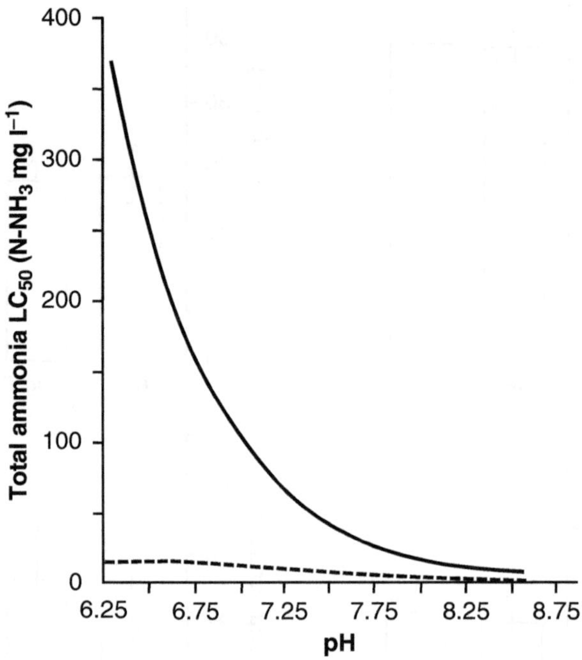

Fig. 6.3 Comparison of total ammonia 96-h LC_{50} for *Lumbriculus variegatus* to pH (*solid line*) and pH water quality criteria for total ammonia (*dashed line*) (From Schubauer-Berigan et al. 1995, Fig. 1, redrawn and reproduced by permission of Wiley & Sons Ltd., conveyed by Copyright Clearance Centre, Inc.)

tables (Fig. 6.3). Escape behaviour of worms to the sediment-water interface occurs if ammonia in pore water is high, and it has been reported for *T. tubifex* (Arrate et al. 2004) and *L. variegatus* (Whiteman et al. 1996).

Differential sensitivity and response of worms of the same species from different cultures or sampled from different localities may result in differences in the endpoints used to assess sediment toxicity. In pairwise experiments, Reynoldson et al. (1996) compared the sensitivity, growth and reproduction endpoints of test populations of *T. tubifex* maintained under the same culture conditions but originating from two different sites, a North American lacustrine population and a Spanish riverine population of animals. The authors reported little differences in performance that could not be accounted for by appropriate calibration. Spanish worms were generally smaller than the Canadian worms, but the size of the mature worms did not appear to influence the number of cocoons or young produced (Table 6.2). However, Canadian worms seemed to be much more sensitive to small changes in temperature with low production of cocoon and young at 20°C, compared with that recorded at 22.5°C.

Although little work has been done with oligochaetes to assess estuarine sediment toxicity, these animals may have some utility particularly for oligohaline and

Table 6.2 Comparison in number of cocoons (CCAD) and juveniles (YGAD) produced in 28 days, at 20°C and 22.5°C, by Spanish and Canadian populations of *Tubifex tubifex*, both cultured in sediment from Long Point (Lake Ontario, Canada) (From Reynoldson et al. 1996, Table 4, with permission of Springer Publ. © conveyed by Copyright Clearance Centre, Inc.)

	20°C		22.5°C	
	CCAA	YGAD	CCAD	YGAD
Tubifex tubifex populations	Mean ± sd	Mean ± sd	Mean ± sd	Mean ± sd
Spanish population				
Experiment 1	7.0 (0.6)	22.3 (3.4)	4.3 (2.2)*	18.8 (9.4)
Experiment 2	6.7 (0.8)	17.2 (8.6)	8.6 (1.0)*	19.0 (4.6)*
Canadian population				
Experiment 1	5.0 (0.4)*	4.6 (2.2)*	10.3 (1.3)	27.4 (5.6)
Experiment 2	4.9 (0.2)*	7.0 (1.2)	10.4 (0.3)	30.6 (2.3)

Marked with * those results where production was significantly lower (Mann-Whitney *U*-test, $p<0.05$) than that obtained in the same experiment by the other population

β-mesohaline sections of estuaries (Rodriguez et al. 2006), where some freshwater tolerant species, such as *Limnodrilus hoffmeisteri*, form substantial populations. This euryhaline freshwater species can be satisfactorily bred under 7–8‰ salinity after a period of progressive acclimation. Salinity should be maintained below 9‰ to avoid potential negative effects on reproduction, although these worms can tolerate as much as 12–13‰ (Brinkhurst and Simmons 1968, in the Bay of San Francisco, USA) or 14‰ in Fraser River Estuary (Chapman and Brinkhurst 1981). Polyhaline and α-mesohaline sections require the development of new bioassays using species with a higher range of tolerance to salinity, *e.g. Heterochaeta costata* or *Monopylephorus* spp. However, there are several methodological issues related to the estuarine ecosystem that requires more research. Many practical decisions used in freshwater bioassays can be applied to estuarine sediment tests, however, criteria should be established to help deciding which salinity should be used to expose animals in estuarine chronic tests. Sites in estuaries exhibit a wide range of salinity conditions depending on the morphology of the estuary, daily and yearly tidal dynamics, and river flow. Water-column salinity is highly variable in estuaries due to the tidal regime but pore water salinity is more stable for infaunal animals and, probably, more relevant.

6.8 Bioaccumulation

There are a few documents that provide general guidance for measuring bioaccumulation of sediment-associated chemicals in freshwater oligochaetes. These provide useful references when making decisions related to test methodology, measurement of chemicals in test organisms, calculation of bioaccumulation factors, or comparison of tissue residues for different compounds (*e.g.* ASTM 2000; USEPA 2000; OECD 2008).

The measurement of bioaccumulation is usually done on a by weight basis of the exposed organisms. Sediment in the gut may account for 10–15% of the worm dry weight (Oliver 1987; Brooke et al. 1996) and bioaccumulation assessment may require either gut clearing or a *post-hoc* correction of the weight of sediment contained into the gut. Some differences exist in the procedures used by different authors with respect to time of purging the worm gut before body-burden analysis (see Sect. 6.4), although the suitability of purging or not should be considered when chemical concentration in the worm is required for studies on trophic transfer, since in the field worms are consumed by prey with the gut full. In bioaccumulation studies, a 24-h gut clearing period is commonly used (Ankley et al. 1992, 1994b; Krantzberg 1994; Brunson et al. 1998; Gillis et al. 2004). However, some toxicants are more rapidly eliminated than others, thus the purging time should be as short as possible to avoid biasing the estimation of bioaccumulation due to depuration of chemicals from tissue during the holding period in clean water, particularly for compounds with log K_{ow} of less than 5 (USEPA 2000). Thus, several authors have recommended shorter depuration periods (*e.g.* 12-h for *L. variegatus,* Ingersoll et al. 2003) for conducting bioaccumulation tests. Kukkonen and Landrum (1995) investigated several approaches for estimating the contribution of the intestinal content to benzo(a)pyrene (BaP) measures of bioaccumulation in *L. variegatus*. They showed that BaP in the gut contributed less than 10% to total body burden at steady state, and suggested that a 10-h water-only clearing was the most appropriate method for eliminating the gut-content influence on the body burden. Working with several PAHs, Sheedy et al. (1998) estimated in *L. variegatus* a loss of 31–68% for fluoranthene, anthracene and fluorene after 24-h depuration period in clean water (although it was zero for pyrene), suggesting the suitability of using shorter periods for gut clearance, prior to the body burden measurement. Mount et al. (1999) demonstrated that 98% of the gut content in *L. variegatus* was purged within 6 h of being placed into clean water without sediment. For non-ionic organic chemicals, the depuration rate was related to K_{ow}, and 6-h purging periods reduced potential errors from chemical depuration, particularly for compounds with log K_{ow} <5. They also proved that, after 24 h, only those chemicals with log K_{ow} >5 would be at more of the 90% of the initial concentration. Standley (1997) used clean sediment to purge the gut of contaminated sediment, leaving the worms overnight (≤24 h) in freshly collected, low organic sediment. However, Ingersoll et al. (1995) argued against this practice because clean sediment may contribute 15–20% to the oligochaete dry weight, thus resulting in a dilution of the body burden of the contaminant.

In an evaluation of purging period for cationic metals, Dawson et al. (2003) also recommended a purging period of 6 h. Probably, the most sophisticated method used to measure the clearing times was applied by Klump et al. (1987) in a study on assimilation efficiencies of organic contaminants. Using dual tracers, these authors calculated a maximum time required for gut clearing of 3.6±3.1 h. Gillis et al. (2004) compared the BSAF calculated for three metals (Cd, Ni and Pb) using tissue metal concentration in undepurated, 24-h water-depurated, and 24-h sediment depurated worms and concluded that BSAF calculations were more accurate in sediment-depurated animals, although this varies with the metal. Calculated BSAFs showed similar values for Pb, and overestimations of 18% for Cd, and 72% for Ni in unpurged worms.

Field tubificids are often covered by a secreted mucous layer which is very difficult to remove and contributes to the total body-burden. Back (1990) recommended keeping the animals in clean tap water for 24 h after exposure to contaminants, or after sampling from a polluted environment, to eliminate metals bound to the mucous surface film on the worm body. After 24 h, no further loss of Pb, Cd and Zn relative to the dry weight of the tubificids was observed, a longer period was not recommended due to worm weight loss due to starvation. However, due to the risk of depuration of chemicals as mentioned above, it seems sensible to use the gut purging period in clean water to also eliminate or reduce the body mucous layer. Preservation in alcohol and formalin has been shown to increase measured tissue metal levels by up to a factor of 3 in some cases, probably due to a reduction of wet weight. Therefore, Chapman et al. (1980) recommended not to use fixatives, but worms may be preserved frozen at −20°C for later analyses.

When working with organisms sampled in the field, the exposure time is assumed to be sufficient for animals to be in steady state However, in laboratory tests the time required to attain steady state depends on the species and on the characteristics of the substance. Some substances reach steady state in a few days, whereas to ensure equilibrium concentrations for superhydrophobic substances (log Kow>7 or 8), more than 30 days may be required (Ankley et al. 1992). The exposure period in standard bioaccumulation experiments is typically of 28 days. Feeding the worms is not recommended (USEPA 2000), but when sublethal endpoints (*e.g.* reproduction) are to be measured together with chemical body-burden, long exposure periods are required, and feeding should be considered.

The density of animals in the experiments may also affect bioaccumulation factors, as demonstrated by Kukkonen and Landrum (1994) who described an inverse relationship between animal density and pyrene body burden. USEPA (2000) has recommended density to be calculated as a ratio of total organic carbon in sediment to dry weight of organisms, with a target value of about 50:1, keeping density into a range of 1–5 g per replicate.

Many other practical issues need to be examined to assure a higher degree of standardisation of bioaccumulation studies with aquatic oligochaetes. Interlaboratory comparison studies have been conducted prior to the development of the OECD guidelines (2008) for *L. variegatus* and *T. tubifex,* and such studies are also necessary for future development of new methods used for bioaccumulation and toxicity tests with aquatic oligochaete species.

References

Abrahamsen G (1973) Biomass and body-surface area of populations of Enchytraeidae and Lumbricidae (Oligochaeta) in Norwegian coniferous forest soils. Pedobiologia 13:28–39

Adreani L, Bonacina C, Bonomi G, Monti C (1984) Cohort cultures of *Psammoryctides barbatus* (Grube) and *Spirosperma ferox* Eisen: a tool for a better understanding of demographic strategies in tubificids. Hydrobiologia 115:113–119

Ahamed MT, Mollah MFA (1992) Effects of various levels of wheat bran and mustard oil cake in the culture media on tubificid production. Aquaculture 107:107–113

Ankley GT, Cook PM, Carlson AR, Call DJ, Swenson JA, Corcoran HF, Hoke RA (1992) Bioaccumulation of PCBs from sediments by oligochaetes and fishes: comparison of laboratory and field studies. Can J Fish Aquat Sci 49:2080–2085

Ankley GT, Benoit DA, Balogh JC, Reynoldson TB, Day KE, Hoke RA (1994a) Evaluation of potential confounding factors in sediment toxicity tests with three freshwater benthic invertebrates. Environ Toxicol Chem 13:627–635

Ankley GT, Leonard EN, Mattson VR (1994b) Prediction of bioaccumulation of metals from contaminated sediments by the oligochaete *Lumbriculus variegatus*. Water Res 28:1071–1076

Anlauf A (1994) Some characteristics of the genetic variants of *Tubifex tubifex* (Müller, 1774) (Oligochaeta, Tubificidae) in laboratory cultures. Hydrobiologia 278:1–6

Anlauf A, Neumann D (1997) The genetic variability of *Tubifex tubifex* (Müller) in 20 populations and its relation to habitat type. Arch Hydrobiol 139:145–162

Arrate JA, Rodriguez P, Martinez-Madrid M (2002) Effects of three chemicals on the survival and reproduction of the oligochaete worm *Enchytraeus coronatus* in chronic toxicity test. Pedobiologia 46:136–149

Arrate JA, Rodriguez P, Martinez-Madrid M (2004) *Tubifex tubifex* chronic toxicity test using artificial sediment: methodological issues. Limnetica 23:25–36

ASTM (1990) Standard guide for collection, storage, characterization and manipulation of sediments for toxicological testing. ASTM E1391-90, Philadelphia

ASTM (2000) Standard guide for determination of bioaccumulation of sediment-associated contaminants by benthic invertebrates. ASTM E-1688. In: Annual book of standards, Vol 11.05, Philadelphia

ASTM (2005) Standard guide for conducting sediment toxicity tests with freshwater invertebrates. ASTM E1706-05, Philadelphia

Aston RJ (1984) The culture of *Branchiura sowerbyi* (Tubificidae, Oligochaeta) using cellulose substrate. Aquaculture 40:89–94

Aston RJ, Milner AGP (1981) Conditions required for the culture of *Branchiura sowerbyi* (Oligochaeta, Tubificidae) in activated sludge. Aquaculture 26:155–160

Aston RJ, Sadler K, Milner AGP (1982) Effects of the temperature on the culture of *Branchiura sowerbyi* (Oligochaeta, Tubificidae) on activated sludge. Aquaculture 29:137–145

Back H (1990) Epidermal uptake of Pb, Cd and Zn in tubificid worms. Oecologia 85:226–232

Bailey HC, Liu DHW (1980) *Lumbriculus variegatus*, a benthic oligochaete, as a bioassay organism. In: Eaton JG, Parrish PR, Hendricks AC (eds) Aquatic toxicology, ASTM Special Technical Publication 707, Philadelphia, pp 205–215

Bailey RC, Norris RH, Reynoldson TB (2004) Bioassessment of freshwater ecosystems. Using the reference condition approach. Kluwer Academic, Boston, 170 pp

Bonacina C, Bonomi G, Monti C (1987) Progress in cohort cultures of aquatic oligochaeta. Hydrobiologia 155:163–169

Bonacina C, Bonomi G, Monti C (1989) Population analysis in mass cultures of *Tubifex tubifex*. Hydrobiologia 180:127–134

Bonacina C, Pasteris A, Bonomi G, Marzuoli D (1994) Quantitative observations on the population ecology of *Branchiura sowerbyi* (Oligochaeta Tubificidae). Hydrobiologia 278:267–274

Bouguenec V (1992) Oligochaetes (Tubificidae and Enchytraeidae) as food in fish rearing: a review and preliminary tests. Aquaculture 102:201–217

Bouguenec V, Giani N (1989) Biological studies upon *Enchytraeus variatus* Bougenec & Giani, 1987 in breeding cultures. Hydrobiologia 180:151–165

Brinkhurst RO (1986) A guide to the freshwater aquatic microdrile oligochaetes of North America. Can Spec Publ Fish Aquat Sci 84:259

Brinkhurst RO (2003) The benthos of lakes. The Blackburn Press, Caldwell, New Jersey, 216 pp

Brinkhurst RO, Simmons ML (1968) The aquatic Oligochaeta of the San Francisco Bay system. Calif Fish Game 54:180–194

Brkovic-Popovic Y, Popovic M (1977) Effects of heavy metals on survival and respiration rate of tubificid worms: Part I- Effects on survival. Environ Pollut 13:65–72

Brooke LT, Ankley GT, Call DJ, Cook PM (1996) Gut content weight and clearance rate for three species of freshwater invertebrates. Environ Toxicol Chem 15:223–228

Brunson EL, Canfield TJ, Dwyer FJ, Ingersoll CG, Kemble NE (1998) Assessing the bioaccumulation of contaminants from sediments of the Upper Mississippi River using field-collected oligochaetes and laboratory-exposed *Lumbriculus variegatus*. Arch Environ Contam Toxicol 35:191–201

Chapman PM (1987) Oligochaete respiration as a measure of sediment toxicity in Puget Sound, Washington. Hydrobiologia 155:249–258

Chapman PM, Brinkhurst RO (1980) Salinity tolerance in some selected aquatic oligochaetes. Int Rev ges Hydrobiol 65:499–505

Chapman PM, Brinkhurst RO (1981) Seasonal changes in interstitial salinities and seasonal movements of subtidal nethic invertebrates in the Fraser River Estuary, B.C. Estuar Coast Mar Sci 12:49–66

Chapman PM, Churchland LM, Thomson PA, Michnowsky E (1980) Heavy metal studies with oligochaetes. In: Brinkhurst RO, Cook DG (eds) Aquatic oligochaete biology. Plenum Press, New York, pp 477–502

Collado R, Schmelz RM (2001) Descriptions of three *Pristina* species (Naididae, Clitellata) from Amazonian forest soils, including *P. marcusi* sp.nov. Hydrobiologia 463:1–11

Dawson TD, Lott KG, Leonard EN, Mount D (2003) Time course of metal loss in *Lumbriculus variegatus* following sediment exposure. Environ Toxicol Chem 22:886–889

Day KE, Kirby RS, Reynoldson TB (1995) The effect of manipulation of freshwater sediments on responses of benthic invertebrates in whole-sediment toxicity tests. Environ Toxicol Chem 14:1333–1343

Downing J (1984) Sampling the benthos of standing waters. In: Downing FA, Rigler FH (eds) A manual on methods for the assessment of secondary productivity in fresh waters, IBP HandBook, vol 17. Blackwell, Oxford, pp 87–130

Drake CM, Elliott JM (1982) A comparative study of three air-lift samplers used for sampling benthic macro-invertebrates in rivers. Freshwater Biol 12:511–533

Dubilier N, Giere O, Grieshaber MK (1994) Concomitant effects of sulfide and hypoxia on the aerobic metabolism of the marine oligochaete *Tubificoides benedii*. J Exp Zool 269:287–297

Egeler P, Römbke J, Meller M, Knacker Th, Franke C, Studinger NR (1997) Bioaccumulation of lindane and hexachlorobenzene by tubificid sludgeworms (Oligochaeta) under standardized laboratory conditions. Chemosphere 35:835–852

Egeler P, Rombke J, Meller M, Knacker Th, Nagel R (1999) Bioaccumulation test with tubificid sludgeworms in artificial media. Hydrobiologia 406:271–280

Finogenova NP, Lobasheva TM (1987) Growth of *Tubifex tubifex* Müller (Oligochaeta, Tubificidae) under various trophic conditions. Int Rev ges Hydrobiol 72:709–726

Giani N, Sambugar B, Rodriguez P, Martinez-Ansemil E (2001) Oligochaetes in southern European groundwater. Hydrobiologia 463:65–74

Gillis PL, Dixon DG, Borgmann U, Reynoldson TB (2004) Uptake and depuration of cadmium, nickel, and lead in laboratory-exposed *Tubifex tubifex* and corresponding changes in the concentration of a metallothionein-like protein. Environ Toxicol Chem 23:76–85

Gnaiger E, Staudigl J (1987) Aerobic metabolism and physiological responses of aquatic oligochaetes to environmental anoxia- Heat dissipation, oxygen consumption, feeding and defecation. Physiol Zool 60:659–678

Hangay G, Dingley M (1985) Biological museum methods. Vol 2. Plants, invertebrates and techniques. Academic Press Australia, Sidney, 323 pp

Harper RM, Fry JC, Learner MA (1981a) A bacteriological investigation to elucidate the feeding biology of *Nais variabilis* (Oligochaeta, Naididae). Freshwater Biol 11:227–237

Harper RM, Fry JC, Learner MA (1981b) Digestion of bacteria by *Nais variabilis* (Oligochaeta) as established by autoradiography. Oikos 36:211–218

Ingersoll CG, Ankley GT, Benoit DA, Bronson EL, Burton GA, Dwyer FJ, Hoke R, Landrum PF, Norberg-King TJ, Winger PV (1995) Toxicity and bioaccumulation of sediment-associated contaminants using freshwater invertebrates. A review of methods and applications. Environ Toxicol Chem 14:1885–1894

Ingersoll CG, Buson EL, Wang N, Dwyer FJ, Ankley GT, Mount DR, Huckins J, Petty J, Landrum PF (2003) Uptake and depuration of nonionic organic contaminants from sediment by the oligochaete, *Lumbriculus variegatus*. Environ Toxicol Chem 22:872–885

Juget J, Lafont M (1982) L'echantillonage de la faune benthique: Revue des techniques de prélévement, d'extraction et de tri- Application aux oligochètes. Rev Sci Eau 1:243–254

Juget J, Goubier V, Barthelemy D (1989) Intrinsic and extrinsic variables controlling the productivity of asexual populations of *Nais* spp (Naididae, Oligochaeta). Hydrobiologia 180:177–184

Kaiser M, Irmer U, Weiler K (1989) Monitoring of water quality: seasonal variation of heavy metals in sediments, suspended particulate water and tubificids of the Elbe River. Environ Technol Lett 10:845–855

Keilty TJ, Landrum PF (1990) Population specific toxicity responses by the freshwater oligochaete *Stylodrilus heringianus* in natural lake Michigan sediments. Environ Toxicol Chem 9:1147–1154

Klerks PL, Bartholomew PR (1991) Cadmium accumulation and detoxification in a Cd-resistant population of the oligochaete *Limnodrilus hoffmeisteri*. Aquat Toxicol 19:97–112

Klump JV, Krezoski JR, Smith ME, Kaster JL (1987) Dual tracer studies of the assimilation of an organic contaminant from sediments by deposit feeding oligochaetes. Can J Fish Aquat Sci 44:1574–1583

Korínková J, Sigmund J (1968) The colouring of bottom-fauna samples before sorting. Acta Societatis Zoologicae Bohemoslovacae 32:300

Kosiorek D (1974) Development of *Tubifex tubifex* Müll. in experimental culture. Pol Arch Hydrobiol 21:411–422

Krantzberg G (1994) Spatial and temporal variability in metal bioavailability and toxicity of sediment from Hamilton Harbour, Lake Ontario. Environ Toxicol Chem 13:1685–1698

Kukkonen J, Landrum PF (1994) Toxicokinetics and toxicity of sediment-associated pyrene to *Lumbriculus variegatus* (Oligochaeta). Environ Toxicol Chem 13:1457–1468

Kukkonen J, Landrum PF (1995) Effects of sediment bound polymethylsiloxane on the bioavailability and distribution of benzo(a)pyrene in lake sediment to *Lumbriculus variegatus*. Environ Toxicol Chem 14:523–531

Lafont M (1983) Annelides oligochètes. Introduction pratique à la systematique des organismes des eaux continentales françaises. Bull Soc Linn Lyon 4:108–135

Learner MA, Edwards RW (1963) The toxicity of some substances to *Nais* (Oligochaeta). Proc Soc Wat Treat Exam 12:161–168

Leppänen MT, Kukkonen JVK (1998) Factors affecting feeding rate, reproduction and growth of an oligochaete *Lumbriculus variegatus*. Hydrobiologia 377:183–194

Lochhead G, Learner MA (1983) The effect of temperature on asexual population growth of three species of Naididae (Oligochaeta). Hydrobiologia 98:107–113

Lundkvist H (1978) A technique for determining individual fresh weights of live small animals, with special reference to Enchytraeidae. Oecologia 35:365–367

Maestre Z, Martinez-Madrid M, Rodriguez P, Reynoldson T (2007) Ecotoxicity assessment of river sediments and a critical evaluation of some of the procedures used in the aquatic oligochaete *Tubifex tubifex* chronic bioassay. Arch Environ Contam Toxicol 53:559–570

Marian MP, Chandrian S, Pandian TJ (1989) A rack culture system for *T. tubifex*. Aquacult Eng 8:329–337

Martinez-Madrid M, Rodriguez P, Perez-Iglesias JI, Navarro E (1999) Sediment toxicity bioassays for assessment of polluted sites in the Nervion River (northern Spain). 2. *Tubifex tubifex* (Oligochaeta) reproduction sediment bioassay. Ecotoxicology 8:111–124

Mount DR, Dawson TD, Burkhard LP (1999) Implications of gut purging for tissue residues determined in bioaccumulation testing of sediment with *Lumbriculus varieg*atus. Environ Toxicol Chem 18:1244–1249

Nijboer R, Wetzel MJ, Verdonschot PFM (2004) Diversity and distribution of Tubificidae, Naididae, and Lumbriculidae (Annelida, Oligochaeta) in the Netherlands: an evaluation of twenty years of monitoring data. Hydrobiologia 520:127–141

OECD (2007) Sediment-water *Lumbriculus* toxicity test using spiked sediment. OECD guideline for the testing of chemicals no 225

OECD (2008) Bioaccumulation in sediment-dwelling benthic oligochaetes. OECD guideline for the testing of chemicals No 315

Oliver BG (1987) Bio-uptake of chlorinated hydrocarbons from laboratory-spiked and field sediments by oligochaete worms. Environ Sci Technol 21:785–790

Paoletti A (1989) Cohort cultures of *Tubifex tubifex* forms. Hydrobiologia 180:143–150

Pasteris A, Bonacina C, Bonomi G (1994) Observations on cohorts of *Tubifex tubifex* cultured at different food levels, using cellulose substrate. Hydrobiologia 278:315–320

Pasteris A, Vecchi M, Bonomi G (1999) A comparison among different population models for *Limnodrilus hoffmeisteri* Claparède (Oligochaeta, Tubificidae). Hydrobiologia 406:183–189

Peckarsky BLO (1984) Sampling the stream benthos. In: Downing FA, Rigler FH (eds) A manual on methods for the assessment of secondary productivity in fresh waters, Chapter 5, 2nd edn. IBP HandBook 17. Blackwell Sci Publ, Oxford, pp 131–160

Phipps GL, Ankley GT, Benoit DA, Mattson VR (1993) Use of the aquatic oligochaete *Lumbriculus variegatus* for assessing the toxicity and bioaccumulation of sediment-associated contaminants. Environ Toxicol Chem 12:269–279

Poddubnaya TL (1980) Characteristics of the life cycle of Tubificidae and Naididae. In: Kothekar VS (ed) Aquatic Oligochaeta worms. Taxonomy, ecology and faunistic studies in the USSR. Amerind Publ Co, New Delhi, pp 97–104

Reynoldson TB (1983) The population dynamics of the aquatic Oligochaeta in the English Lake District. PhD thesis, University Lancaster, 245 pp

Reynoldson TB (1987) The role of environmental factors in the ecology of tubificid oligochaetes-an experimental study. Holarct Ecol 10:241–248

Reynoldson TB, Day KE (1998) Biological guidelines for the assessment of sediment quality in the Laurentian Great Lakes. National Water Research Institute, Canada, Report: 498–232

Reynoldson TB, Thompson SP, Bamsey JL (1991) A sediment bioassay using the tubificid oligochaete worm *Tubifex tubifex*. Environ Toxicol Chem 10:1061–1072

Reynoldson TB, Day KE, Clarke C, Milani D (1994) Effects of indigenous animals on chronic endpoints in freshwater sediment toxicity tests. Environ Toxicol Chem 13:973–977

Reynoldson TB, Rosa F, Mudroch A, Day KE (1995) Investigation of the Spanish river area of the north channel of Lake Huron. II. Benthic invertebrate community structure and sediment toxicity, with reference to biological sediment guidelines. NWRI report no 95–14

Reynoldson TB, Rodriguez P, Martinez-Madrid M (1996) A comparison of reproduction, growth and acute toxicity in two populations of *Tubifex tubifex* (Müller, 1774) from the North America Great lakes and northern Spain. Hydrobiologia 334:199–206

Rodriguez P, Reynoldson TB (1999) Laboratory methods and criteria for sediment bioassessment. In: Mudroch A, Azcue J, Mudroch P (eds) Manual of bioassessment of aquatic sediments quality. Chapter 3. Lewis Publisher, Boca Raton, pp 83–133

Rodriguez P, Arrate J, Martinez-Madrid M, Reynoldson TB, Schumacher V, Viguri J (2006) Toxicity of Santander Bay sediments to the euryhaline freshwater oligochaete *Limnodrilus hoffmeisteri*. Hydrobiologia 564:157–169

Rosenberg DM, Resh VH (1993) Freshwater biomonitoring and benthic invertebrates. Chapman & Hall, New York, 488 pp

Sambugar B, Martinez-Ansemil E, Giani N (2005) Oligochaetes from springs in southern Europe. Bullettino del Museo Civico di Storia Naturale di Verona, Botanica Zoologia 29:93–106

Schubauer-Berigan MK, Monson PD, Ankley GT (1995) Influence of pH on the toxicity of ammonia to *Chironomus tentans* and *Lumbriculus variegatus*. Environ Toxicol Chem 14:713–717

Sheedy B, Mattson VR, Cox JS, Kosian PA, Phipps GL, Ankley GT (1998) Bioconcentration of polycyclic aromatic hydrocarbons by the freshwater oligochaete *Lumbriculus variegatus*. Chemosphere 36:3061–3070

Smit H, Dudok E, Wiersma S (1993) Biovolume as a tool in biomass determination of Oligochaeta and Chironomidae. Freshwater Biol 29:37–46

Smith DP, Kennedy JH, Dickson KL (1991) An evaluation of a naidid oligochaete as a toxicity test organism. Environ Toxicol Chem 10:1459–1465

Standley LJ (1997) Effect of sediment organic matter composition on the partitioning and bioavailability of dieldrin to the oligochaete *Lumbriculus variegatus*. Environ Sci Technol 31:2577–2583

Steinlechner R (1987) Identification of immature tubificids (Oligochaeta) of Lake Constance and its influence on the evaluation of species distribution. Hydrobiologia 155:57–63

Sturmbauer C, Opadiya GB, Niederstätter H, Riedmann A, Dallinger R (1999) Mitochondrial DNA reveals cryptic oligochaete species differing in cadmium resistance. Mol Biol Evol 16:967–974

Timm T (1980) Culture methods for aquatic Oligochaeta. In: Kothekar VS (ed) Aquatic Oligochaeta worms: taxonomy, ecology and faunistic studies in the USSR. Proceedings of the symposium on aquatic Oligochaeta, Tartu, 1967 (trans: Russian). Amerind Publ Co, New Delhi, pp 119–131

USEPA (2000) Methods for measuring the toxicity and bioaccumulation of sediment-associated contaminants with freshwater invertebrates, 2nd edn. EPA 600/R-99/064

Verdonschot PFM (2001) Hydrology and substrates: determinants of oligochaete distribution in lowland streams (The Netherlands). Hydrobiologia 463:249–262

Westheide W, Bethke-Beilfuss D (1991) The sublethal enchytraeid test system: guidelines and some results. In: Esser G, Overdieck D (eds) Modern ecology: basic and applied aspects, Chapter 23. Elsevier, Amsterdam, pp 497–508

Whiteman FW, Kahl MD, Rau DM, Balcer MD, Ankley GT (1996) Evaluation of interstitial water as a route of exposure for ammonia in sediment tests with benthic macroinvertebrates. Environ Toxicol Chem 15:794–801

Chapter 7
Appendices

P. Rodriguez and T.B. Reynoldson, *The Pollution Biology of Aquatic Oligochaetes*,
DOI 10.1007/978-94-007-1718-3_7, © Springer Science+Business Media B.V. 2011

Appendix 1

LC50 values (or the equivalent EC50 values for movement inhibition) of several chemical substances for oligochaete species, in water and sediment short-term exposures (References in Chap. 4)

Taxon	Chemical and or environmental factors	Water LC_{50} ($mg\ l^{-1}$) 48 h	Water LC_{50} ($mg\ l^{-1}$) 96 h	Sediment LC_{50} ($mg\ g^{-1}$) 48 h	Sediment LC_{50} ($mg\ g^{-1}$) 96 h	Source
Enchytraeidae						
Achaeta cf. affinis	Na PCP	0.18				Römbke and Knacker (1989)
	2,4,5 Trichloro phenoxyacetic acid (2,4,5 T)	569				
Cognettia sphagnetorum	NaPCP	0.6				Römbke and Knacker (1989)
	2,4,5 T	708				
Enchytraeus albidus	NaPCP	0.3				Römbke and Knascker (1989)
	Cd	6.4				
	Benomyl	0.2				
	Parathion	7.1				
	2,4,5 T	1,585				
	Chloracetamide	8.9				
	Tetrapropylene benzol sulphonate	1.0				
	Potassium dichromate	1.9				
Enchytraeus cf. *bucholzi*	NaPCP	0.20 (0.18 young)				Römbke and Knacker (1989)
	Potassium dichromate	4.42 (2.34 young)				
	Cu	0.41				Didden and Römbke (2001)
	Zn	57				

Marionina cambrensis	NaPCP	0.21			Römbke and Knacker (1989)
	2,4,5 T	269			
Tubificidae					
Branchiura sowerbyi	Aniline		586		Bhunia et al. (2003)
	Cadmium:				Dutta and Kaviraj (1996)
	$Ca(OH)_2 = 0$ mg l^{-1}		55		
	$Ca(OH)_2 = 200$ mg l^{-1}		93		
	NaPCP		0.28	0.56	Chapman et al. (1982b)
	Black liquor(%)		0.79	1.4	
	Hg		0.08	3.2	
	Cd		0.24	5.7	
	Sewage %		2.5	7.6	
	Low pH		3.7	2.5	
	High pH		10.5	11.3	
	Salinity (‰)		7.5	12	
	Temperature (°C)		35	35	
	Linear alkylbenzene sulphonate (LAS):				Casellato and Negrisolo (1989)
	(10°C)		4.38	10.8	
	(20°C)		4.82		
	NaPCP		2.20		Inoue and Kondo (1962)
	Parathion		3.50		
	EPN		2.06		
	Diazinon		4.95		
	DDT		19.91		

(continued)

(continued)

Taxon	Chemical and or environmental factors	Water LC_{50} (mg l^{-1}) 48 h	Water LC_{50} (mg l^{-1}) 96 h	Sediment LC_{50} (mg g^{-1}) 48 h	Sediment LC_{50} (mg g^{-1}) 96 h	Source
	Lindane		11.60			
	Endrin		7.66			
	Dieldrin		4.12			
	Rotenone		0.25			
	Pyrethrins		0.56			
Ilyodrilus frantzi	NaPCP		0.31			Chapman and Mitchell (1986)
	Hg		0.29			
	Low pH		3.5			
	High pH		10.5			
	Salinity (‰)		11.5			
	Temperature (°C)		24			
Limnodrilus hoffmeisteri	NaPCP		0.33		1.25	Chapman et al. (1982b)
	Black liquor(%)		0.58		1.35	
	Hg		0.18		3.2	
	Cd		0.17		3.5	
	Sewage (%)		5.5		12.0	
	Low pH		3.5		2.6	
	High pH		10.5		11.0	
	Salinity (‰)		10.0		14.5	
	Temperature (°C)		34		35	
	Linear alkyl benzene sulfonate (LAS) (10°C)		1.98		7.8	Casellato and Negrisolo (1989)

	Pesticides:				Kabir and Katoon (1980) (mixture of *Limnodrilus* spp., 80% *L. hoffmeisteri*)
	Nogos	5.62			
	Diazinon	6.21			
	Carbicron	6.56			
	Endrin			2.0–5.6 (4.0 mixed with *S. heringianus*)	Keilty et al. (1988a)
	Phenol	870	780		Green et al. (1985)
	NaPCP		0.60		Inoue and Kondo (1962) (as *L. socialis*)
	Parathion		5.05		
	EPN		2.46		
	Diazinon		0.73		
	DDT		19.25		
	Lindane		5.39		
	Endrin		0.78		
	Dieldrin		4.53		
	Rotenone		0.82		
	Pyrethrins		1.00		
	Furadan	14	11		Dad et al. (1982)
	Malataf	18	16		
Limnodrilus udekemianus	Parathion		22.60		Inoue and Kondo (1962) (as *L. willeyi*)
	EPN		28.12		
	Diazinon		10.02		
	DDT		73.45		
	Lindane		21.86		
	Endrin		17.64		
	Dieldrin		15.31		
	Rotenone		1.39		
	Pyrethrins		21.45		

(continued)

(continued)

Taxon	Chemical and or environmental factors	Water LC_{50} (mg l^{-1}) 48 h	Water LC_{50} (mg l^{-1}) 96 h	Sediment LC_{50} (mg g^{-1}) 48 h	Sediment LC_{50} (mg g^{-1}) 96 h	Source
Monopylephorus cuticulatus	NaPCP		0.55		1.3	Chapman et al. (1982b)
	Black liquor (%)		6.0		12.0	
	Hg		0.23		1.8	
	Cd		135		135	
	Sewage (%)		18		35	
	Low pH		3.5		2.5	
	High pH		10.5		10.5	
	Salinity (‰)		No lethality in freshwater		No lethality in freshwater	
	Temperature (°C)		31		34	
Quistadrilus multisetosus	NaPCP		0.57		0.92	Chapman et al. (1982b)
	Black liquor (%)		0.75		3.6	
	Hg		0.25		6.0	
	Cd		0.32		7.4	
	Sewage (%)		2.7		7.2	
	Low pH		3.5		2.5	
	High pH		9.6		11.5	
	Salinity (‰)		7.5		14	
	Temperature (°C)		32		35	
Rhyacodrilus montana	NaPCP		0.75			Chapman et al. (1982b)
	Black liquor (%)		1.8			
	Hg		0.24			
	Cd		0.63			
	Sewage (%)		1.8		5.5	
	Low pH		2.5			

	High pH	9.5		
	Salinity (‰)	7.5		
	Temperature (°C)	25.0		
Spirosperma ferox	NaPCP	0.43		Chapman et.al. (1982b)
	Black liquor (%)	0.37		
	Hg	0.33		
	Cd	0.35		
	Sewage (%)	1.1		
	Low pH	3.8		
	High pH	9.6		
	Salinity (‰)	7.5		
	Temperature (°C)	35.0		
Spirosperma nikolskyi	NaPCP	0.98	3.6	Chapman et al. (1982b)
	Black liquor (%)	0.37	5.0	
	Hg	0.50	7.5	
	Cd	0.45	12.0	
	Sewage (%)	1.3	5.6	
	Low pH	3.5	2.3	
	High pH	9.8	11.0	
	Salinity (‰)	7.5	12	
	Temperature (°C)	25.0	25.0	

(continued)

(continued)

Taxon	Chemical and or environmental factors	Water LC_{50} ($mg\ l^{-1}$) 48 h	Water LC_{50} ($mg\ l^{-1}$) 96 h	Sediment LC_{50} ($mg\ g^{-1}$) 48 h	Sediment LC_{50} ($mg\ g^{-1}$) 96 h	Source
Tectidrilus verrucosus (as *Limnodriloides verrucosus*)	NaPCP		0.25		0.98	Chapman et al. (1982b)
	Black liquor (%)		0.36		0.55	
	Hg		0.12		1.1	
	Cd		10		18	
	Sewage (%)		2.7		6.4	
	Low pH		4.5		3.8	
	High pH		10		10.2	
	Salinity (‰)		9		7	
	Temperature (°C)		19		22	
Tubifex tubifex and Limnodrilus hoffmeisteri	NaPCP		0.58			Chapman et al. (1982a)
	Hg		0.23			
	Cd		0.58			
	Low pH		3.5			
	High pH		10.5			
	Salinity (‰)		8.8			
	Temperature (°C)		34.5			
	Malathion	20.7	16.7			Whitten and Goodnight (1966)
	Parathion	6.3	5.23			
	DDT	>100	>100			
	Dieldrin	13.1	6.71			
	BHC	5.66	3.15			
	Sevin	>50	>50			
	NaPCP	1.0				Slooff (1983)
	Hg	0.18				
	Cd	6.5				

	n Propanol	9,200			
	n Heptanol	26			
	Ethyl acetate	760			
	Ethyl proprionate	180			
	Acetone	15,000			
	Trichlorethylene	132			
	Allylamene	18			
	Anilene	450			
	Benzene	>320			
	Pyridine	1300			
	o-Creosol	165			
	Salicyl aldehyde	1.8			
Tubifex tubifex	NaPCP	0.38*	0.38	0.82	Chapman et al. (1982b) [* Chapman et al.(1982a)]
	Black liquor (%)		0.64	3.5	
	Hg	0.14*	0.14	1.25	
	Cd	0.9*	0.32	3.8	
	Sewage (%)		3.0	8.6	
	Low pH		3.6	2.5	
	High pH		10.5	10.7	
	Salinity (‰)		9	14	
	Temperature (°C)		35	35	
	Temperature (°C)		34		Birtwell and Arthur (1980)

(continued)

(continued)

Taxon	Chemical and or environmental factors	Water LC_{50} (mg l^{-1}) 48 h	Water LC_{50} (mg l^{-1}) 96 h	Sediment LC_{50} (mg g^{-1}) 48 h	Sediment LC_{50} (mg g^{-1}) 96 h	Source
	Cu	0.89				Brkovic-Popovic and Popovic (1977a) (Test in tap water, distilled water, and dilution water for BOD test)
	Cd	0.72				
	Hg	0.1				
	Zn	60.2				
	Cr	4.57				
	Ni	61.4				
	Furadan	18	14			Dad et al. (1982)
	Malataf	32	28			
	As^{5+}	190.54	127.36			Fargasová (1994) (25°C)
	Pb^{2+}	19.498	14.62			
	Cr^{6+}	6.45, 3.16	3.31, 2.91			
	Hg^{2+}	0.38	0.28			
	Cd^{2+}	1.462	1.032			
	11 Organotin compounds	0.0055–0.0089	0.0001–0.0055			Fragasova and Kizlink (1996) (20°C)
	Plant growth regulators:					Fargasová (1997) (25 ± 1°C)
	IAA	97.50				
	2,4-D	161.17				
	IPA	123.0				
	MCPA	171.0				
	NAA	78.60				

	Cu^{2+}	0.005	0.002		Fargasová (1998) (20°C, 80 mg l^{-1} $CaCO_3$)
	Cu^{+}	0.012	0.006		
	Mn^{2+}	0.429	0.295		
	Mo^{6+}	14.096	4.563		
	Ni^{2+}	0.629	0.537		
	V^{5+}	0.250	0.211		
	300 pesticides			0.02– >10,000	Högger and Ammon (1994) 2–7 d exposure with sand-soil sediment
	32 metals:				Khangarot (1991) (30°C, 400 mg $CaCO_3$ l^{-1}, 5.8 mg O_2 l^{-1})
	Os	0.009	0.0067		
	Ag	0.039	0.031		
	Pt	0.086	0.061		
	Cu	0.282	0.158		
	Cr	0.196	0.190		
	Cd	59.43	47.53		
	Zn	21.13	17.78		
	Ni	96.38	66.75		
	Pb	0.142	0.042		
	Hg	0.121	0.051		
	As	8.87	8.87		

(continued)

(continued)

Taxon	Chemical and or environmental factors	Water LC_{50} (mg l^{-1}) 48 h	Water LC_{50} (mg l^{-1}) 96 h	Sediment LC_{50} (mg g^{-1}) 48 h	Sediment LC_{50} (mg g^{-1}) 96 h	Source
	Cu in the presence of 24 amino acids (2–20 mM)		0.084 (Cu alone) 0.164–11.80 in the presence of aminoacids			Khangarot and Rathore (2004) (20°C)
	Cu (distilled water)		0.014			Lucan-Bouché et al. (1997)
	Cu (well water)		0.25			
	Pb (distilled water)		$0.2 < LC_{50} < 0.5$			
	Pb (well water) (300 mg l^{-1} Ca CO_3)		*ca.* 0.25			
	Cd	0.064	0.030			Lucan-Bouché et al. (1999a), Bouché et al. (2000)
	Cd	70.02	5.91			Rathore and Khangarot (2002) (20°C) (Data for 15°C, 20°C, 25°C and 30°C)
	Cr	3.2	2.72			
	Co	202.99	179.71			
	Cu	0.154	0.092			
	Fe	157.82	125.42			
	Pb	581.64	514.19			
	Mn	350.16	275.70			
	Hg	0.051	0.048			
	Ni	29.59	18.97			
	Zn	20.81	14.74			

	$CaCO_3$ *mg l*$^{-1}$		$CaCO_3$ *mg l*$^{-1}$		
	12	*300*	*12*	*300*	
Cd	1.0	14.92	0.13	8.50	Rathore and Khangarot (2003) (20°C)
Cr	8.84	22.93	2.81	14.31	
Co	374.1	>1,800	128.6	565.8	
Cu	0.157	0.940	0.097	0.615	
Pb	4.47	95.41	3.5	60.82	
Mn	89.95	>1,800	26.8	464.75	
Hg	0.065	0.025	0.065	0.013	
Ni	236.8	337.0	15.31	99.9	
Zn	4.49	27.20	0.76	17.89	
Cd	22.47		14.75		Redeker and Blust (2004)
Cu	0.18, 0.26		0.07, 0.09		Reynoldson et al. (1996)
Cd	0.9–6.5		0.4–3.2		
Cr	33.2–95.5		9.8–38.1		
Lindane			3.5, 3.9		
Cd			0.35–2.44		Sturmbauer et al. (1999) (range for 4 genetic lineages)

(continued)

(continued)

Taxon	Chemical and or environmental factors	Water LC_{50} (mg l^{-1}) 48 h	Water LC_{50} (mg l^{-1}) 96 h	Sediment LC_{50} (mg g^{-1}) 48 h	Sediment LC_{50} (mg g^{-1}) 96 h	Source
Tubificoides fraseri	NaPCP		0.46		0.7	Chapman et al. (1982b, c) (as *T. gabriellae*)
	Black liquor (%)		0.67		0.98	
	Hg		0.098		1.2	
	Cd		24		62	
	Sewage (%)		7.5		29	
	Low pH		4.5		2.7	
	High pH		10.6		10.6	
	Salinity (‰)		2		1	
	Temperature (°C)		27		32	
Varichaetadrilus pacificus	NaPCP		0.105			Chapman et al. (1982b)
	Black liquor (%)		0.60			
	Hg		0.10			
	Cd		0.38			
	Low pH		3.5			
	High pH		10.0			
	Salinity (‰)		7.5			
	Temperature (°C)		25.0			
Naididae						
Nais communis	NaPCP		0.11			Chapman and Mitchell (1986)
	Hg		0.16			
	Low pH		3.9			
	High pH		9.9			
	Salinity (‰)		6.0			
	Temperature (°C)		22.0			

Dero digitata	Carbendazim	0.980		Van Wijngaarden et al. (1998)
	Formalin (µl l^{-1})	22.4		Mischke et al. (2001)
	Chloramine-T	28.1		
	Sodium chloride	6,700		
	Potassium permanganate	5.7		
	Copper sulfate	1.8		
	Hydrogen peroxide (µl l^{-1})	13.2		
	Rotenone ® (µl l^{-1})	0.21		
	Bayluscide®	0.24		
Pristina longiseta (as P. leidyi)	Cadmium (95 mg l^{-1} $CaCO_3$)	0.215		Smith et al. (1991)
	Vanadium	30.830		
Stylaria lacustris	Carbendazim	2.035	0.821	Van Wijngaarden et al. (1998)
Lumbriculidae				
Lumbriculus variegatus	TNT	4.9		Bailey and Liu (1980)
	TNT/ROX	8.8		
	Nitroaromatics	24.5		
	Cu	0.23	0.15	
	Zn	8.1	6.3	
	Cd	0.12	0.074	
	Pb	3.4	1.8	
	Cr	25.3	13.3	
	Hg	0.11	0.10	
	Terbutryn		23.7	Brust et al. (2001)

(continued)

(continued)

Taxon	Chemical and or environmental factors	Water LC_{50} ($mg\ l^{-1}$) 48 h	Water LC_{50} ($mg\ l^{-1}$) 96 h	Sediment LC_{50} ($mg\ g^{-1}$) 48 h	Sediment LC_{50} ($mg\ g^{-1}$) 96 h	Source
	Allyl alcohol		0.32			Ewell et al. (1986)
	Pentachlorophenol		3.2			
	Thymol		3.2			
	12 organics		>100			
	6 inorganics		>100			
	Cu		0.32			
	Ni, Cr, Zn		32.0			
	NaCN		21.0			
	NaClO		3.2			
	Hg		0.044			McCrary and Heagler (1997)
	Phenol		35.6			Hickey and Martin (1995)
	PCP		0.69			
	Cd (II)		0.15			
	Pt^{4+} (distilled water)		0.397			Veltz et al. (1996)
	Pt^{4+} (hard water)		30			
	Cu (distilled water)		0.060			Veltz-Balatre (2000)
	Cu (hard water)		0.186			
	Pb (distilled water)		0.225			

Stylodrilus heringianus	NaPCP	0.63	1.35	Chapman et al. (1982b)
	Black liquor(%)	0.83	2.3	
	Hg	0.14	6.0	
	Cd	0.55	7.5	
	Sewage (%)	2.8	5.9	
	Low pH	3.7	2.6	
	High pH	10.3	10.5	
	Salinity (‰)	7.5	11	
	Temperature (°C)	25.0	25.0	
	Endrin		1.4–5.4 (4.0 mixed with *L. hoffmeisteri*)	Keilty et al. (1988a) (10°C)

Appendix 2 Data on bioaccumulation in oligochaete species exposure to several chemical substances. Abbreviations: k_u, uptake constant, usually from dissolved phase, but also used by some authors as a generic uptake constant; k_s, uptake constant from sediment; k_e, elimination constant; Partition coefficients: K_{ow} octanol-water, K_b biota-environment, K_{oc} sedimentary organic carbon, K_{doc} dissolved and colloidal organic carbon in pore water (References in Chap. 5)

Chemical	Exposure conditions (t)	Measurements (other than tissue residue)	Bioaccumulation data	Species	Reference
Metals					
Ag	Bioassay in silver sulphide spiked sediment (28 days) (Sediment TOC approx. 2%)	Accumulation factor (AF)	Highest tissue concentration = 80.3 mg Ag kg^{-1} at highest sediment concentration (444 mg Ag kg^{-1}dw). AF = 0.18	*L. variegatus*	Hirsch (1998)
		Survival and reproduction	Survival and reproduction were not affected even at highest concentrations		
Cd	Spiked sediment bioassays (10 days)	Influence of AVS on bioavailability	Tissue concentration = 240–690 μg Cd g^{-1}(including gut content), similar to the maximum residue of 670 μg Cd g^{-1}measured in water-only bioassays	*L. variegatus*	Carlson et al. (1991)
Cd	Microcosms (56 days)	Toxicokinetics bioturbation	50 μg g^{-1} dw (not purged worms), that is about 6 times the concentration in the higher 5-cm layer of sediment. No effects in bioturbation	Tubificids	Ciutat et al. (2005b)
Cd	Bioassays in cadmium chloride spiked sediment (28 days)	Toxicity	Threshold tissue concentration (significantly higher than control) = 2.26 μmol g^{-1} dw	*T. tubifex*	Gillis et al. (2002)
		MTLP	Tissue residue corresponding to the lowest significant reduction in young per adult = 30.38 μmol g^{-1} dw		
			MTLP and tissue Cd were significantly elevated above control levels after exposure to the 0.67 μmol Cd g-1 t dw.		

Cd	Aqueous test (6 days) spiked with ^{109}Cd	Resistance	Aqueous test: 1,1010 μmol g^{-1}ww (control worms); 2,232 μmol g^{-1}ww (polluted site worms); 2,040 μmol g^{-1}ww (second generation)	*L. hoffmeisteri*	Klerks and Bartholomew (1991)
	Field Cd- polluted sediments (28 days)	Metallothionein-like protein	Sediment tests (5,400–34,000 μg Cd g^{-1}dw): up to ca. 3 μmol Cd g^{-1}ww in worms from control site, and ca. 15 μmol Cd g^{-1}dw in worms from polluted site		
Cd	Water-only tests (7 days)	Bioconcentration (BCF)	Tissue concentration (exposed to Cd concentrations of 5, 20 and 200 μg l^{-1}) = 0.6, 2.2, and 30.3 μg g^{-1} ww (0.1 μg g^{-1} ww in control worms)	*L. variegatus*	Ng and Wood (2008)
			BCF = 134.9 l kg^{-1}		
		Trophic transfer to fish	Diet-borne Cd reduced trout growth by 50% in fish exposed to highest dose, during 1 month		
Cd	Toxicokinetic tests: 0.1 μM Cd solution spiked with ^{109}Cd (4-d uptake + 8-d elimination)	LC_{50}, Critical Body Concentration (CBC)	Average tissue residue associated to 50% mortality = 0.37 μmol g^{-1}ww in worms exposed to 0.1–10 μM Cd.	*T. tubifex*	Redeker and Blust (2004)
		Toxicokinetics			
	Bioaccumulation tests: aqueous tests (15 days)	Biological half-life	Tissue concentration corresponding to LC_{50}, CBC_{50} = 0.32 μmol g^{-1}ww		
			CBC_{50} = 0.38 μmol g^{-1}ww calculated by a sigmoidal function of mortality % on tissue concentration for all exposure times and concentrations		
			k_u = 0.117 day^{-1} (during first 8 h). Overall k_e = 14.57 day^{-1}		
			Half-life = approx.15 days		

(continued)

(continued)

Chemical	Exposure conditions (t)	Measurements (other than tissue residue)	Bioaccumulation data	Species	Reference
Cd	Aqueous tests spiked with ^{109}Cd and stable Cd, 0.52 μg l^{-1} (10-d uptake + 10-d elimination)	Toxicokinetics Influence of water hardness	k_u = 0.058 l $g^{-1}day^{-1}$ k_e = 0.002 day^{-1} k_u is a function of the log ([Cd^{2+}]/[Ca^{2+}]) and water hardness, and it shows exponential decrease with increased hardness	*L. variegatus*	Xie et al. (2008)
Cd, Ni, Pb	Spiked sediment (6 and 14-week uptake +: 10-week depuration in sediment)	Toxicokinetics, MTLP, BSAF Influence of gut purging in BSAF calculation	k_u (h^{-1}) = 0.003 and 0.0175 (Cd), 0.1727 (Ni), 0.0192 (Pb) k_e (h^{-1}) = 0.0008 and 0.0013 (Cd), 0.0755(Ni), 0.0001 (Pb) BSAF (6 weeks) = 21.9 (Cd), 2.3 (Ni), 192.0 (Pb) Cd-BSAF (10 weeks) = 19.4 MTLP reached a maximum after 96-h exposure to Cd	*T. tubifex*	Gillis et al. (2004b)
Cd, Hg, Pb	Field sediments and tubificids from polluted sites	Seasonal variations	Cd: 0.3–0.7 mg kg^{-1}; Hg: 0.2–1.2 mg kg^{-1} Pb: 8.0-23.0 mg kg^{-1} No correlation of tissue concentration and sediment concentration of metals. No seasonal variations detected	Field tubificids	Kaiser et al. (1989)
Cu, Cd	Artificial silica-sand sediment bioassay (14 days)	BAF (%) and linear relationship with asexual reproduction	Cu tissue concentration = up to 171.3 mg Cu kg^{-1} tissue ww (0.83 in control) Cd tissue concentration = up to 47.2 mg Cd kg^{-1} tissue ww ((0.14 in control)	*L. variegatus*	Chapman et al. (1999)
Cu, Pb	Field tubificids sampled from two polluted sites	Seasonal variations	Cu tissue concentration: higher in winter (31.6 and 53.7 μg g^{-1}dw) Pb tissue concentration: higher in winter (17.9 and 21.3 μg g^{-1}dw) Minimal tissue concentration in summer	Field tubificids	Hernandez et al. (1988)

Zn	Aqueous bioassay, in a $ZnCl_2$ solution containing ^{65}Zn 6-h to 17-day exposure period	Toxicokinetics Body wall adsorption and total tissue concentration	Zn adsorption (μg g^{-1}dw) = 0.552 $x^{0.691}$ Zn total uptake (μg g^{-1}dw) = 4.159 $x^{0.771}$, where x is Zn concentration in mg l^{-1} Adsorption represents ca. 10% total Zn accumulated, after 6-h exposure to 10 days	*T. tubifex*	Fleming and Richards (1982)
Several metals	Field worms and sediments	Bioaccumulation factor (BAF) Zinc localisation by X-ray diffraction	Maximum tissue concentration: 6,296 (Zn), 132 (Mn), 197.3 (Cu), 1,017 (Cd) μg g^{-1}dw Maximum BAF = 39.3 (Zn), 0.150 (Mn), 0.731 (Cu), 9.01 (Cd) g g^{-1}dw Zn represents 0.5–1% of the dry tissue in the gut and chloragogen cells, and up to 4% at some tissue portions	Field tubificids	Say and Giani (1981)
Several metals	Tissue analysis from field organisms	Metal availability in laboratory and field organisms Seasonal variation	Metal bioaccumulation in oligochaetes is different than in fishes. Pb, Cu, Cd, Fe and Mn body concentration is higher in tubificids than in fishes	Field tubificids	Krantzberg (1994)
Several metals	Tissue analysis from field organisms used in aquaculture	Tissue concentration in worms form 6 sites polluted by sewage and industrial wastes	Cd: 2.38–7.21 mg kg^{-1} dw Fe: 671.9–5,738 mg kg^{-1} dw Pb: 14.95–33.49 mg kg^{-1} dw Zn: 60.20–166.60 mg kg^{-1} dw Cu: 29.38–108.90 mg kg^{-1} dw	Red worms (*Tubifex* spp.) Probably a mixture of tubificid species	Singh et al. (2007)

(continued)

(continued)

Chemical	Exposure conditions (t)	Measurements (other than tissue residue)	Bioaccumulation data	Species	Reference
Several metals	Field sediment bioassays (28 days)	Bioaccumulation Biomagnification risk	Tissue concentration (mg kg^{-1}dw) after exposure to dredged sediments (tissue concentration in the culture in parentheses): As: 5.6–26.7 (0.7), Cd: 0.2–0.5 (0.1), Cr: 0–3.3 (0.0), Cu: 17.6–79.2 (9.2), Hg: 0.01–0.17 (0.08), Mn: 41–113 (3), Mo: 0.2–0.7 (0.1), Ni: 0–1.6 (0.7); Pb: 3.4–6.6 (1.4), Se: 2.6–5.1 (1.4), Zn: 430–956 (266)	*L. variegatus*	Winger et al. (2000)
Organic compounds (miscellanea)					
Benzo(a) pyrene (BaP) and PCB-52	Field sediments with radiolabelled contaminants in sediment, algae and bacteria (7 days)	Bioaccumulation factors Toxicokinetics Absorption efficiency	Maximum BAF – BaP=13.1–25.0 $g_{food}\ g^{-1}{}_{tissue}$, Maximum BAF- PCB-52=2.9–6.8 (12.2) $g_{food}\ g^{-1}{}_{tissue}$ BaP Absorption efficiencies=38 and 50.7 % (from labelled bacteria), and 22.9 and 74.8% (from labelled sediment)	*L. variegatus*	Bott and Standley (2000)
2,4,6-trinitrotoluene (TNT) and 4 metabolites	Aqueous test, using ^{14}C-labelled TNT (up to 96-h uptake+53-h elimination)	Toxicokinetics Bioconcentration factors of TNT and metabolites	Steady state of all absorbed compounds reached at 1 h. BCF-TNT=2.53, BCF- (TNT+all metabolites)=12.25. BCFs of detectable compounds were lineally related to log K_{ow} 82% absorbed TNT was metabolized to amino-dinitrotoluenes (ADNTs). All absorbed compounds were eliminated in 1–3 h	*T. tubifex*	Conder et al. (2004)
Dichloro-phenol (DCP) and pyrene	Water only and water+sand (<100 μm) tests with [^{14}C]DCP Sediment spiked with pyrene	DCP uptake (5 days) Pyrene uptake (12 days) Bioaccumulation factors Exposure routes	DCP uptake only from water Pyrene uptake via sediment ingestion represents approx. 20% body burden in feeding worms	*L. variegatus*	Conrad et al. (2000)

4-nonyl-phenol (4-NP)	Spiked and field sediment bioassays (56 days)	Toxicokinetics BSAF	Steady state was not reached in 56 days Tissue residue: up to 4.4 μg g^{-1}ww (in spiked sediment) 56-d BSAF=24 g_{OC} g^{-1}_{lipid} (*L. variegatus*) 14-d BSAF=6 g_{OC} g^{-1}_{lipid} (*L. variegatus*) 56-d BSAF=39–55 g_{OC} g^{-1}_{lipid} (field worms)	*L. variegatus* and field oligochaetes	Croce et al. (2005)
Perfluoro-chemicals (PFCs)	Filed and spiked sediment bioassays (56 days) Depuration (28-d uptake+42–49-d elimination)	Toxicokinetics Bioaccumulation factors Biotransformation	Tissue concentration reached to 65–270 ng g^{-1} ww (in 56 days) k_s (× 10^4)=10.9–19.3 g_{oc} g^{-1}ww h^{-1} k_e (× 10^4)=6.8–36.8 h^{-1} BSAF=7–55 g_{oc} $g^{-1}_{lipid,}$ based in kinetic constants BAF=0.02–1.22 g_{oc} g^{-1}ww (non-lipid normalised). BAF provides accurate predictions of the bioaccumulation of PFCs from field sediments	*L. variegatus*	Higgins et al. (2007)
DDT, DDD, DDE, PAHs	Sediment bioassays using field contaminated sediments (56 d and 28-d uptake+7-d depuration)	Toxicokinetics Steady-state BSAF	Body burden in field oligochaetes was similar to steady-state concentration estimated in *L. variegatus*. DDT and its metabolites-BSAF range 1–8 (from day 14 to 56). Depuration is very low in water and gradual along 7 days in sediment Most PAHs with K_{ow}<5.6 peaked at day 3, with a plateau from day 7 to 56. BSAF range 0.7–3.0 (from day 14–56). Depuration in water or sediment: 53% and 23% reduction during the first 12 h, respectively PAHs with K_{ow}>5.6 reached steady state by day 14. BSAF range 0.4–2.0 (from day 14 to 56). Depuration in water is very low. 18% and 23% reduction during the first 12 h in water or sediment, respectively	*L. variegatus*	Ingersoll et al. (2003)

(continued)

(continued)

Chemical	Exposure conditions (t)	Measurements (other than tissue residue)	Bioaccumulation data	Species	Reference
Benzo(a) pyrene (BaP) and polydimethylsiloane (PDMS)	Sediment bioassay spiked with [^{3}H]BaP and [^{14}C]PDMS (7 days)	Bioaccumulation factors Toxicokinetics Bioavailability Biological half-life	Steady state achieved in within 96–168 h Reference sediments: BaP – k_s=0.0688 g_{sed} g_{worm}^{-1} $h^{-1}$and 0.00029 g_{OC} g_{worm}^{-1} h^{-1} ; BAP – k_e=0.023 h^{-1}(depuration in sediment) and 0.0004 h^{-1} (depuration in water) In PDMS polluted sediments, k_s slightly decreases BaP: BAF=up to ca. 1.5 g_{sed} g_{worm}^{-1} PDMS: BAF=0.08-0.11 g_{sed} g_{worm}^{-1} BAP bioavailability was reduced by PDMS BAP Half-life: 30.3 h in sediment	*L. variegatus*	Kukkonen and Landrum (1995b)
Methylmercury	Sediment bioassay, spiked with labeled [^{14}C]MeHg (2 weeks)	Toxicokinetics Selenium and OC influence on bioaccumulation	Steady state was not reached. Lake Höytiäinen-k_u=0.0089 g_{sed} g^{-1} h^{-1}(7.1% TOC; 90 ng ^{14}C-MeHg g^{-1} dw) (LOI=7%) Lake Mekrijärvi-k_u=0.0032 g_{sed} g^{-1} h^{-1}(17.8% TOC; 106 ng ^{14}C-MeHg g^{-1} dw) (LOI=18%) Bioaccumulation decreased by 75% and 86% as compared to the reference, when 15 and 50 mg Se kg^{-1}, respectively, were added to sediment.	*L. variegatus*	Nuutinen and Kukkonen (1998)

Polybrominated diphenylethers (Tetrabromo and Pentabromo diphenil ethers, TeBDE and PeBDE)	Field sediment bioassay, using [14C]TeBDE and PeBDE (3, 4 week-exposure Depuration study: 3-week uptake + 4-week elimination in clean sediment	Toxicokinetics Bioaccumulation factor	Steady-state tissue concentration: 38.5–44.0 nmol g_{lip}^{-1} k_s = 0.066–0.251 (feeding worms), 0.010 (worms not ingesting sediment) $g_{OC}\ g_{lipid}^{-1}\ h^{-1}$ k_e = 0.022–0.071 (feeding worms), 0.006 (worms not ingesting sediment) h^{-1} BSAF = 3.0–3.7 (log K_{ow} = 6.01–6.77 TeBDE, and 6.53–7.66 PeBDE)	*L. variegatus*	Leppänen and Kukkonen (2004)
17α-ethinyl estradiol (EE_2)	Artificial sediment, bioassay, spiked with [14C]EE_2 (35-d uptake + 10-d elimination)	Toxicokinetics Steady state not reached	Accumulation factor (AF, 35 days) = 75 $g_{lip}\ g_{oc}^{-1}$ AF (estimated at steady state) = 191 $g_{lip}\ g_{oc}^{-1}$	*L. variegatus*	Leibig et al. (2005)
Surfactants: 4-nonylphenol (4-NP) and LAS	Water-only (48 h) and spiked sediment bioassays (11 days), using [14C]-labelled substances, in 3 sediments of different %LOI (3.15–41.39%)	Toxicokinetcs Lethal body residues (LBR_{50}) BAF (k_s/k_e), BSAF	4-NP LBR_{50}: 11.55 µmol g^{-1}ww (48-h LC_{50} = 6.26 µmol l^{-1}) LAS LBR_{50}: 14.26 µmol g^{-1}ww (48-h LC_{50} = 5.65 µmol l^{-1}) 4-NP: BAF = 1.8–33.6, BSAF = 14.1–55.4 LAS: BAF = 0.5–4.0, BSAF = 5.7–35.5	*L. variegatus*	Mäenpää and Kukkonen (2006)

(continued)

(continued)

Chemical	Exposure conditions (t)	Measurements (other than tissue residue)	Bioaccumulation data	Species	Reference
Wood preservatives chlorophenols (CP), organo-halogens (EOX), As, Cu, Cr,	Sediment bioassay (28 days)	Tissue concentration	Ky-5CP: up to 337 ng g^{-1}dw Other CP: up to 288 ng g^{-1}dw As: up to 362 μg g^{-1}dw (good correlation with pore water concentration) Cu: up to 52 μg g^{-1}dw Cr: < detection limit	*L. variegatus*	Lyytikäinen et al. (2001)
PAHs					
Benzo(a) pyrene (BaP)	Desorbed sediments, spiked with BaP, supplemented with [^{3}H]BaP (38 days)	Toxicokinetics, BSAF, biotransfor-mation, Assimilation efficiency (AE)	BSAF = 1.3 g_{oc} g_{lip}^{-1} (at steady state) Steady state was reached in about one month Major uptake route by ingestion of sediment particles. k_s = 0.032 and 0.035 g_{sed} g_{worm}^{-1} h^{-1}. Low elimination rate k_e = 0.003 h^{-1}. Negligible biotransformation AE = 80%, during a single gut passage	*Ilyodrilus templetoni*	Lu et al. (2004)
Fluoranthene	Transference to shrimps through diet (worms exposed to spiked sediments) (5-d feeding + 3-d depuration)	Bioaccumulation Trophic transfer coefficient (TTC) Influence of piperonyl butoxide (PBO), inhibitor of cytochrome P-450	Tissue residue: 671.6 μg g_{worms}^{-1}dw (day 5) Trophic transfer very low TTC = 0.02 (presence of PBO) TTC = 0.01 (absence of PPO)	*M. rubroniveus*	Filipowicz et al. (2007)

Fluoranthene	Water-only test (72-h uptake without UV radiation + 96 h with UV radiation)	Bioconcentration factor Salinity influence (7.3, 13.5, 20.6 and 27 ‰)	BCF = 19,802–502.9 salinity ‰, $r^2 = 0.655$ Tissue residue μg g^{-1}dw = 168.0 - 4.5 salinity ‰ (worms exposed to 7.7 μg l^{-1})	*M. rubroniveus*	Weinstein (2003)
Fluoranthene	Water-only tests (3 days) Spiked sediment tests (10 days)	Bioconcentration and bioaccumulation factors Effects of UV radiation	Mean BCF = 10,893 ± 2,828 (on a dw basis) across all treatments 72-h LC_{50} > 120.4 μg l^{-1} 72-h LD_{50} >1289.1 μg g^{-1} dw 10-d LC_{50} >191,765 μg $g^{-1}{}_{OC}$ 10-d tissue residue: 5.13–6.43 mg g^{-1} dw BAF = up to 23.0 (on a dw basis)	*M. rubroniveus*	Weinstein et al. (2003)
Anthracene, pyrene, fluorene	Water-only bioassays (96-h uptake and 48-h depuration in the dark) Effects of UV radiation in exposed worms (96 h)	Bioconcentration factor Time-dependent mortality (LT_{50}) Light intensity * initial tissue residue	96-h BCF = 1370 l g^{-1} (anthracene), 1720 l g^{-1} (pyrene), 400 l g^{-1} (fluorene) k_e : 0.20 day^{-1} (anthracene), 0.14 $days^{-1}$ (pyrene), ND (fluorene) Linear regression of the LT_{50} on the product (Light intensity * initial tissue residue) for anthracene and pyrene, with some deviation at longer exposures. Fluorene was not phototoxic.	*L. variegatus*	Ankley et al. (1997)

(continued)

(continued)

Chemical	Exposure conditions (t)	Measurements (other than tissue residue)	Bioaccumulation data	Species	Reference
Fluorene, anthracene, benzo(*k*) fluoranthene, benzo(*a*) pyrene, indeno (*1,2,3 cd*) pyrene, dibenzo (*g,h,i*) perylene	Creosote contaminated sediment bioassay (28 days)	Bioaccumulation factors and K_{ow}	BSAFs of individual PAHs range 1.2–5.7 Individual PAH concentrations in sediment and in the tissue were correlated (r=0.96) Log K_{ow} and BSAFs correlated negatively (r = –0.75). K_{ow}=4.38–7.04, for the 6 studied PAHs	*L. variegatus*	Hyötiläinen and Oikari (2004)
Fluoranthene, anthracene, fluorene, pyrene	Water-only test (96-h uptake+96-h depuration)	Bioconcentration factor	96-h mean BCF (l g^{-1}, on a ww basis)=2390 (fluoranthene); 1210 (anthracene), 452 (fluorene); 1920 (pyrene) By 24-h depuration period, losses of 0–68% of the four PAHs were reported.	*L. variegatus*	Sheedy et al. (1998)
Pyrene	Spiked sediment bioassays (10 days)	BSAF (dw / ww) and Faecal-sediment accumulation factor (FSAF, dw basis, normalised for OC) Biotransformation	Body residue up to 4.4 µmol g^{-1} ww, which did not cause significant mortality Low biotransformation: mean 14%, independent of time or pyrene concentration BSAF negatively related to sediment concentration, may be related with selection Faecal pyrene concentrations increased with pyrene levels in sediment. FSAF values suggest removal of pyrene during gut passage: 1 (on day 5)–0.26 (on day 10)	*Limnodrilus hoffmeisteri*	Millward et al. (2001)

Pyrene	Sediment bioassay, spiked with pyrene, and using [^{3}H]pyrene (14 and 58 days) Elimination test (7-d uptake + 4-d elimination)	Toxicokinetics (dw / ww) Effects of density on bioaccumulation Avoidance behaviour Biological half-life	k_s = 0.039–0.132 g_{sed} g^{-1} h^{-1}. k_s decreased with increasing sediment concentration, which was attributed mainly to selective feeding on particle size k_e = 0.0256 h^{-1}(in sediment), 0.0043 h^{-1}(in water) Steady state was achieved within 2–7 days, except for 132 µg g^{-1} ww Half-life: 27 h in sediment, 161 h in water Bioaccumulation declined as density decreased	*L. variegatus*	Kukkonen and Landrum (1994)
Pyrene, *trans*-chlordane, benzo(a)pyrene,	Sediment bioassay (7 days)	Toxicokinetics: Uptake rate coefficients for elutriate, pore water and whole sediment	K_{uoc} (µg OC cleared g^{-1}worm ww h^{-1}): – *Trans*-chlordane: 95.35 (elutriate), 60.4 (pore water), 356.1 (whole sediment) – Benzo(a)pyrene: 8.1 (elutriate), 30.4 (pore water), 53.5 (whole sediment) – Pyrene: 88.75 (elutriate), 72.85 (pore water), 169.2 (whole sediment)	*L. variegatus*	Harkey et al. (1994)
Several PAHs	Field worms and sediment bioassays (28 days)	Tissue concentration comparisons between field and laboratory exposed oligochaetes BASF	Naphtalenes (log K_{ow} < 4.5) body burden similar in laboratory in field worms. PAHs of log K_{ow} > 5.1 with body burden lower in field worms Typically, concentrations in the laboratory were within a factor of 3 between laboratory and field collected worms, higher in laboratory worms as PAHs molecular weight increases BSAF = 0.97 (benzo(a)anthracene); 5.3 (naphthalene), for laboratory-exposed worms BSAF = 1.1 (chrysene); 8.8 (naphthalene), for field-exposed worms	*L. variegatus* and field oligochaetes	Brunson et al. (1998)

(continued)

(continued)

Chemical	Exposure conditions (t)	Measurements (other than tissue residue)	Bioaccumulation data	Species	Reference
Several PAHs	Sediment bioassay (4 weeks) Depuration experiment (6–20 h uptake phase + up to 120, 1 40 and 256 h) using radiolabelled PAHs	Toxicokinetics	Lower-molecular-weight compounds (pyrene and fluoranthene) have higher uptake constants and body concentration peaked between 48–96 h Higher-molecular-weight compounds (benzo (g) perylene and inden(1,2,3-cd)pyrene reached to an apparent steady-state after 100–336 h exposure Elimination is most rapid for phenanthrene (k_e = 0.0399 h^{-1}). Ln $k_e = -1.2807\ \chi^v + 2.807$, where χ^v is the first-order valence index of the PAH. Estimated k_e values ranged 0.0015–0.0198 h^1	*L. variegatus*	Harkey et al. (1995)
Chlorinated contaminants					
Chlorinated hydrocarbons	Field sediment bioassays (110 days)	Toxicokinetics K_b relationships with porewater and sediment concentration	$\log K_b = 6.25 - 0.75\ (\log K_{ow} - 6.84)^2$, where extrapolated max. values of $\log K_b$ (6.29) correspond to a $\log K_{ow} = 7.52$ $\log K_b = 1.03 \log K_{ow} + \log C_b - \log C_s + \log f_{oc} - 0.18$ Linear fit for compounds with $\log K_{ow}$ 4.4–6.4: $\log K_b = 2.30 \log K_{ow} - 7.95$	*T. tubifex and L. hoffmeisteri*	Connell et al. (1988)

Chlorinated hydrocarbons (37 chemicals)	Spiked sediment slurry (ca. 20% solids) bioassay Sediment TOC = 4.6% (79 d uptake + 84 d depuration)	Bioaccumulation factors (BCF, BAF, on dw basis) Toxicokinetics Biological half-life	11-d body burden = 130–9000 ng g^{-1}dw 79-d body burden = 40–6400 ng g^{-1}dw 11-d BCF = 1900–34000 $l_{porewater}$ kg_{worm}^{-1} dw Half-life < 5–200 days Field BAF for 7 compounds = 0.34–5.6 g_{sed} g_{worm}^{-1} dw. Good agreement for the most persistent compounds	Field tubificids (mainly *T. tubifex* and *L. hoffmeisteri*)	Oliver (1987)
Chloro-phenols (2,4,5-TCP and PCP)	Water-only (6, 7 days) and spiked sediment bioassays (7 days), using ^{14}C labelled substances	Toxicokinetics Bioaccumulation factors (BCF, BAF) Critical body residue (LBR_{50}) Influence of sediment avoidance behaviour	PCP-BCF = 1955 ml g^{-1} ww PCP-BAF = 26–32 g_{sed} dw g^{-1}_{worm} ww PCP- k_u = 34.0 ml g^{-1} ww h^{-1}, 0.005–0.012 g_{OC} g^{-1}_{worm} ww h^{-1} PCP-LBR_{50} = 0.83 µmol g^{-1} ww (in water), 0.43 and 0.62 µmol g^{-1} ww (in sediment) 2,4,5-TCP-BAF = 8 g_{sed} dw g^{-1}_{worm} ww (no apparent steady state was reached in water) 2,4,5-TCP- k_u = 14.6 ml g^{-1} ww h^{-1}, 0.004 and 0.009 g_{OC} g^{-1}_{worm} ww h^{-1} 2,4,5-TCP-LBR_{50} = 1.18 µmol g^{-1} ww (in water), 1.07 and 1.10 µmol g^{-1} ww (in sediment)	*L. variegatus*	Nikkilä et al. (2003)

(continued)

(continued)

Chemical	Exposure conditions (t)	Measurements (other than tissue residue)	Bioaccumulation data	Species	Reference
Dieldrin	Spiked sediment bioassays (7 days)	Bioaccumulation factors Bioavailability and partition coefficients	log K_{oc}=4.38–5.01 log K_{doc}=3.22–3.97 Body burden: 2.4 (2 days) 2.3 (7 days) μg g^{-1}(on a ww basis) Dissolved OC was 1 order of magnitude less sorptive to dieldrin than particulate OC BSAF correlate negatively with log K_{oc} and positively with log K_{doc}	*L. variegatus*	Standley (1997)
Endrin	Microcosms spiked with a mixture of [^{12}C] and [^{14}C] endrin (four experiments of 980–1312 h)	Bioaccumulation factors Reworking rates Survival and final biomass with single and mixed species	In single species tests, bioaccumulation factor in *L. hoffmeisteri* range 4.5–31.3, and 17.6–29.0 in *S. heringianus* (on a dw basis) In mixed species tests, *L. hoffmeisteri* mean bioaccumulation factor range 5.2–13.6, and *S. heringianus* 2.4–43.3 Highest bioaccumulation factors occurred at the intermediate concentrations (6300 ng g^{-1})	*S. heringianus* *L. hoffmeisteri*	Keilty et al. (1988a)
Endrin	Microcosms spiked with [^{14}C] endrin(1300 h)	Bioaccumulation Reworking rates Survival and final biomass	Worms bioaccumulate endrin, 34–67 times the sediment concentration (on a dry weight basis)	*S. heringianus*	Keilty et al. (1988b)

Hexa chlorobenzene (HCB), PCBs, penta chlorotoluene (PCT), hexa chlorobutadiene (HCBD), octa chloroestyrene (OCS), Mirex, *p,p*'-DDE	Field polluted sediment bioassays (33, 52 and 110 days)	Bioaccumulation factor (BAF, on a dry weight basis corrected for contaminant levels in control worms)	Maximum 110-day BAF=5.1 (PCBs); 0.63 (HCB), 0.17 (HCBD), 6.7 (OCS),1.5 (PCT), 4.7 (Mirex), 2.6 (DDE) Relationships of BAFs and K_{ow} was complex (non linear), with a maximum BAF values for log K_{ow} values of about 6	Field oligochaetes (mainly *L. hoffmeisteri* and *T. tubifex*)	Oliver (1984)
Herbicides: ioxynil, pendimethalin, bentazone	Water-only toxicity tests (48 h) Sediment bioaccumulation tests, using [^{14}C]-labelled substances (10 days)	BSAF CBR_{50}: tissue concentration at LC_{50} Toxicokinetics	Ioxynil: 48-h $LC_{50}=1.79$ mg l^{-1}, BSAF=15.7–101.7, water exposure-$CBR_{50}=267.1$ µmol kg^{-1}ww Bentazone 48-h $LC_{50}=79.11$ mg l^{-1}, BASF=2.9–14.9, water-exposure-CBR_{50}: 2849 µmol kg^{-1}ww Pendimethalin. Non toxic. BSAF=2.2–4.9	*L. variegatus*	Mäenpää et al. (2003)
Herbicide: Isoproturon	Water-only tests, spiked with 0.1 to 10 mg l^{-1} a.i. (not feeding) (7 days)	Metallothionein (MT) Tissue residues of parent substance and metabolites	MT=up to 121.42 (formulated substance), 91.62 µg g^{-1} ww (pure substance), at day 4 ($MT_{control}=46.85$–48.84 µg g^{-1} ww) Tissue residue: No parent substance was found, only Isoproturon metabolites. Up to 23.20 (day 2) and 18.20 µg g^{-1} ww (at day 7) (initial concentration 10 mg l^{-1})	*T. tubifex*	Mosleh et al. (2005)
Hexahydro-1,3,5-trnitro-1,3,5-triazine (RDX)	Water-only tests (16-h uptake + 9.1-h elimination in water)	Toxicokinetics Biological half-life Trophic transfer to fish	BCF≈2.4 ml g^{-1}(on a ww basis), at steady-state $k_u=5.12$ ml g^{-1} h^{-1}; $k_e=2.51$ h^{-1} BCF=2.1 Half-life=0.28 h RDX dietary uptake by fish is minimal	*L. variegatus*	Belden et al. (2005)

(continued)

(continued)

Chemical	Exposure conditions (t)	Measurements (other than tissue residue)	Bioaccumulation data	Species	Reference
Lindane, hexachlorobenzene (HCB)	Artificial sediment spiked using [^{14}C]-labelled substances (12-d uptake + 12-d elimination)	Toxicokinetics Bioaccumulation factors (BAF on a ww basis, BAF_{pc}: based onparent compound, and BSAF) Depuration	Lindane-BAF: 4.6, 4.8 (*T.t*), 4.1, 5.1 (*Lh*). CHB-BAF: 4.8, 7.6 (*Tt*), 6.7, 7.3 (*Lh*) Lindane-BSAF: 2.81, 2.93 (*T.t*), 1.49, 1.82 (*Lh*). CHB-BSAF: 3.13, 4.96 (*Tt*), 2.41, 2.66 (*Lh*) Lindane-k_s (h^{-1}): 3.31, 3.57 (*Tt*), 3.1, 16.1 (*Lh*). HCB-k_s (h^{-1}): 0.17, 0.47 (*Tt*), 0.46, 0.59 (*Lh*) Lindane-k_e (h^{-1}): 0.35 (*Tt*), 0.195 (*Lh*). HCB-k_e (h^{-1}): 0.037 (*Tt*), 0.047 (*Lh*) Depuration (% of steady state concentration after 12-d elimination): Lindane = 11% (*Tt*), 7% (*Lh*). HCB = 4% (*Tt*), 1% (*Lh*)	*L. hoffmeisteri* (*Lh*) *and T. tubifex* (*Tt*)	Egeler et al. (1997)
Lindane, hexachlorobenzene (HCB) and 3,4-dichloroaniline (DCA)	Artificial spiked sediment using [^{14}C]-labelled substances Lindane and HCB: 6-d uptake and DCA: 12-d uptake + 12-d elimination	Toxicokinetics Bioaccumulation factors on ww basis and ^{14}C activity (BAF and BAF_{pc}: corrected for content of parent compound)	Lindane-BAF: 4.7 (*Tt*), 4.6 (*Lh*). CHB-BAF: 6.2 (*Tt*), 7.0 (*Lh*). DCA-BAF: 13.2 (*Tt*) Lindane-BAF_{pc}: 4.37 (*Tt*), 4.52 (*Lh*). CHB- BAF_{pc}: 6.16 (*Tt*), 6.92 (*Lh*). DCA-BAF_{pc}: no steady-state (*Tt*) Tissue residue (μg kg^{-1}): 127 (Lindane), 1,360 (HCB), 342 (DCA)	*L. hoffmeisteri* (*Lh*) *and T. tubifex* (*Tt*)	Egeler et al. (1999)

Hexa-chlorobenzene (HCB)	Artificial spiked sediment using [^{14}C]HCB (0.26 μg g^{-1} nominal concentration) (63 day)	Toxicokinetics Accumulation factor (AF, on a ww basis) Biomagnification	The steady state at day 10. BAF = 7.8 ± 0.4 (SE) from kinetic parameters Tissue residue: 3.51 μg g^{-1} ww, 25.07 μg g^{-1}dw, 129.11 μg g^{-1}lipid $AF_{fish/worm}$ = 1.3 on ww, 0.5 based on dw and lipid Combined exposure of fish to spiked sediment and contaminated worms: $AF_{fish/worm}$ = 3.2 on ww, 1.3 based on dw and lipid.	*T. tubifex*	Egeler et al. (2001)
PCBs (total and 9 homologues)	Sediment bioassays (30 days) Bioaccumulation in field oligochaetes	Bioaccumulation factor (BSAF)	Total-PCB tissue concentration: In field oligochaetes = 35.890 mg g_{lipid}^{-1} *L. variegatus* = 32.950 mg g_{lipid}^{-1} BSAF: In field oligochaetes = 0.87 g OC g_{lipid}^{-1} *L. variegatus* = 0.84 g OC g_{lipid}^{-1} Limited capability to metabolize PCBs	*L. variegatus* and field oligochaetes (primarily *Limnodrilus* sp)	Ankley et al. (1992)
PCBs	Field sediment % PCBs = 183–2948 ng g^{-1}dw, and 10,182–42, 205 ng g^{-1}OC)	Bioaccumulation factor (BASF)	Total PCB tissue concentration in field oligochaetes = 2.6-6.2 μg g_{lipid}^{-1} Log BASF= −2 to 0 over the log K_{ow} range (5–8), lower than for fishes	Field oligochaetes	Bremle and Ewald (1995)

(continued)

(continued)

Chemical	Exposure conditions (t)	Measurements (other than tissue residue)	Bioaccumulation data	Species	Reference
PCBs (MCNP, CBP, HCBP) and DDE	Aqueous bioassays using [^{14}C]-labelled substances dosed in algae solutions, and used as food for worms (10-d acute and 35-d chronic test)	Toxicokinetics Tissue concentration on living and dead individuals. Critical Body residues (CBR) for lethal and sub-lethal endpoints (biomass and reproduction)	k_s = 105.4–137.4 ml g^{-1} h^{-1} k_e = 0.0013–0.22 h^{-1} Tissue concentration (max. values, ww): 1.038 nmol DDE kg^{-1}; 0.343 nmol MCBP kg^{-1}; 0.658 nmol TCBP kg^{-1}; 1.340 nmol DCBP kg^{-1}; 0.35 nmol HCBP kg^{-1} Lethal body residues: 0.88–1.35 mmol kg^{-1} (no mortality for MCNP) Sublethal CBR (ww): 0.34–0.56 mmol kg^{-1}	*L. variegatus*	Fisher et al. (1999)
PCBs and organochlorine pesticides (p,p'-DDE and HCB)	Field sediment bioassays	Bioconcentration factor (C_{lipid} / $C_{porewater}$) Biomagnification	Tissue concentration (mg kg^{-1}dw) = 3.71 (Total PCBs), 0.002 (p,p'-DDE), 0.002 (HCB) BCFs agree with values predicted from log K_{ow}, except for HCB.	*Field oligochaetes*	Galassi et al. (1994)
Tetrachlorobiphenyl (TCBP)	Sediment bioassay using labelled [^{14}C] TCBP	Bioaccumulation factor	BAF declined with increasing concentrations: from 3.8 to 0.9 (22°C) and from 2.1 to 1.1 (10°C)	*L. variegatus*	Landrum et al. (2004a)
	Reworking study: 24 days at 22°C and 26 days at 10°C	Burial rate (W_b) using ^{137}Cs Biological diffusion rate (D_b)	Body residue of 96 nmol g^{-1} (22°C) and 124 nmol g^{-1} (10°C) were estimated to reduce 50% burial rates, from linear regression models		

Tetrachloro-biphenyl (TCBP)	Sediment bioassay using labelled [^{14}C] TCBP, at 10 and 22°C (up to 30 days)	Feeding rates (faeces production) and burial rates BAF (10 days)	Tissue concentration: Total TCBP = 4–7 (at 10°C), 18–35 nmol g^{-1}ww (at 22°C). It does not relate to feeding rate Feeding and burial rates relative to control vs TCBP tissue concentrations fit a regression model, which estimated a body residue of 88 nmol TCBP g^{-1}ww to reduce by 50% the biological response	*L. variegatus*	Landrum et al. (2004b)
Tetrachlorodibenzo-*p*-dioxin (TCDD) Octachlorodi-benzo-*p*-dioxin (OCDD)	Sediment and water + sand tests spiked with [^{3}H]2,3,7,8-TCDD and [^{14}C]OCDD (28 days) Elimination test (20 day)	Toxicokinetics, BAF (lipid weight based) in sand + overlying water BSAF (worm ww / sediment dw)	k_u (from water) = 1,700 (TCDD) and 180 (OCDD) ml g^{-1} day^{-1} k_s (from sediment) = 0.49 (TCDD) and 0.006 (OCDD) g g^{-1} day^{-1} k_e = 0.175 (TCDD) and 0.049 (OCDD) day^{-1} log BAF (l kg^{-1}): 5.9 (for TCDD) and 5.5 (for OCDD) 28-d BSAF: 1.6 (for TCDD) and 0.07 (for OCDD) Half-life: 4.0 days (TCDD) and 14 days (OCDD)	*L. variegatus*	Loonen et al. (1997)
2,3,7,8 – Tetrachlorodi-benzo-*p*-dioxin (TCDD)	Water-only test (fed with spiked trout chow: 30–3,000 ng [^{3}H]TCDD $g_{food\ TOC}^{-1}$(28 days)	Toxicokinetics Biological half-life	No toxic effects were observed in 28-d tests. Body residue up to 9,533 ng TCDD g_{lipid}^{-1} k_e = 0.0014–0.0022 h^{-1} Half-lives ranged 315–495 h for elimination of total compound	*L. variegatus*	West et al. (1997)

Units used in different papers are not readily comparable. Unless stated otherwise bioaccumulation factors are calculated as the ratio between concentration in the worm by concentration in the environment (water, sediment or porewater).

Index

P. Rodriguez and T.B. Reynoldson, *The Pollution Biology of Aquatic Oligochaetes*,
DOI 10.1007/978-94-007-1718-3, © Springer Science+Business Media B.V. 2011

MIX
Papier aus verantwortungsvollen Quellen
Paper from responsible sources
FSC® C105338

If you have any concerns about our products,
you can contact us on
ProductSafety@springernature.com

In case Publisher is established outside the EU,
the EU authorized representative is:
Springer Nature Customer Service Center GmbH
Europaplatz 3, 69115 Heidelberg, Germany

Printed by Libri Plureos GmbH
in Hamburg, Germany